D1271491

Chromatographic Separations Based on Molecular Recognition

543.089
C 4672j

Chromatographic Separations Based on Molecular Recognition

EDITED BY

Kiyokatsu Jinno

WITHDRAWN

⊗ WILEY-VCH

NEW YORK • CHICHESTER • WEINHEIM • BRISBANE • SINGAPORE • TORONTO

Kiyokatsu Jinno
School of Materials Science
Toyohashi University of Technology
Tempaku-cho, Toyohashi
441 Japan

This book is printed on acid-free paper.∞

Library of Congress Cataloging-in-Publication Data

Chromatographic separations based on molecular recognition / edited by Kiyokatsu Jinno.
   p. cm.
   Includes bibliographical references and index.
   ISBN 0-471-18894-8 (alk. paper)
   1. Chromatographic analysis.   2. Molecular recognition.
   I. Jinno, Kiyokatsu, 1945– .
   QD79.C4C482   1996
   543′.0892—dc20

                                        96-23299
                                            CIP

Copyright © 1997 Wiley-VCH, Inc.

All rights reserved. Published simultaneously in Canada.

Reproduction or translation of any part of this work
beyond that permitted by Section 107 or 108 of the
1976 United States Copyright Act without the permission
of the copyright owner is unlawful. Requests for
permission or further information should be addressed to
the Permissions Department, John Wiley & Sons, Inc.,
605 Third Avenue, New York, NY 10158-0012.

Printed in the United States of America

ISBN 0-471-18894-8 Wiley-VCH, Inc.

10 9 8 7 6 5 4 3 2 1

Preface

Chromatography is the technique of separating the various components of complex mixtures on the basis of differences at the molecular level of interactions among solutes. It is a biphasic process in which solids or liquids, the so-called stationary phase, are borne on a carrier, called the mobile phase. Depending on the kind of carrier, we categorize the process as gas chromatography (GC) when the mobile phase is a gas, liquid chromatography (LC) when the mobile phase is a liquid, and supercritical fluid chromatography (SFC) when the mobile phase is a supercritical fluid. Of course other modes of chromatography are also available, such as thin-layer chromatography (TLC) and paper chromatography (PC), although those two are less popular than the first three methods.

Because the mobile phase gas is inert, GC separations generally show the difference in elution time as a function of the strength of the interaction between the solute and the stationary phase. In LC and SFC, however, the interactions among the solute, the stationary phase, and the mobile phase are all dominant factors, combining to control the elution time (retention). Various methods have been applied to promote an understanding of the contribution of solute–stationary phase, solute–mobile phase, and stationary phase–mobile phase interactions at the molecular level. Spectroscopic instruments are used for the interpretation of the stationary phase structures. Computer calculations for molecular modeling usually are used in interpreting solute structures. And several spectroscopic and solvatochromic methods have been proposed for assessing the contribution of the mobile phase to the solute and stationary phase structures. Upon gathering such information about chromatographic retention mechanisms, one can study the separation process theoretically and experimentally.

v

ALLEGHENY COLLEGE LIBRARY

The retention mechanisms in chromatography have not been fully elucidated. Yet to design the best separation system for a particular problem, the mechanisms should be known, for otherwise it is difficult to interpret the interactions at the molecular level, particularly in LC systems, where interactions are extremely complex. Indeed, in LC setups, the part of the interaction that controls retention is sometimes dominant and sometimes nondominant. In such cases one can model the interaction in chromatographic separation processes by using information about the solute and stationary phase structures. Molecular recognition is one important source of information about separation. A molecule can be recognized by its stationary phase, and the separation process can be induced by different molecular level interactions between the stationary phase and the molecules that differ from each other in shape and size. This book, which examines various aspects of molecular recognition in chromatographic separation, is intended to emphasize that such ideas will help in the design of new separation systems for many difficult separation problems.

I thank the contributors for useful and interesting discussions of their works and the stimulation it has provided for our own studies. I also thank my research students at Toyohashi University of Technology, who have contributed to our own studies in this field. I mention as well H. Ohta, who works with me in my laboratory and spent much time in editing and organizing whole chapters to be compatible with Macintosh computers. I hope that this book will help readers to arrive at a better understanding of the retention process in chromatography, especially the gas and liquid forms.

Kiyokatsu Jinno
Toyohashi, Japan

Contents

3. Molecular Recognition for Fullerenes in Liquid Chromatography   147
Kiyokatsu Jinno

4. Chromatographic Enantiomer Separation on Chiral Polymers   239
Yoshio Okamoto and Eiji Yashima

7. Molecular Recognition in Complexation Gas Chromatography   371
Volker Schurig

Contributors

KEN HOSOYA. Department of Polymer Science, Kyoto Institute of Technology, Matsugasaki, Sakyo-ku, Kyoto 606, Japan

KIYOKATSU JINNO. School of Materials Science, Toyohashi University of Technology, Tempaku-cho, Toyohashi 441, Japan

ROMAN KALISZAN. Department of Biopharmaceutics and Pharmacodynamics, Medical University of Gdansk, and Department of Biotechnology, University of Gdansk, Gen. J. Hallera 107, 80-416 Gdansk, Poland

YOSHIO OKAMOTO. Department of Applied Chemistry, School of Engineering, Nagoya University, Furo-Cho, Chukasa-Ku, Nagoya 464-01, Japan

LANE C. SANDER. Analytical Chemistry Division, National Institute of Standards and Technology, Gaithersburg, MD 20899, USA

VOLKER SCHURIG. Institut für Organische Chemie der Universität, Auf der Morgenstelle 18, D-72076 Tübingen, Germany

IRVING W. WAINER. Pharmacokinetics Division, Department of Oncology, McGill University, and the Montreal General Hospital, 1650 Cedar Avenue, Room B7113, Montreal, Quebec H3G 1A4, Canada

STEPHEN A. WISE. Analytical Chemistry Division, National Institute of Standards and Technology, Gaithersburg, MD 20899, USA

EIJI YASHIMA. Department of Applied Chemistry, School of Engineering, Nagoya University, Furo-cho, Chukasa-ku, Nagoya 464-01, Japan

CHAPTER

1

Molecular Shape Recognition for Polycyclic Aromatic Hydrocarbons in Reversed-Phase Liquid Chromatography

Stephen A. Wise and Lane C. Sander

1.1 Introduction*

For the past 20 years we have been involved in investigations of the chromatographic and physical characteristics of alkyl bonded stationary phases used in reversed-phase liquid chromatography (LC) to gain a better understanding of the mechanisms responsible for retention and selectivity. These investigations were initiated at the National Institute of Standards and Technology (NIST) as part of LC methods development for the separation of polycyclic aromatic hydrocarbons (PAHs). Since its inception in the early 1970s, high performance LC has been used for the separation of PAHs. In 1971 Schmit et al. [1] first described the separation of PAHs using a chemically bonded octadecylsilane (C_{18}) stationary phase, and since that time reversed-phase LC on C_{18} phases has become the most popular LC mode for the separation of PAHs [2–6]. The popularity of reversed-phase LC for PAH separations is due, in part, to its excellent selectivity for the separation of PAH isomers.

In the late 1970s, studies at NIST (then the National Bureau of Standards), were focused on the retention of PAHs on various stationary phases in both normal

---

* Certain commercial equipment, instruments, or materials are identified in this chapter to specify adequately the experimental procedure. Such identification does not imply recommendation or endorsement by the National Institute of Standards and Technology, nor does it imply that the materials or equipment identified are necessarily the best available for the purpose.

1

and reversed-phase LC [7,8]. Confronted with the need to compare retention on different C_{18} columns, investigators reported retention data by means of a retention index system described by Popl et al. [9] for normal phase separations of PAHs (see discussion below). However, we observed that the retention indices for PAHs in reversed-phase LC often varied significantly among the different columns. These studies indicated that even though reversed-phase LC on C_{18} stationary phases provides excellent separations of PAHs, not all C_{18} stationary phases provide the same selectivity (i.e., relative separation) for PAHs. In the early 1980s, studies at NIST [7,10] and other laboratories [11–14] compared different commercial C_{18} columns from various manufacturers for the separation of PAHs with particular emphasis on the separation of the 16 PAHs identified by the U.S. Environmental Protection Agency (EPA) as priority pollutants. These studies found that even though all the different columns were "generically" C_{18} phases, some provided significantly enhanced selectivity for the separation of these 16 PAHs. During these early studies, it became evident that such investigations were limited because the exact details concerning the silica substrate and the bonded phase syntheses were difficult to obtain from the LC column manufacturers. As a result, investigations were initiated at NIST to understand more fully the influence on selectivity of PAH separations in reversed-phase LC of factors such as bonded-phase type, silica substrate characteristics, alkyl chain length, and C_{18} ligand density.

During these early comparisons of different C_{18} phases, we also observed that for the separation of PAH isomers, the shape of the isomer was an important parameter that influenced both retention and selectivity. A shape parameter, defined as the length-to-breadth (L/B) ratio, was found to correlate with the reversed-phase LC retention of PAH isomers [10]. This observation led to more detailed investigations on retention and selectivity in reversed-phase LC of the influence of various PAH molecular descriptors such as L/B ratio and dihedral angle of distortion (to describe nonplanarity). The L/B ratio has since become one of the most widely used descriptors for correlation with chromatographic retention, particularly LC retention, for PAHs [15–22].

The influence of both the stationary phase and solute shape parameters in the separation of PAHs has been described in a number of research papers published by the NIST group during the past 20 years [4,10,22–35]. In addition, several review papers have appeared summarizing these investigations and our understanding of the mechanism of retention of PAHs in reversed-phase LC [36–42]. However, these reviews have focused primarily on the influence of the stationary phase parameters rather than the PAH solute. This chapter describes briefly the results of our stationary phase investigations; in addition, our studies concerning the solute characteristics and their effect on selectivity for PAH separations are revisited and discussed in detail, particularly in the context of recently proposed models for retention in LC. All the reversed-phase LC retention data reported in earlier investigations are combined to provide a retention and solute shape parameter database for more than 200 PAHs and alkyl-substituted PAHs.

1.2 Stationary Phase Characteristics Affecting Shape Selectivity

1.2.1 Stationary Phase Chemistry

The ability of a stationary phase to discriminate among PAH isomers (and other classes of compounds) on the basis of the molecular structure is described as "shape selectivity." Shape recognition in LC has been the subject of extensive reviews [38,39]. Perhaps the single most important parameter affecting stationary phase shape selectivity toward PAHs is a property involving phase structure that has been all but ignored over the past 20 years, namely, "phase type" (i.e., whether a phase was prepared by monomeric or polymeric synthesis chemistry) [35,41]. Bonded phases are prepared through the reaction of silica with chloro- or alkoxysilanes as illustrated in Figure 1.1. Most reversed-phase C_{18} stationary phases are prepared by reaction of dimethylchlorooctadecylsilane with silica. The reaction, when carried out with an excess of the silane reagent, is limited by the number of available silanols at the silica surface. This type of phase is termed "monomeric" because only bonds between the silane reagent and the silica silanols are possible. In contrast, reaction of silica with octadecyltrichlorosilane in the presence of water results in bonds between silane reagent molecules as well as with the silica. This type of phase is termed "polymeric," and the extent of the reaction is controlled by the amount of water added, though the pore diameter of the silica also affects phase loading, as described later [32,33].

Selectivity differences between monomeric and polymeric C_{18} phases are clearly evident for the separation of PAHs. In 1979–1981 several studies [7,10–14] compared the differences in selectivity for commercial C_{18} stationary phases from various manufacturers. In 1981 Wise et al. [10] compared the retention behavior of more than 100 PAHs on a monomeric and a polymeric C_{18} column. In all these studies it was found that only one particular stationary phase material, Vydac 201TP/HC-ODS, was successful in separating all 16 of the priority pollutant PAHs listed by the U.S. Environmental Protection Agency (EPA). Because of the unique selectivity of this particular material for the separation of PAHs, reversed-phase LC using this C_{18} column was specified in EPA Method 610 for the determination of PAHs in aqueous effluents [43]. However, none of these early studies recognized the significance of the bonded phase synthesis on selectivity. Only after a series of investigations in the early 1980s was the importance of the polymeric phase synthesis established [30–33].

Today, the enhanced selectivity of polymeric phase chemistry is widely recognized by analysts involved in the measurement of PAHs. However, such differences in phase selectivity are perhaps just as widely ignored outside this specialty because variations in retention behavior between monomeric and polymeric C_{18} phases are less dramatic for solutes without rigid structure (i.e., molecules with free rotation about a single bond). As a result, the vast majority of stationary phases currently used in reversed-phase LC are prepared using monomeric surface modification

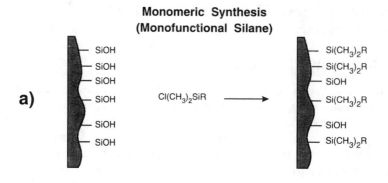

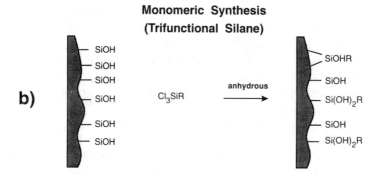

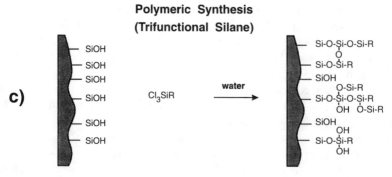

Figure 1.1 Synthesis schemes for monomeric and polymeric alkyl stationary phases.

chemistry. The relatively limited number of commercial sources of polymeric C_{18} phase columns has hindered appreciation of the unique shape recognition properties of these phases. A prejudice still exists against polymeric phases, based in part on the assumption that the phases exhibit high silanol activity and low efficiency. In fact, the LC column manufacturers that do produce polymeric phases generally do not promote their product as a polymeric phase but instead target their application to PAH separations to express the novel selectivity differences inherent with these phases.

The selectivity differences for monomeric and polymeric C_{18} phases for the separation of the 16 EPA priority pollutants are shown in Figure 1.2. Even though these 16 PAHs were the focus of many of the early investigations in phase selectivity, the differences in the separations are not overwhelming. The main difference is the lack of separation of chrysene and benz[a]anthracene (isomers) and the difference in elution order of benzo[ghi]perylene and indeno[1,2,3-cd]pyrene (isomers). However, these seemingly minor differences are indicators of the significant differences between these two phase types for the separation of PAH isomers. The enhanced shape recognition inherent to the polymeric C_{18} phases is dramatically illustrated in Figure 1.3 for the separation of PAH isomers of molecular mass 302. Very little separation of these isomers could be achieved on a monomeric C_{18} column; however, all 12 isomers were resolved with a heavily loaded polymeric C_{18} phase using the same mobile phase conditions [44]. Similar differences in column selectivity for monomeric and polymeric phases are apparent for the separation of 11 PAH isomers of molecular mass 278, as shown in Figure 1.4. Again under the same mobile phase conditions, the isomers were not completely resolved on the monomeric phase, and they eluted in a much narrower band than on the polymeric phase. This difference is not a consequence of column strength or absolute retention. Attempts to improve such separations with monomeric stationary phases typically fail, even with application of weak mobile phase or lengthy gradient elution conditions. The enhanced shape recognition ability illustrated in Figures 1.3 and 1.4 is a general property observed for polymeric phases. Other phase parameters such as the extent of polymerization (i.e., phase loading), ligand length, and substrate properties also affect shape selectivity, although monomeric phases never exhibit the degree of shape discrimination possible with polymeric phases.

1.2.2 Assessing Stationary Phase Shape Selectivity

Because column manufacturers are often reluctant to detail their column synthesis procedures, it is difficult to predict whether a given column will have monomeric- or polymeric-like selectivity. When studies were made of the retention behavior of more than 100 PAH standards on monomeric and polymeric C_{18} phases [10,30], several of the compounds were observed to exhibit marked changes in relative retention on the two phase types, and the relative retention of these particular compounds varied as the surface coverage or ligand density of the polymeric phase varied [30]. Based on these observations, a simple empirical test was developed to

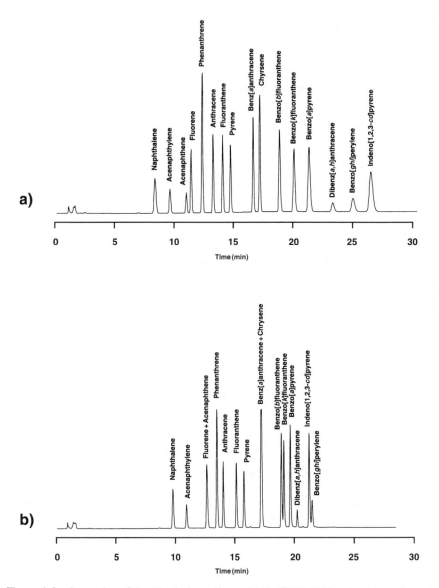

Figure 1.2 Separation of the 16 priority pollutant PAHs (SRM 1647) using (a) a polymeric C_{18} column and (b) a monomeric C_{18} column.

assess stationary phase shape selectivity toward PAHs [27,39,45,46]. This test is based on the relative retention of three PAH solutes (see Figure 1.5 for structures), two of which have nonplanar conformations. (The importance of the nonplanar structure in assessing the phase structure is discussed later.) A solution containing these components is available as Standard Reference Material (SRM) 869, "Column

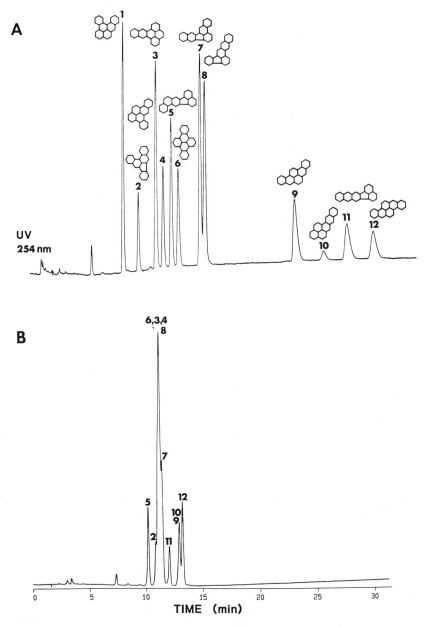

Figure 1.3 Separation of PAH isomers of molecular mass 302 on (a) a polymeric C_{18} column and (b) a monomeric C_{18} column. Peaks: 1, dibenzo[a,l]pyrene; 2, dibenzo[b,e]fluoranthene; 3, naphtho[2,3-e]pyrene; 4, dibenzo[a,e]pyrene; 5, naphtho[1,2-k]fluoranthene; 6, dibenzo[e,l]pyrene; 7, dibenzo[b,k,]fluoranthene; 8, naptho[2,3-b]fluoranthene; 9, dibenzo[a,i]pyrene; 10, naphtho[2,3-a]pyrene; 11, naphtho[2,3-k]fluoranthene; 12, dibenzo[a,h]pyrene.

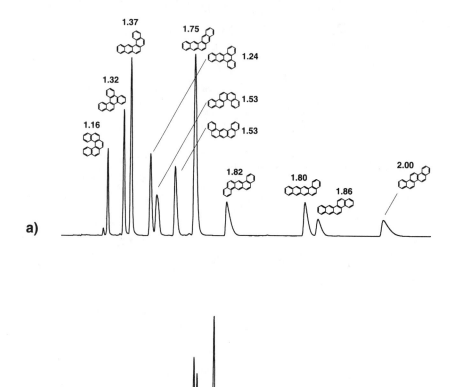

Figure 1.4   Separation of molecular mass 278 PAH isomers on (a) a polymeric C_{18} column and (b) a monomeric C_{18} column; numbers refer to L/B values.

Selectivity Test Mixture for Liquid Chromatography (Polycyclic Aromatic Hydrocarbons)."*

The retention of benzo[a]pyrene (BaP, planar conformation) relative to 1,2:3,4:5,6:7,8-tetrabenzonaphthalene (TBN, nonplanar conformation, alternate name dibenzo[g,p]chrysene) and phenanthro[3,4-c]phenanthrene (PhPh, nonplanar conformation) provides a sensitive measure of the polymeric or monomeric character of the phase. Phases prepared using monomeric surface modification chemistry

---

* Standard Reference Materials Program, NIST, Gaithersburg, MD 20899.

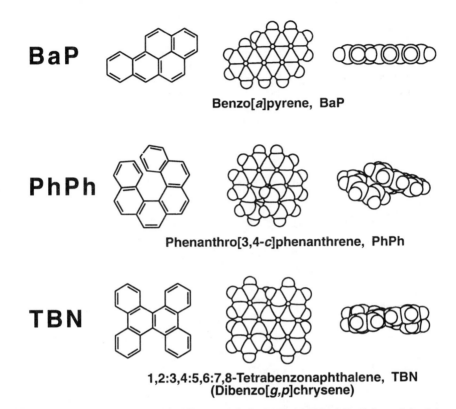

BaP

Benzo[a]pyrene, BaP

PhPh

Phenanthro[3,4-c]phenanthrene, PhPh

TBN

**1,2:3,4:5,6:7,8-Tetrabenzonaphthalene, TBN
(Dibenzo[g,p]chrysene)**

Figure 1.5 Structures and space-filling models for PAHs in SRM 869, Column Selectivity Test Mixture.

give the elution order BaP ≤ PhPh < TBN, while phases prepared using polymeric surface modification chemistry give the order PhPh < TBN ≤ BaP. The first category has been described as "monomeric-like selectivity" and the second, "polymeric-like selectivity." Phases with intermediate properties such as densely loaded monomeric C_{18} phases, or lightly loaded polymeric C_{18} phases are indicated by the elution order PhPh < BaP < TBN and are considered to have "intermediate selectivity." It should be emphasized that shape selectivity does not correlate with absolute retention. High carbon loading monomeric C_{18} phases have high absolute retention but have little shape recognition ability, whereas polymeric phases with low carbon loading have high shape selectivity. Figure 1.6 illustrates separations of SRM 869 for several commercial C_{18} columns. The dramatic differences in these separations show the variations in selectivity that exist among columns, and serve to emphasize the importance of appropriate stationary phase selection in method development for PAH separations.

Using SRM 869, a quantitative measure of phase shape selectivity can be calculated to enable relative comparisons between different C_{18} phases. The shape selec-

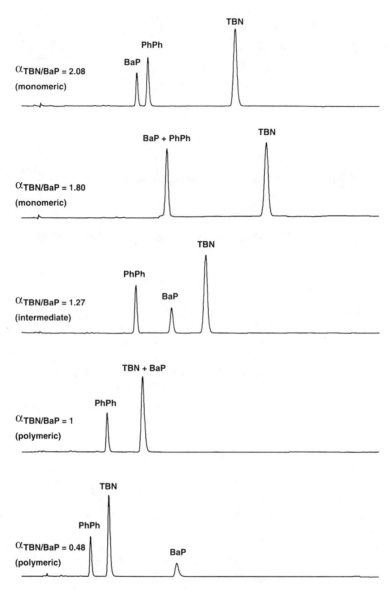

Figure 1.6 Separation of SRM 869, Column Selectivity Test Mixture, on various monomeric and polymeric C_{18} columns: BaP, benzo[*a*]pyrene; PhPh, phenanthro[3,4-*c*]phenanthrene; TBN, 1,2:3,4:5,6:7,8-tetrabenzonaphthalene.

tivity factor $\alpha_{TBN/BaP}$ (defined as k'_{TBN}/k'_{BaP}) has been shown to correlate with retention behavior for PAHs and phase type [27,31,32,41]. A classification scheme has been proposed based on measurement of $\alpha_{TBN/BaP}$ values for experimental and commercial C_{18} columns to assist in selection of the appropriate column [27].

Values for $\alpha_{TBN/BaP} \leq 1$ reflect moderately loaded polymeric C_{18} phases, and values for $\alpha_{TBN/BaP} \geq 1.7$ reflect monomeric C_{18} phases. For values $1 < \alpha_{TBN/BaP} < 1.7$, the synthesis scheme is less certain and may indicate light polymerization with di- or trifunctional reagents, or densely loaded monomeric phases. Thus, the following classification scheme has been proposed:

$\alpha_{TBN/BaP} \leq 1$, "polymeric-like"
$1 < \alpha_{TBN/BaP} < 1.7$, "intermediate"
$\alpha_{TBN/BaP} \geq 1.7$, "monomeric-like"

More than 40 commercial C_{18} columns from different manufacturers were evaluated using SRM 869 and grouped according to this classification scheme [39]. Among commercial C_{18} columns, a wide range of $\alpha_{TBN/BaP}$ values are observed from about 0.6 to 2.2 [39]. It should be emphasized that significant variations in selectivity are possible within each of the classification groups. Experience indicates that these variations are greatest for polymeric phases, and changes in elution order have been observed among polymeric C_{18} columns with different phase loadings [22,39].

SRM 869 serves several purposes, including column classification and selectivity assessment to aid method development, and stationary phase characterization in quality control during the production of columns. Several column manufacturers currently use SRM 869 as part of their quality control criteria to assure that phase selectivity is reproducible from one column lot to another [47–49]. We have used SRM 869 extensively to investigate the factors that affect stationary phase selectivity, such as phase synthesis, substrate pore size, alkyl chain length, mobile phase composition, and temperature, as discussed in the sections that follow.

1.2.3 Substrate Pore Size Effects

The possible influence of pore size on retention is often neglected for low molecular mass solutes such as PAHs because their molecular dimensions are small compared to the pore diameter of most reversed-phase substrates. For example, benzo[a]pyrene has a molecular length of approximately 14 Å, which is small compared with pore diameters of 60–300 Å for common commercial phases. Despite this assumption, changes in column selectivity have been observed for C_{18} phases prepared on different pore diameter silicas. The effect of pore size on column selectivity was studied for four substrates with pore diameters of 60, 100, 150, and 300 Å [33]. Monomeric and polymeric C_{18} phases were synthesized on each substrate, and the selectivity of the phases evaluated using SRM 869 and the mixture of the 16 EPA priority pollutant PAHs. All four monomeric C_{18} phases exhibited similar chromatographic behavior, except for variations in absolute retention. Differences in absolute retention are expected because the surface area (and thus the phase loading) of the small pore substrates was significantly greater than that for the wide pore substrates. The effect of pore size on selectivity was pronounced for the polymeric C_{18} phases. Polymeric phases prepared on narrow pore substrates exhibited retention behavior more like commercial monomeric C_{18} phases than polymeric phases. Thus, separation of the 16 priority pollutant PAHs, which is normally possible only with certain polymeric C_{18} phases, could be achieved only for the 150 and 300 Å polymeric C_{18} columns. In an examination of polymeric C_{18} phases prepared on many dissimilar silicas of differing pore sizes, narrow pore substrates consistently yielded phases with monomeric-like selectivity. Typically, substrates with pore diameters of 120–150 Å or larger produced phases with enhanced "polymeric-like" selectivity for PAH isomers.

Because monomeric C_{18} phases did not exhibit this variation in selectivity with pore size, size exclusion effects were ruled out as the source of the differences. We envision a size exclusion effect that influences the extent of polymeric surface modification during the phase synthesis. Polymeric reaction schemes utilizing trichlorosilanes and water form silane polymers in solution. These reactive polymers may have an appreciable molecular mass, and consequently surface modification is affected by pore size. An excess of the trichlorosilane is used, and monomer coexists with silane polymer in solution. Monomer and polymer molecules compete for reaction at the silica surface. Because the smaller silane monomers can diffuse into narrow pores more easily than large silane polymer molecules, more monomer molecules reach the surface of narrow pore substrates, resulting in a high percentage of bound monomers. In the large pore substrates, however, the silane polymers compete with silane monomer molecules more effectively and the resulting phases have overall polymeric-like character.

1.2.4 Bonding Density

Studies of differences in PAH selectivity among different C_{18} phases [7,11–14] indicated that bonded phase loading (surface coverage or density) played a role in PAH selectivity. In an early investigation [30] using seven lots of commercially prepared polymeric C_{18} phases that differed in C_{18} surface coverage (as determined from carbon loading and surface area measurements on the C_{18} modified silica), selectivity for PAHs was observed to vary in a regular manner with phase loading. Another study [31] using the same lots of polymeric C_{18} materials demonstrated an increase in shape recognition for PAH isomers as the C_{18} surface coverage increased. These two studies led to more detailed investigations using phases prepared under known conditions to better understand the effects of phase loading.

The effect of bonded phase loading density on selectivity has been studied by a number of research groups, particularly for monomeric C_{18} phases [50–55]. Phase density can be regulated by altering the synthesis reaction conditions. For monomeric C_{18} phases, surface coverage values typically range from about 3.0 to 3.5 $\mu mol/m^2$. Phases with low alkyl chain density are easily prepared by reducing the concentration of the reactive silane or by reducing the reaction time. High density monomeric phases with surface coverage values of 4 $\mu mol/m^2$ or greater have been prepared by Sentell, Barnes, and Dorsey using a reaction procedure under ultrasound conditions [56]. High density phases have also been prepared by Szabo and co-workers using a novel dimethylaminodimethyloctadecylsilane modifying reagent [57]. In general, retention is observed to increase with percent carbon loading of the phase. These changes are as expected (based on increases in the phase ratio), and selectivity toward PAHs remains relatively constant, at least for monomeric phases with low to normal phase density. Anomalous shifts in retention and selectivity have been reported for high density phases. Sentell and Dorsey have observed that retention is greatest for a bonded phase density of about 3.1 μmol/m^2, and at higher monomeric phase densities retention actually decreases [53]. It should be noted that column shape selectivity also changes for high density monomeric phases. Using SRM 869, Sentell and Dorsey found that the high density phases were more selective toward shape, with a shift in selectivity toward that normally associated with polymeric C$_{18}$ phases [54].

A detailed investigation of the influence of stationary phase chemistry and bonding density was recently carried out by our research group [23]. Various monomeric and polymeric surface modification schemes were used to prepare C$_{18}$ stationary phases on a single lot of silica. Surface coverage values ranged from 2.0 μmol/m^2 for a monomeric phase, to 7.4 μmol/m^2 for a surface polymerized stationary phase. Column selectivity was studied through the use of various PAH probes selected based on differences in molecular shape. Thus, the separation of PAHs including triptycene (rigid, nonplanar), pyrene (planar), o-terphenyl (nonrigid, nonplanar), tetraphenylmethane (globular), and p-terphenyl and 1,6-diphenylhexatriene (linear, extended) indicated differences in selectivity among the columns that can be attributed to differences in molecular shape. Molecular mass 278 PAH isomers were also utilized in this evaluation. In general, shape recognition was observed to increase with increasing bonding density regardless of the way in which the density was achieved.

Two stationary phases were an exception, and these are particularly noteworthy. These phases were prepared by dissimilar synthetic approaches, but had similar bonding densities (5.25 vs. 5.34 μmol/m^2). Significantly different chromatographic properties were observed for these two columns, and the differences were attributed to local order and homogeneity (i.e., interchain spacing) for the bonded surface. A comparison of the absolute retention for various solutes on the different columns provides further insight into solute retention mechanisms. A plot of k' versus surface coverage is shown in Figure 1.7. The trends indicated in the figure are similar to those observed by Sentell and Dorsey [53], and predicted by Dill [58], in that a retention maximum is observed as a function of bonded phase loading. An additional observation can be made. If the curves in Figure 1.7 are examined carefully, it is apparent that maximum retention occurs at different surface coverage values for different solutes. Several solutes exhibit maximum retention at or near 4 μmol/m^2, whereas other solutes exhibit a maximum at much higher phase loadings. Bulky solutes such as triphenylbenzene, tetraphenylmethane, and o-terphenyl are examples of solutes for which maximum retention is observed at lower surface coverages. Extended solutes such as 1,6-diphenylhexatriene and p-terphenyl exhibit maximum retention at much higher phase loadings approaching 7 μmol/m^2.

The retention behavior just described might be expected based on interchain distances. Bulky solutes should require larger interchain spacing for partitioning within the stationary phase compared with planar or linear solutes [58]. The reten-

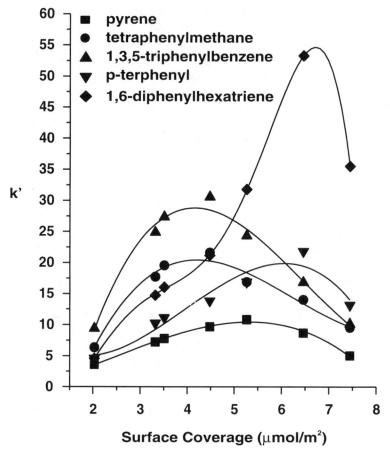

Figure 1.7 Retention k' plotted as a function of surface coverage for various C_{18} columns and selectivity probes.

tion of bulky molecules is reduced for stationary phases with high surface coverages because the energetic cost of cavity creation is higher than that associated with planar or linear molecules. As the space between bonded alkyl chains decreases, solute–stationary phase interactions are favored for extended and linear solutes.

Polymeric C_{18} phases typically have phase densities nearly twice that of monomeric C_{18} phases (5–6 μmol/m²). Because these phases are prepared through polymerization of a trifunctional silane, phase structure at the molecular level is difficult to conceptualize. One possibility is that the additional phase loading results from branched structures extending away from the silica surface. Another possibility is that phase structure is like a monomeric phase, but with substantially increased phase density. We favor the latter model for polymeric phases. Studies utilizing small angle neutron scattering (SANS) have been carried out on both monomeric and polymeric C_{18} phases [59,60]. Bonded phase thicknesses were evaluated, and although the value for the polymeric C_{18} phase was greater than for the monomeric phase (21 Å versus 17 Å, respectively), the difference was small, and not sufficient to account for the differences in phase loading. Bonded phase density was also obtained from the SANS measurements, and the polymeric phase was found to be significantly denser than the monomeric phase. At the molecular level, the two phase types do not appear to be fundamentally different; instead the differences appear to be a matter of degree and the result of alkyl chain packing density. This view of bonded phase structure should be used with caution because the covalent linkages for polymeric C_{18} phases are known to differ from monomeric C_{18} phases, and silanol density is also known to differ between the two phase types.

Among phases prepared using polymeric phase syntheses, phase density can be altered by changing the quantity of water added during the reaction [32]. Phases with loadings as high as 7.4 μmol/m^2 have been prepared [23]. Such heavily loaded polymeric phases exhibit even greater shape selectivity, as is evidenced through isomer separations. The effect of phase loading on selectivity is shown in Figure 1.8 for a separation of 278 molecular mass PAH isomers. The four phases were prepared using monomeric and polymeric surface modification approaches, and for polymeric (and intermediate) syntheses, different amounts of water were added to produce different phase loadings. Using SRM 869, the selectivity factor $\alpha_{TBN/BaP}$ decreases with increasing phase loading, and as can be seen in Figure 1.8, the ability to resolve PAH isomers (i.e., shape selectivity) increases. To date, the smallest $\alpha_{TBN/BaP}$ value observed was approximately 0.2, for an experimental lot of a commercial polymeric C_{18} column.

1.2.5 Bonded Phase Length

The effect of alkyl phase length on retention and selectivity has been studied extensively over many years. For nonrigid, nonpolar solutes, retention is observed to increase with increasing bonded phase length, either through a linear or a logarithmic relationship [61–63]. Berendsen and de Galan observed a critical chain length behavior for solute retention such that k' increased with alkyl phase length up to a point and then leveled off [64]. The chain length at which this leveling off of retention occurs is called the critical chain length. Others have observed a similar phenomenon related to changes in selectivity [65].

Despite the existence of this considerable body of research, little effort has been expended in examining possible relationships between phase length and shape selectivity. We studied this effect by preparing monomeric and polymeric phases with alkyl ligands ranging from C_8 to C_{30} [29]. Shape selectivity was probed using SRM 869. For the monomeric phases, selectivity changed little among the C_8, C_{12}, and C_{18} phases, but $\alpha_{TBN/BaP}$ was observed to decrease significantly (i.e., shape selectivity increased) for the C_{22} and C_{30} phases. The decrease in $\alpha_{TBN/BaP}$ is indicative of phase selectivity similar to that expected for C_{18} polymeric phases. Figure 1.9 plots $\alpha_{TBN/BaP}$ versus chain length of the phase. For the polymeric phases of various alkyl lengths, selectivity is observed to change significantly for the shorter chain

ALLEGHENY COLLEGE LIBRARY

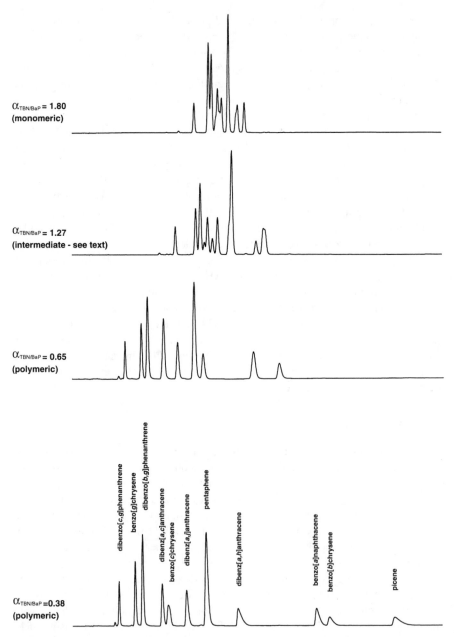

Figure 1.8   Separation of PAH isomers of molecular mass 278 on polymeric C_{18} columns of different phase loadings. Increasing phase loading is indicated by decreasing $\alpha_{TBN/BaP}$ value.

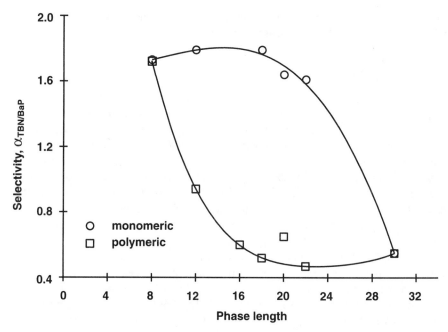

Figure 1.9 Variations in column shape selectivity ($\alpha_{TBN/BaP}$) as a function of alkyl phase length for monomeric, and polymeric phases.

lengths. Values for $\alpha_{TBN/BaP}$ increase for the shorter alkyl phases. The monomeric phases take on "polymeric-like" selectivity for long alkyl lengths, and polymeric phases become "monomeric-like" for short chain lengths (12 carbons and below).

1.2.6 Mobile Phase Composition

In reversed-phase LC, the most important consideration for controlling absolute solute retention is mobile phase composition. Numerous studies have examined the relationship between the composition of the mobile phase and retention, and most of the data indicate linear free energy relationships: retention decreases with increasing organic composition of the mobile phase, and plots of the natural logarithm k' versus the percentage of organic modifier are linear. Selectivity is also observed to vary as a function of mobile phase composition, although trends are not easily predicted, especially among dissimilar solutes. We have examined changes in selectivity as a function of mobile phase composition using SRM 869 for various mobile phase systems [37]. Although changes in the shape selectivity factor $\alpha_{TBN/BaP}$ were small, in each case $\alpha_{TBN/BaP}$ decreased with increasing organic content in the mobile phase, indicating that shape recognition increases with increasing organic mobile phase composition. It is doubtful that these changes are large enough to be of practical significance in method development. The change in selectivity with mobile phase composition is of concern, however, in the reporting of retention indices, and this problem is discussed in a later section.

1.2.7 Temperature

Unlike gas chromatography, LC rarely uses temperature as a separation parameter. When column temperature is employed, it is typically controlled to improve retention reproducibility or efficiency. Solute retention decreases with increasing temperature, and occasionally this property is exploited to adjust retention when the mobile phase composition is held constant. In our laboratory, we have observed dramatic changes in column selectivity toward planar and nonplanar solutes with changes in column temperature [26]. These changes occur continuously over the useful temperature range for both monomeric and polymeric C_{18} columns. Figure 1.10 plots $\alpha_{TBN/BaP}$ versus temperature for a monomeric and a polymeric C_{18} column. The selectivity factor $\alpha_{TBN/BaP}$ decreases at subambient temperatures and increases at elevated temperatures. Low values for $\alpha_{TBN/BaP}$ (i.e., $\alpha_{TBN/BaP} < 1$ at ambient temperature) typically indicate polymeric phase selectivity behavior; thus the observation that $\alpha_{TBN/BaP}$ decreases with temperature suggests that "polymeric-like" selectivity can be temperature induced. Likewise, increases in temperature produce increases in $\alpha_{TBN/BaP}$ and "monomeric-like" selectivity. Temperature-induced selectivity changes have been used to provide "polymeric-like" separations of the 16 priority pollutants on a monomeric C_{18} phase at subambient temperatures ($-8°C$) and, more importantly for practical applications, to provide enhanced shape

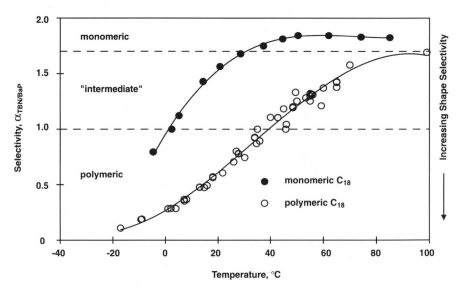

Figure 1.10   Column shape selectivity ($\alpha_{TBN/BaP}$) plotted as a function of temperature for monomeric and polymeric C_{18} columns.

selectivity for difficult isomer separations by using a polymeric phase operating at subambient temperatures [26].

The increased ability of a column to separate solutes on the basis of shape with reductions in temperature is a general trend that is observed for all C_{18} columns. At a given subambient temperature, the relative degree of shape discrimination depends on column selectivity observed at ambient temperature. For example, although enhanced shape recognition is displayed at subambient temperatures for both monomeric and polymeric C_{18} phases (compared to the same columns at ambient temperature), greater shape recognition is possible with the polymeric C_{18} phase at low temperatures because the column exhibits significant shape selectivity at ambient temperature. This is evident from the plots of $\alpha_{TBN/BaP}$ in Figure 1.10 for which $\alpha_{TBN/BaP(polymeric)}$ is less than $\alpha_{TBN/BaP(monomeric)}$ at any given temperature.

These differences in selectivity due to temperature are the result of changes in alkyl bonded phase structure that occur with temperature. Sander and co-workers [66] used Fourier transform infrared (FTIR) spectra from alkyl bonded phases at various temperatures to identify changes in the conformation of the bonded C_{18} chain. Transitions in the IR spectra corresponding to bends and kinks in the alkyl chain were observed to decrease with decreasing temperature, but did not exhibit any evidence of first-order phase transition of the type found when normal paraffins are frozen. These results indicated that the bonded alkyl ligands become straighter and more extended at low temperatures but do not associate to undergo a phase transition. The inability of the bound chains to associate in a crystalline lattice is not unexpected because the movement of the ligands is restricted by the covalent linkage. Thus, phase ordering results not from enthalpic contributions from lattice formation, but instead from entropic effects of individual chains as conformational defects are eliminated at low temperatures. The structure of this ordered, bonded phase can be conceptualized as a liquid crystalline system of monolayer thickness. Solute interaction with such an ordered phase can be expected to involve molecular shape; solute shape should be a less important factor for interaction with randomly oriented alkyl chains. It can be argued that bonded phases at ambient temperature possess more order than true liquids because the covalently bound ligands have restricted degrees of freedom and necessarily extend away from the silica surface.

1.2.8 Controlling Shape Selectivity

As indicated by the preceding discussion, a number of factors influence column selectivity for PAH separations, and different parameters can be altered to achieve the same results. The various parameters for increasing shape selectivity are summarized in Figure 1.11. Thus, column selectivity can be controlled at several levels. At the stationary phase preparation level, column manufacturers can select the substrate, reaction conditions, and silane reagents to alter phase type, phase density, and alkyl chain length. For example, the degree of phase loading with polymeric C_{18} phases can be varied by controlling the quantity of water introduced in the synthesis. An alternative approach to tailoring selectivity is to blend stationary phase materials with different selectivities. Thus, to achieve a column with $\alpha_{TBN/BaP}$

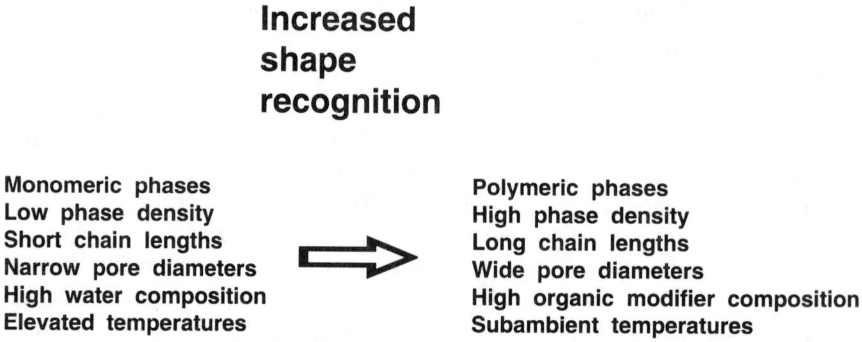

**Increased
shape
recognition**

Monomeric phases
Low phase density
Short chain lengths
Narrow pore diameters
High water composition
Elevated temperatures

Polymeric phases
High phase density
Long chain lengths
Wide pore diameters
High organic modifier composition
Subambient temperatures

Figure 1.11   Trends in shape selectivity in reversed-phase LC.

$= 0.65$, a heavily loaded polymeric phase (e.g., $\alpha_{TBN/BaP} = 0.4$) could be blended with a monomeric phase (e.g., $\alpha_{TBN/BaP} = 1.8$) in fractions appropriate to yield the desired selectivity. Wise et al. [34] have shown that columns prepared with blended silicas have retention behavior very similar to that of columns prepared with un-blended bonded silica with comparable selectivity. Unfortunately, neither of these approaches is readily accessible to practicing chromatographers for routine applications. A third, more easily utilized approach is column coupling [34]. When dissimilar columns are joined together, the selectivity is intermediate to the selectivity of individual columns. The resulting retention behavior is most easily predicted for isocratic separations. With gradient elution separations, solute retention is affected by column order, since the analytes will interact with each stationary phase under different mobile phase conditions. However, the most useful approach to controlling column selectivity for PAHs probably is to adjust the temperature at which the separation is performed. When using polymeric C_{18} phases for PAHs separations, small changes in temperature can have dramatic effects.

1.3 Retention Indices for PAHs

The reporting of LC retention data in a meaningful form remains a difficult problem. Because absolute retention is dependent on a variety of stationary phase, mobile phase, and operational parameters, capacity factors have limited utility in the expression of solute retention. However, if all chromatographic conditions are held constant including mobile phase composition and temperature, then a comparison of k' values for different solutes is valid for a specific column. If these criteria are met, a tabulation of k' values for a series of solutes should provide an indication of relative elution order and degree of separation. A better approach, however, is the normalization of retention relative to a standard(s) and the expression as a retention index. The retention index approach was popularized for GC with the introduction in 1958 of the Kovats retention index system, in which a series of homologous normal hydrocarbons served as the retention index standards [67].

An LC retention index system for PAHs, which was similar to Kovats indices for GC, was described by Popl et al. in 1974 for use in reporting normal phase LC retention data [68]. In this system, retention is expressed relative to PAH standards containing from one to five aromatic rings and assigned the following retention indices: benzene 10, naphthalene 100, phenanthrene 1000, benz[*a*]anthracene 10,000, and benzo[*b*]chrysene 100,000. As illustrated in Figure 1.12, each successive PAH standard represents the addition of a catacondensed aromatic ring in a regular manner. Popl et al. [9,68] reported the retention indices (*I*) as log (*I*) values using the following expression:

$$\log I_x = \frac{\log R_{x'} - \log R_{n'}}{\log R_{n'+1} - \log R_{n'}} \tag{1.1}$$

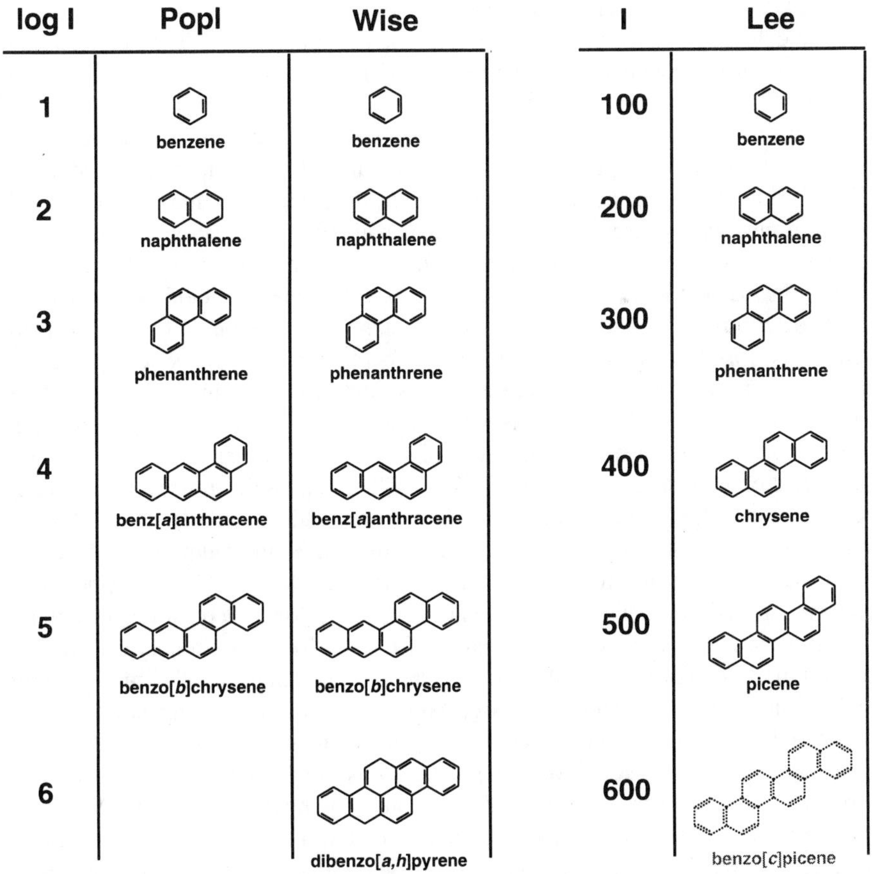

log I	Popl	Wise	I	Lee
1	benzene	benzene	100	benzene
2	naphthalene	naphthalene	200	naphthalene
3	phenanthrene	phenanthrene	300	phenanthrene
4	benz[*a*]anthracene	benz[*a*]anthracene	400	chrysene
5	benzo[*b*]chrysene	benzo[*b*]chrysene	500	picene
6		dibenzo[*a,h*]pyrene	600	benzo[*c*]picene

Figure 1.12 PAHs used as retention index standards and the assigned retention index *I* values.

where R is the corrected retention volume, x represents the solute, and n and $n + 1$ represent the lower and higher eluting PAH standards. Capacity factors can be used in place of corrected retention volumes with no change in the retention index. The five retention standards, benzene through benzo[b]chrysene, are assigned log I values of 1 through 5. Using this system, a retention index of log $I = 4.5$ indicates elution intermediate to benz[a]anthracene and benzo[b]chrysene.

In 1979 Lee et al. [69] developed a similar retention index system based on PAH standards for reporting GC data. Using the Kovats retention index system, Lee and co-workers found that the retention index for PAH solutes varied by as much as 20 index units when two columns with different stationary phase film thicknesses were compared [69]. To minimize these differences, Lee suggested the use of the angular catacondensed PAH series (see Figure 1.12) of benzene, naphthalene, phenanthrene, chrysene, and picene as the retention standards rather than the n-alkanes used in the Kovats system. These standards were assigned retention index values of 100 (benzene), 200 (naphthalene), 300 (phenanthrene), 400 (chrysene), and 500 (picene). Using this system, differences of only 0.14 index unit were observed on the two columns with different stationary phase film thicknesses. The Lee retention index system has found widespread use in GC for the reporting of PAH retention data [69–71].

As mentioned in Section 1.1, we have investigated the use of the retention indices described by Popl et al. [9] for comparing LC data on different columns [7,10]. This work, carried out on a number of different C_{18} columns, indicated a problem inherent to reversed-phase LC data, namely, that retention indices in this system vary with the type of column utilized. The primary utility of retention indices in LC is retention normalization. Retention indices should be similar for PAHs determined on columns with dissimilar absolute retention, but similar overall selectivity for the compounds of interest. For example, monomeric C_{18} phases prepared with low and high surface area substrates will differ in absolute retention behavior, but selectivity for PAHs will remain nearly constant. Thus, retention indices for PAHs determined with each column will be similar, even though absolute retention may differ considerably. Conversely, perhaps the greatest limitation in the use of retention indices in LC for PAHs is the tendency of variations to occur with changes in column selectivity. Because differences in column selectivity can result in changes in the relative retention of various classes of solutes (including changes in elution order), retention indices will necessarily depend on selectivity.

As discussed earlier, column selectivity for PAHs is most strongly affected by phase type (monomeric vs. polymeric), column temperature, bonded phase length and density, pore size, and to a lesser extent, mobile phase composition. Thus, retention indices can be compared only for solutes run on similar columns under the same chromatographic conditions. In practice, it is usually sufficient to specify C_{18} monomeric or polymeric phase chemistry, column temperature, and mobile phase composition to obtain somewhat consistent retention index values; however, more accurate comparisons could be obtained by comparing columns with similar shape selectivity (i.e., $\alpha_{TBN/BaP}$) values.

Ideally, the series of PAHs selected as the retention standards should provide a linear relationship between log retention and number of rings in the PAH standard. In normal phase LC both the Popl and the Lee standards provide nearly linear plots of log retention versus number of rings. The same is true in reversed-phase LC on monomeric C_{18} phases. However, on polymeric C_{18} phases both series of standards produce nonlinear plots with the Lee standards providing the most curvature. Therefore, we adopted the retention index system used by Popl et al. [9] to report retention data for PAHs [7]. In the original work of Popl et al. [9] a six-aromatic-ring PAH retention standard was not specified; however, we adopted the use of dibenzo[a,h]pyrene for this purpose. Obviously, dibenzo[a,h]pyrene does not represent the next member of the series of catacondensed PAHs used as retention standards. Because the choice of retention standards is somewhat arbitrary, however, dibenzo[a,h]pyrene was selected as the longest retained PAH of the six-ring peri-condensed PAHs studied [44]. Structures of the various standards used in Popl, Lee, and Wise retention index measurements are shown in Figure 1.12. The Lee retention index standards follow a logical progression of ring addition, and though not specified, the six-ring standard would be benzo[c]picene. For the retention data reported in this chapter, we employed the standards suggested by Popl, with the addition of dibenzo[a,h]pyrene for the six-ring standard. (At the time of these studies, benzo[c]picene was not available for use as the six-ring standard.) Interconversion of retention indices based on different standards is possible; however, absolute retention data for the individual standards must be available.

The data presented in this work are for typical monomeric and polymeric C_{18} columns operated at room temperature ($\approx 25°C$), with approximate selectivity values $\alpha_{TBN/BaP}$ of 1.7 and 0.7, respectively. Because of the wide range of compounds studied, retention was measured under different mobile phase conditions. Some of the compounds were chromatographed at two or more compositions and an average value for log I_x was calculated, although this value is less reliable than retention indices measured at specific mobile phase compositions.

1.4 Solute Shape Parameter

1.4.1 Development of the Length-to-Breadth (L/B) Ratio

The length-to-breadth parameter provides an indication of the overall two-dimensional shape of a molecule, and good correlations have been reported between L/B and both GC and LC retention for PAHs. One of the earliest observations concerning solute shape and chromatographic retention for PAHs was made by Janini and co-workers [72]. Using only a few solutes, these investigators observed that for GC on a liquid crystalline stationary phase, retention among isomeric PAHs increased with increasing ratio of the length to the width of the molecule as determined by a very rough estimate of these dimensions; that is, the more rodlike solutes were retained longer. Later Radecki et al. [73] expanded on the GC observation of Janini et al. [74] for the liquid crystalline phase and proposed a formal shape parameter defined as the ratio of the length to the breadth (L/B) of a box of minimum area drawn to enclose the atoms of the PAH molecule. Wise and co-workers [10] first reported the effect of PAH isomer shape on retention in reversed-phase LC. Wise et al. [10] not only extended the approach of Radecki et al. [73] to reversed-phase LC, but calculated L/B in a slightly different way: specifically, for a given PAH solute, the box was drawn about a planar representation of the molecule such that the ratio of the sides of the rectangle was maximized (see Figure 1.13a). Wise et al. [10] demonstrated that a better correlation of L/B with LC retention was obtained using a maximized L/B rather than a "minimized area" L/B are proposed by Radecki et al. [73], particularly for methyl-substituted PAHs (see reference 10 for examples of the differences between the two approaches).

In the initial paper by Wise et al. [10] the L/B ratio was calculated from a planar

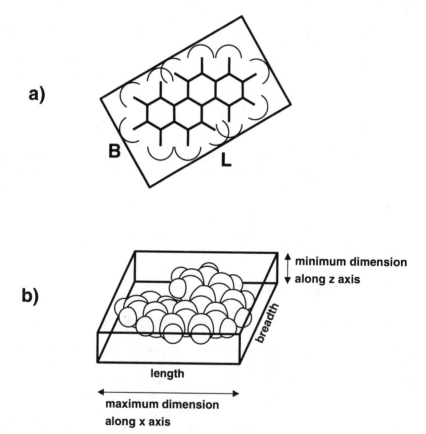

Figure 1.13 Approaches for specification of the length-to-breadth molecular descriptor. (a) The box defining the L/B ratio is drawn to enclose the van der Waals radii of the atoms, such that a maximum value for L/B is obtained (Original approach as described in reference 10.) (b) Calculation of L/B for nonplanar molecules. The molecule is oriented such that the minimum dimension (the thickness) is aligned with the z axis, and a box of maximum length is drawn to enclose the xy projection of the molecule (including van der Waals radii).

representation of the molecule using 120° bond angles and the following approximate bond lengths: C—C aromatic (1.4 Å), C—C aromatic to aliphatic (1.54 Å), and C—H (1.1 Å). The van der Waals radius of the hydrogen atom was 1.2 Å. This approach is illustrated in Figure 1.13a. Later compilations of L/B ratios were calculated using molecular dimensions generated from molecular modeling programs as described in the next section.

1.4.2 Current Method of Calculation of L/B

Over the past 15 years, we have published L/B values for PAHs calculated by various approaches. As the algorithm for calculation of L/B has been refined to accommodate three-dimensional shape, some of these values have changed slightly. Initial L/B values were based on simplistic two-dimensional structures with fixed bond angles and bond lengths as described in the preceding section. In a later approach, molecular structure was modeled with a commercial program (XICAMM, Xiris Co., New Monmouth, NJ), and two-dimensional projections of atom coordinates (with van der Waals radii) were used in calculation of L/B. Three-dimensional molecular orientation was set by the modeling program; however, this procedure was haphazard at best.

In our current implementation for the calculation of the L/B ratio, the molecular structure of each compound is generated using commercial molecular modeling programs (PC-Model and MMX, Serena Software, Bloomington, IN). For planar PAHs, L/B can be determined from a two-dimensional representation of the molecule. The structure is rotated and various trial values for L/B are calculated until a maximum value for L/B is determined. For nonplanar molecules, however, different values for L/B can result, depending on the initial orientation. If the molecule is only slightly nonplanar, this effect may be insignificant. For globular molecules (e.g., phenanthro[3,4-c]phenanthrene), however, suitable orientation is less straightforward and ambiguities are possible. Therefore, for the calculation of L/B for nonplanar PAHs, an iterative approach is used.

The algorithm starts with an arbitrary molecular orientation. The molecule is sequentially rotated about x, y, and z axes and xy, yz, and xz projections are determined with each rotation. An orientation is set such that when a box is drawn about the molecule to enclose the van der Waals surface, the minimum dimension is aligned with the z axis and the maximum dimension is aligned with the x axis (Figure 1.13b). L/B is then calculated from the xy projection as if the molecule were planar. This algorithm is also useful for planar molecules, since the final orientation will be with the plane of the molecule aligned with the xy plane. The z dimension is the "molecular thickness," and this parameter can be utilized as a descriptor for solute nonplanarity.

Using the current approach, the calculated L/B values are remarkably similar to those obtained in the original work, particularly for planar solutes; however, the L/B ratios calculated for PAHs with some nonplanarity are slightly higher. For example, the L/B value for benzo[c]phenanthrene increased from 1.22 [10] to 1.28; the value for dibenzo[c,g]phenanthrene increased from 1.12 [31] to 1.16; the value for dibenzo[b,g]phenanthrene increased from 1.33 [31] to 1.37; and the value for naphtho[2,3-j]fluoranthene increased from 1.55 [44] to 1.58. Recently, we compiled a comprehensive PAH structure index containing the structures of more than 600 PAHs with the calculated L/B ratios using the current approach [75]. The L/B values in the structure index [75] and this chapter represent the current approach and supersede all L/B values published earlier by our group [4,10,22,28,31,36–39,41,44,76].

1.5   Results and Discussion

1.5.1   *L/B* Ratio and Retention in LC

1.5.1.1   Unsubstituted PAHs

The retention indices on both polymeric and monomeric C_{18} phases and various molecular descriptors including L/B values are reported in Table 1.1 for 67 unsubstituted PAHs. The measurement uncertainty for log I values determined for a single column is estimated to be less than 0.02 unit; however, difference in log I values between columns can be expected to be larger. In Table 1.1 the PAHs are grouped according to molecular mass, and the structures are identified as catacondensed (carbon atom shared by only two aromatic rings) or pericondensed (carbon atoms shared by three aromatic rings). Within molecular mass groups for pericondensed structures, PAHs are subdivided into groups with only six-membered rings, PAHs containing a five-membered ring with five shared carbon atoms (fluoranthene structures), and PAHs with a five-membered ring with three shared carbon atoms (see the isomers of molecular mass 252 for an example of these subdivisions). Within each group, the PAHs are listed in order of increasing L/B values. The PAHs are grouped according to molecular mass because the correlation between the L/B ratio and the retention is valid only for isomeric PAHs. Table 1.1 also includes values for the molecular dimensions of each PAH (i.e., length and width) used to calculate the L/B value, and the thickness, which is a measure of the nonplanarity (see discussion below).

As discussed earlier, the polymeric C_{18} phase exhibits enhanced shape recognition compared to the monomeric C_{18} phase. Thus, the discussions that follow focus primarily on the retention behavior of the PAH isomers on polymeric C_{18} phases. In the original study of Wise et al. [10], LC retention and L/B values were correlated for 31 unsubstituted PAHs in nine isomeric groups with the largest groups containing five or six isomers. Later studies expanded the number of isomers to include all 12 possible catacondensed five-ring isomers of molecular mass 278 [31] and 19 pericondensed six-ring isomers of molecular mass 302 [44].

The correlation between L/B ratio and LC retention was first observed for the catacondensed three-ring, four-ring, and five-ring isomers of molecular mass 178, 228, and 278, respectively. The PAH isomer retention was found to increase as the L/B ratio increased (i.e., as the solutes become more rodlike). For the three-ring isomers, phenanthrene elutes prior to anthracene as predicted by the L/B ratios. The

Table 1.1 RETENTION INDICES AND SHAPE PARAMETERS FOR UNSUBSTITUTED PAHs

| Compound | Molecular Mass | Structure[a] | Molecular Dimensions (Å) | | | | Log I | |
			Length	Width	Thickness	L/B	Polymeric Phase[b]	Monomeric Phase[c]
Two Rings								
Naphthalene	128	C6	9.19	7.43	3.88	1.24	2.00	2.00
Three Rings								
Phenanthrene	178	C6	11.75	8.03	3.89	1.46	3.00	3.00
Anthracene	178	C6	11.65	7.44	3.89	1.57	3.16	3.11
Four Rings								
Pyrene	202	P6	11.66	9.28	3.89	1.26	3.55	3.63
Fluoranthene	202	PF5	11.16	9.24	3.88	1.21	3.39	3.43
Acephenanthrylene	202	PF4	11.70	9.07	3.89	1.29	3.37	3.38
Aceanthrylene	202	PF4	11.70	8.64	3.88	1.35	3.39	3.39
7H-Benzo[c]fluorene	216	CF4	12.65	9.14	4.24	1.38	3.49	3.64
11H-Benzo[a]fluorene	216	CF4	13.39	8.03	4.24	1.67	3.81	3.75
11H-Benzo[b]fluorene	216	CF4	13.76	7.78	4.24	1.77	3.82	3.78
Triphenylene	228	C6	11.68	10.44	4.37	1.12	3.75	3.82
Benzo[c]phenanthrene	228	C6	11.91	9.32	4.99	1.28	3.69	3.91
Benz[a]anthracene	228	C6	13.94	8.72	3.89	1.60	4.00	4.00
Chrysene	228	C6	13.94	8.04	3.93	1.73	4.06	3.97
Naphthacene	228	C6	14.12	7.45	3.88	1.90		
Five Rings								
Benzo[g,h,i]fluoranthene	226	PF5	11.40	9.66	3.88	1.18	3.95	4.07
Cyclopenta[cd]pyrene	226	PF4	11.52	9.60	3.89	1.20	3.94	3.95
Benzo[e]pyrene	252	P6	11.76	10.52	3.89	1.12	4.29	4.51

(Continued)

Table 1.1 (*Continued*)

Compound	Molecular Mass	Structure[a]	Molecular Dimensions (Å)			L/B	Log I	
			Length	Width	Thickness		Polymeric Phase[b]	Monomeric Phase[c]
Perylene	252	P6	11.80	9.25	3.91	1.28	4.33	4.52
Benzo[a]pyrene	252	P6	13.88	9.30	3.89	1.49	4.51	4.68
Benz[a]aceanthrylene	252	PF5	12.72	10.77	3.89	1.18	4.24	4.45
Benz[e]acephenanthrylene	252	PF5	13.82	9.96	3.88	1.39	4.29	4.44
Benzo[j]fluoranthene	252	PF5	13.41	9.58	3.94	1.40	4.26	4.37
Benzo[k]fluoranthene	252	PF5	13.62	9.24	3.89	1.47	4.38	4.50
Benz[e]aceanthrylene	252	PF4	13.97	9.78	3.88	1.43	4.25	4.34
Benz[l]aceanthrylene	252	PF4	13.57	9.17	4.53	1.48	4.26	4.38
Benz[j]aceanthrylene	252	PF4	13.88	9.11	3.89	1.52	4.26	4.29
Benz[k]acephenanthrylene	252	PF4	14.03	9.15	3.90	1.53	4.39	4.43
13H-Dibenzo[a,g]fluorene	266	CF4	14.97	9.16	4.29	1.63	4.53	4.62
11H-Indeno[2,1-a]phenanthrene	266	CF4	15.56	8.05	4.24	1.93	4.91	4.74
13H-Dibenzo[a,h]fluorene	266	CF4	15.83	8.03	4.24	1.97	4.96	4.77
Dibenzo[c,g]phenanthrene	278	C6	11.82	10.15	6.24	1.16	4.07	4.51
(Dibenz[a,c]anthracene	278	C6	13.92	11.25	3.89	1.24	4.40	4.73
Benzo[g]chrysene	278	C6	13.78	10.49	5.32	1.31	4.27	4.71
Dibenzo[b,g]phenanthrene	278	C6	13.91	10.13	5.23	1.37	4.33	4.8
Benzo[c]chrysene	278	C6	14.19	9.34	5.39	1.53	4.45	4.85
Dibenz[a,j]anthracene	278	C6	14.54	9.50	3.89	1.53	4.56	4.84
Pentaphene	278	C6	16.09	9.21	3.89	1.75	4.67	4.96
Benzo[a]naphthacene	278	C6	16.22	9.00	3.89	1.80	4.99	4.5
Dibenz[a,h]anthracene	278	C6	15.90	8.73	3.89	1.82	4.73	4.86
Benzo[b]chrysene	278	C6	16.26	8.76	3.89	1.86	5.00	5.00
Picene	278	C6	16.11	8.04	3.90	2.00	5.18	5.02

Six Rings

Compound	Mass	Structure						
Benzo[ghi]perylene	276	P6	11.78	10.48	3.89	1.12	4.76	5.36
Dibenzo[def,mno]chrysene	276	P6	12.90	9.59	3.88	1.35	5.08	5.61
Indeno[1,2,3-cd]pyrene	276	PF5	13.78	9.93	3.88	1.39	4.84	5.23
Indeno[1,2,3-cd]fluoranthene	276	PF5	14.75	9.12	3.89	1.62	4.93	5.05
Dibenzo[def,p]chrysene	302	P6	13.69	11.70	5.17	1.17	4.65	5.57
Benzo[a]perylene	302	P6	13.65	11.59	5.47	1.18	4.93	5.54
Naphtho[1,2,3,4-def]chrysene	302	P6	13.90	11.21	4.19	1.24	4.97	5.56
Dibenzo[de,qr]naphthacene	302	P6	14.37	11.16	4.41	1.29	4.91	5.53
Dibenzo[fq,op]naphthacene	302	P6	13.98	10.51	3.89	1.33	5.04	5.52
Benzo[b]perylene	302	P6	14.38	10.30	4.84	1.40	5.04	5.56
Naphtho[2,1,8-qra]naphthacene	302	P6	16.23	9.59	3.89	1.69	5.86	5.92
Benzo[rst]pentaphene	302	P6	16.09	9.29	3.89	1.73	5.74	5.93
Dibenzo[a,h]pyrene	302	P6	16.08	9.29	3.90	1.73	6.00	6.00
Dibenzo[j,l]fluoranthene	302	PF5	13.13	11.53	4.41	1.14	4.79	5.35
Dibenzo[b,e]fluoranthene	302	PF5	13.76	12.09	4.64	1.14	4.80	5.48
Dibenz[a,e]aceanthrylene	302	PF5	13.97	12.14	3.90	1.15	4.90	5.50
Naphth[2,3-a]aceanthrylene	302	PF5	14.62	11.51	3.89	1.27	4.91	5.51
Idenol[1,2,4-fg]naphthacene	302	PF5	14.11	11.13	3.97	1.27	5.07	5.71
Dibenzo[b,k]fluoranthene	302	PF5	16.16	10.31	3.90	1.57	5.27	5.64
Naphtho[2,3-j]fluoranthene	302	PF5	15.51	9.81	3.93	1.58	4.98	5.40
Dibenz[e,k]acephenanthrylene	302	PF5	16.20	10.07	3.89	1.61	5.28	5.59
Naphtho[1,2-k]fluoranthene	302	PF5	15.52	9.47	3.91	1.64	5	5.34
Naphtho[2,3-k]fluoranthene	302	PF5	16.07	9.24	3.90	1.74	5.92	5.79
Phenanthro[3,4-c]phenanthrene	328	C6	11.86	11.05	7.41	1.07	4.21	4.77
Dibenzo[g,p]chrysene	328	C6	13.55	12.38	6.15	1.09	4.45	5.53

[a] C6, catacondensed structure with only six-membered rings; P6, pericondensed structure with only six-membered rings; PF4, pericondensed structure containing a five-membered ring with four shared carbons; PF5, pericondensed structure containing a five-membered ring with five shared carbons (i.e., a fluoranthenic structure); CF4, catacondensed structure with a five-membered ring with four shared carbons.

[b] Polymeric C_{18} phase with $\alpha_{TBN/BaP} \approx 0.7$.

[c] Monomeric C_{18} phase with $\alpha_{TBN/BaP} \approx 1.7$.

four-ring isomers elute in order of L/B ratio with the exception of benzo[c]phenanthrene, which elutes earlier than expected as a result of nonplanarity (see discussion below).

The most impressive correlation of shape and elution order is observed for the five-ring catacondensed isomers of molecular mass 278 as shown in Figure 1.4. This group is the largest group of unsubstituted catacondensed PAH isomers investigated and also represents the isomer group with the largest range of L/B values (1.16–2.00). (Pentacene, the remaining isomer in this group, was not included in the chromatogram because its solubility in the mobile phase is limited; however, it does elute significantly later than picene on a polymeric C_{18} phase.) For the 11 isomers, only dibenz[a,c]anthracene ($L/B = 1.24$) appears to elute out of order based on L/B ratio. Good correlation between L/B values and elution order is observed for the two remaining groups of catacondensed isomers (molecular mass 216 and 266), which both contain a fluorene structure; however, only three isomers were available in each group.

For the pericondensed PAH isomers the correlation between retention (elution order) and shape (L/B ratio) is strongest when the isomers are subdivided into groups based on whether a five-membered ring is present. The isomers of molecular mass 252 can be divided into three subgroups based on whether only six-membered rings are present (3 isomers), a five-membered ring with five shared carbons (a fluoranthene structure) is present (4 isomers), or a five-membered ring with three shared carbons is present (4 isomers). When each subgroup is considered separately, the elution order correlates well with increasing L/B values.

A similar situation is observed for the 19 pericondensed six-ring isomers of molecular mass 302, which can be divided into two groups: isomers containing only six-membered rings (9 dibenzopyrene and benzoperylene isomers) and isomers containing a five-membered ring in a fluoranthene structure (10 dibenzofluoranthene isomers). Figure 1.14 gives two plots of L/B value versus Log I for the dibenzopyrene/benzoperylene isomers and the dibenzofluoranthene isomers. Correlation coefficients of 0.972 and 0.746 were found for the dibenzopyrenes/benzoperylenes and the dibenzofluoranthenes, respectively, compared to a correlation coefficient of 0.846 for all 19 isomers combined.

1.5.1.2 Methyl-Substituted PAHs

Methyl-substituted PAHs provide large groups of isomers for investigating the relationship of shape and chromatographic retention. The position of the methyl group on the PAH can significantly change the L/B value. In fact, it was the observation of the excellent selectivity of the polymeric C_{18} phases for the separation of five of the six possible methylchrysene isomers (see Figure 1.15) that initiated our investigations of the effect of shape on LC retention. The L/B values and the LC retention indices are summarized in Table 1.2 for 84 methyl-substituted anthracenes, phenanthrenes, fluoranthenes, pyrenes, benz[a]anthracenes, benzo[c]phenanthrenes, chrysenes, triphenylenes, benzo[a]pyrenes, benzo[b]fluoranthenes, perylenes, and picenes (see Figure 1.16 for structures and substitution numbering).

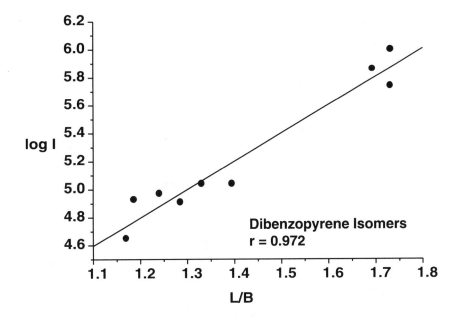

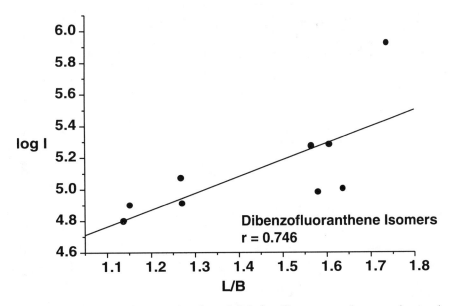

Figure 1.14   Log I plotted as a function of L/B for dibenzopyrene/benzoperylene and dibenzofluoranthene isomers of molecular mass 302.

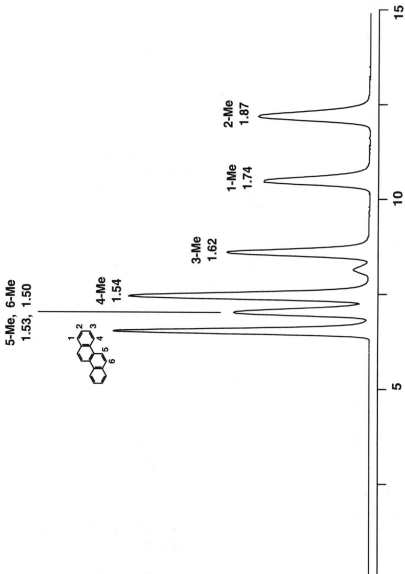

Figure 1.15 Reversed-phase LC separation of methylchrysene isomers on polymeric C_{18} phase; L/B values for methyl isomers are indicated above each peak.

Table 1.2 RETENTION INDICES AND SHAPE PARAMETERS FOR SELECTED METHYL-SUBSTITUTED PAHs

Compound	Molecular Mass	Dihedral Angle (deg)	Molecular Dimensions (Å)				Log I	
			Length	Width	Thickness	L/B	Polymeric Phase[a]	Monomeric Phase[b]
Anthracene	178						3.16	3.11
9-Methylanthracene	192		11.70	8.58	4.14	1.36	3.41	3.52
1-Methylanthracene	192		11.99	8.40	4.21	1.43	3.43	3.57
2-Methylanthracene	192		12.74	7.45	4.20	1.71	3.69	3.69
Phenanthrene	178	*def*					3.00	3.00
9-Methylphenanthrene	192	0	11.74	9.28	4.22	1.27	3.38	3.51
4-Methylphenanthrene	192	−19.5	11.70	9.07	4.69	1.29	3.26	3.40
3-Methylphenanthrene	192	0.3	11.82	8.70	4.20	1.36	3.34	3.47
1-Methylphenanthrene	192	0	11.83	8.03	4.22	1.47	3.40	3.50
2-Methylphenanthrene	192	0	12.86	8.02	4.20	1.60	3.68	3.71
Fluoranthene	202	*ghi* *mno*					3.39	3.43
1-Methylfluoranthene	216	−0.1	11.46	10.13	4.22	1.13	3.73	3.87
7-Methylfluoranthene	216	0	11.20	9.24	4.22	1.21	3.80	3.91
3-Methylfluoranthene	216	0	12.29	9.25	4.20	1.33	3.86	3.91
8-Methylfluoranthene	216	0	12.38	9.23	4.20	1.34	3.85	3.95
Pyrene	202						3.53	3.63
4-Methylpyrene	216		11.65	10.53	4.21	1.11	3.98	4.13
1-Methylpyrene	216		11.66	9.29	4.20	1.25	3.98	4.15
2-Methylpyrene	216		12.77	9.27	4.20	1.38	4.04	4.21
Benz[a]anthracene	228	*pqr*					4.00	4.00
6-Methylbenz[a]anthracene	242	6.9	13.92	9.85	4.21	1.41	4.15	4.41
5-Methylbenz[a]anthracene	242	0.2	13.93	9.59	4.22	1.45	4.28	4.48
11-Methylbenz[a]anthracene	242	0	13.61	9.21	4.22	1.48	4.17	4.36
2-Methylbenz[a]anthracene	242	0	13.92	9.19	4.21	1.51	4.14	4.43
1-Methylbenz[a]anthracene	242	−24.7	13.76	9.01	5.04	1.53	4.18	4.39

(Continued)

Table 1.2 (*Continued*)

Compound	Molecular Mass	Dihedral Angle (deg)			Molecular Dimensions (Å)				Log I	
					Length	Width	Thickness	L/B	Polymeric Phase[a]	Monomeric Phase[b]
7-Methylbenz[a]anthracene	242	11.2			13.90	9.00	4.57	1.54	4.17	4.36
12-Methylbenz[a]anthracene	242	−26.7			13.82	8.96	5.16	1.54	4.14	4.37
10-Methylbenz[a]anthracene	242	−0.5			14.28	8.97	4.20	1.59	4.18	4.42
8-Methylbenz[a]anthracene	242	0.5			13.94	8.71	4.21	1.60	4.21	4.40
4-Methylbenz[a]anthracene	242	0.6			14.41	8.71	4.22	1.65	4.33	4.44
3-Methylbenz[a]anthracene	242	0.1			15.05	8.71	4.21	1.73	4.39	4.51
9-Methylbenz[a]anthracene	242	0.1			15.04	8.70	4.20	1.73	4.37	4.52
Benzo[c]phenanthrene	228	opq	pqr						3.69	3.91
6-Methylbenzo[c]phenanthrene	242	−25.6	−24.6		11.79	10.32	5.05	1.14	4.17	4.37
2-Methylbenzo[c]phenanthrene	242	−23.6	−24		11.97	9.97	5.18	1.20	3.94	4.29
1-Methylbenzo[c]phenanthrene	242	−22.9	−32		11.71	9.28	5.94	1.26	3.73	4.10
5-Methylbenzo[c]phenanthrene	242	−23.2	−25.2		11.84	9.47	5.28	1.25	4.04	4.37
4-Methylbenzo[c]phenanthrene	242	−22.6	−26.2		12.87	9.34	5.39	1.38	4.04	4.37
3-Methylbenzo[c]phenanthrene	242	−23.8	−23.9		13.04	9.36	5.00	1.39	4.09	4.41
Chrysene	228	def	mno						4.06	3.97
6-Methylchrysene	242	−0.6	−0.8		13.92	9.27	4.22	1.50	4.17	4.35
5-Methylchrysene	242	−24	−13.5		13.86	9.05	4.97	1.53	4.17	4.36
4-Methylchrysene	242	−23.4	−13.1		13.72	8.91	5.01	1.54	4.20	4.36
3-Methylchrysene	242	−0.7	−0.7		14.13	8.73	4.20	1.62	4.28	4.41
1-Methylchrysene	242	−0.7	−0.7		14.00	8.03	4.22	1.74	4.39	4.46
2-Methylchrysene	242	−0.6	−0.7		15.05	8.03	4.20	1.87	4.49	4.54
Triphenylene	228	def	jkl	pqr					3.75	3.82
1-Methyltriphenylene	242	15.5	9.6	27.4	11.53	10.40	5.19	1.11	3.88	4.19
Benzo[a]pyrene	252	mno							4.51	4.68
12-Methylbenzo[a]pyrene	266	0			13.86	10.52	4.21	1.32	4.61	5.14
5-Methylbenzo[a]pyrene	266	0			13.88	10.54	4.22	1.32	4.64	5.15

						cde	ijk	qrs	
5.26	4.84	1.34	4.21	9.68	12.97	0			266	4-Methylbenzo[a]pyrene
5.14	4.66	1.35	4.89	9.97	13.46	−18.4			266	11-Methylbenzo[a]pyrene
5.16	4.73	1.43	4.98	9.56	13.65	−18.1			266	10-Methylbenzo[a]pyrene
5.18	4.78	1.47	4.20	9.57	14.11	0			266	9-Methylbenzo[a]pyrene
5.11	4.73	1.50	4.36	9.27	13.86	1.8			266	6-Methylbenzo[a]pyrene
5.24	4.83	1.49	4.20	9.28	13.86	0.1			266	1-Methylbenzo[a]pyrene
5.13	4.74	1.50	4.22	9.29	13.91	0			266	7-Methylbenzo[a]pyrene
5.25	4.90	1.51	4.22	9.29	14.03	0			266	3-Methylbenzo[a]pyrene
5.33	4.94	1.61	4.21	9.29	14.99	0			266	2-Methylbenzo[a]pyrene
5.33	4.96	1.61	4.20	9.29	15.00	0			266	8-Methylbenzo[a]pyrene
5.54	4.90	1.33	4.20	10.45	13.86	−2.8			290	4,5-Dimethylbenzo[a]pyrene
5.77	5.08	1.33	4.22	10.53	13.99	0.5			290	1,4-Dimethylbenzo[a]pyrene
5.62	4.94	1.34	4.22	10.75	14.43	0			290	3,12-Dimethylbenzo[a]pyrene
5.51	4.82	1.45	5.23	9.43	13.65	−21			290	7,10-Dimethylbenzo[a]pyrene
5.68	5.01	1.48	4.92	9.55	14.10	19.3			290	3,11-Dimethylbenzo[a]pyrene
5.63	5.09	1.51	4.46	9.27	14.00	2.3			290	3,6-Dimethylbenzo[a]pyrene
5.75	5.26	1.51	4.22	9.29	13.99	−0.5			290	1,3-Dimethylbenzo[a]pyrene
5.65	5.10	1.51	4.51	9.27	13.99	−4.5			290	1,6-Dimethylbenzo[a]pyrene
5.73	5.21	1.61	4.21	9.29	15.00	0.8			290	1,2-Dimethylbenzo[a]pyrene
5.74	5.37	1.62	4.21	9.29	15.01	0.5			290	2,3-Dimethylbenzo[a]pyrene
4.44	4.29								252	**Benzo[b]fluoranthene**
	4.48	1.25	4.21	11.08	13.82	0	0	0	266	2-Methylbenzo[b]fluoranthene
	4.42	1.30	4.22	10.41	13.53	0	0	0	266	9-Methylbenzo[b]fluoranthene
		1.30	4.21	10.44	13.56	0.5	−6.9	−1.2	266	8-Methylbenzo[b]fluoranthene
	4.50	1.32	4.22	10.42	13.72	0	−0.4	−0.2	266	7-Methylbenzo[b]fluoranthene
	4.64	1.36	4.22	10.17	13.80	0.3	0	−0.1	266	3-Methylbenzo[b]fluoranthene
	4.61	1.37	4.33	10.06	13.81	0	0.9	9.7	266	1-Methylbenzo[b]fluoranthene
		1.38	4.22	9.96	13.80	−0.4	−0.1	0.1	266	4-Methylbenzo[b]fluoranthene
	4.60	1.39	4.61	9.96	13.82	1.3	−2.2	−11.1	266	12-Methylbenzo[b]fluoranthene
		1.40	4.21	10.21	14.35	0	0	0	266	6-Methylbenzo[b]fluoranthene
	4.63	1.44	4.21	10.08	14.51	0	−0.1	0	266	10-Methylbenzo[b]fluoranthene
		1.50	4.20	9.95	14.92	0	0	0	266	5-Methylbenzo[b]fluoranthene
	4.63	1.50	4.21	9.96	14.90	0	0	0	266	11-Methylbenzo[b]fluoranthene

(Continued)

Table 1.2 (Continued)

Compound	Molecular Mass	Dihedral Angle (deg)			Molecular Dimensions (Å)				Log l	
		ghi	pqr	tuv	Length	Width	Thickness	L/B	Polymeric Phase[a]	Monomeric Phase[b]
Perylene	252	*ghi*	*pqr*						4.33	4.52
1-Methylperylene	266	17.6	29		11.78	10.18	5.27	1.16	4.28	4.85
2-Methylperylene	266	−8.6	−8.5		12.46	10.10	4.46	1.23	4.50	5.03
3-Methylperylene	266	−10.9	−10.5		12.90	9.26	4.44	1.40	4.69	4.57
Picene	278	*ghi*	*pqr*	*tuv*					5.18	5.02
5-Methylpicene	292	21	12	15.5	16.05	9.17	5.09	1.75	4[c]	
6-Methylpicene	292	−31.3	−15.5	−16.3	15.95	8.98	5.56	1.78	2	
13-Methylpicene	292	17	28	−6.4	15.92	8.96	5.42	1.78	1	
1-Methylpicene	292	−15.4	7.3	−28.6	15.90	9.00	5.39	1.77	3	
2-Methylpicene	292	0	0	0	16.05	8.91	4.20	1.80	5	
4-Methylpicene	292	−14.6	−15.2	6.6	16.06	7.98	4.51	2.01	6	
3-Methylpicene	292	−0.1	0	−0.1	17.22	8.03	4.20	2.15	7	

[a] Polymeric C_{18} phase with $\alpha_{TBN/BaP} \approx 0.7$.

[b] Monomeric C_{18} phase with $\alpha_{TBN/BaP} \approx 1.7$.

[c] Elution order of methylpicenes on polymeric C_{18} phase.

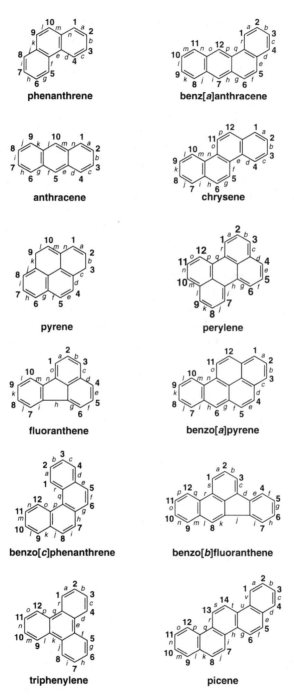

Figure 1.16 Structures and substitution position numbering for methyl-substituted PAHs investigated.

For the methyl-substituted anthracene, phenanthrene, fluoranthene, and pyrene isomers (3–5 isomers per set), the elution order generally follows increasing L/B ratio with 3-methylphenanthrene as the only major exception (see Table 1.2). The retention (expressed as log I) of the 25 methyl-substituted four-ring isomers (benzo[c]phenanthrene, benz[a]anthracene, triphenylene, and chrysene) is strongly correlated with L/B ratio, as illustrated in Figure 1.17 (correlation coefficient 0.848). The six methylchrysene isomers represent one of the best correlations of elution order and L/B ratio as shown in Figure 1.15 (correlation coefficient 0.995).

For the methylbenz[a]anthracenes, the correlation between log I and L/B ratio is less than the other PAHs studied (correlation coefficient 0.745); however, the five isomers with the largest L/B values do have the greatest retention. Only the elution of the 5-methyl isomer differs significantly from that predicted by the L/B ratio. A close examination of the data in Figure 1.17 indicates that most of the points that do not fit the linear relation between retention and L/B represent nonplanar isomers (see later discussion on nonplanarity effects). In addition, for the methyl-substituted PAHs, the range of L/B ratios for each compound is relatively small (e.g., 0.24–0.37 for the molecular mass 228 isomers).

The retention indices and the L/B values for the 12 isomers of methylbenzo[a]pyrene and 12 isomers of methylbenzo[b]fluoranthene are summarized in Table 1.2, and separations of these isomers are shown in Figures 1.18 and 1.19, respectively. The correlation coefficient for the retention indices versus the L/B ratio for the 12 methylbenzo[a]pyrene isomers is 0.806, and for the methylbenzo[b]fluoranthene isomers, the correlation coefficient is 0.784.

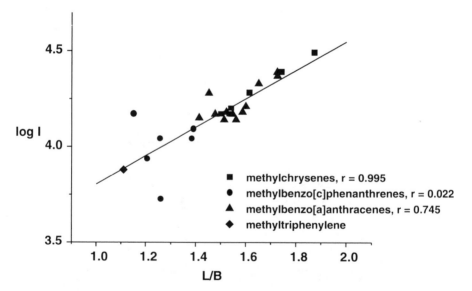

Figure 1.17 Log I plotted as a function of L/B for methylbenzo[c]phenanthrenes, methyltriphenylenes, methylbenz[a]anthracenes, and methylchrysenes.

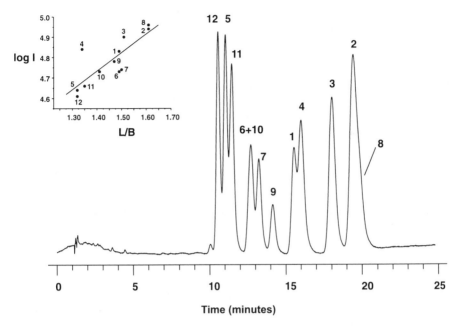

Figure 1.18 Reversed-phase LC separation of methylbenzo[*a*]pyrene isomers on polymeric C_{18} phase; position of methyl substitution is indicated above each peak.

1.5.1.3 Sulfur Heterocycles

Polycyclic aromatic heterocycles are fused aromatic ring compounds closely related to PAHs, but with one or more heteroatoms in a ring. Common classes of heterocycles include compounds with an oxygen atom (furans), compounds with a nitrogen atom (quinolines, carbazoles, and pyridines), and compounds containing a sulfur atom (thiophenes). As with PAHs and alkyl-substituted PAHs, heterocycles constitute classes of compounds with considerable isomeric complexity. The total number of possible isomeric structures for heterocycles is greatly increased compared with the corresponding hydrocarbon analogues because both ring arrangement and position of heteroatom substitution within the rings give rise to unique isomers. In practice, only a small fraction of these compounds are found in environmental samples; however, sufficient complexity exists to make separations challenging.

As with PAHs and methyl-substituted PAHs, heterocyclic isomers provide an opportunity to study structure–retention relations in LC. Wise et al. [77] investigated the retention behavior of about 30 different polycyclic aromatic sulfur heterocycles (PASHs) on monomeric and polymeric C_{18} columns, as well as an aminopropyl column operated in the normal phase mode. Three isomer sets were studied: catacondensed four-ring compounds (molecular mass 234), pericondensed five-ring compounds (molecular mass 258), and catacondensed five-ring com-

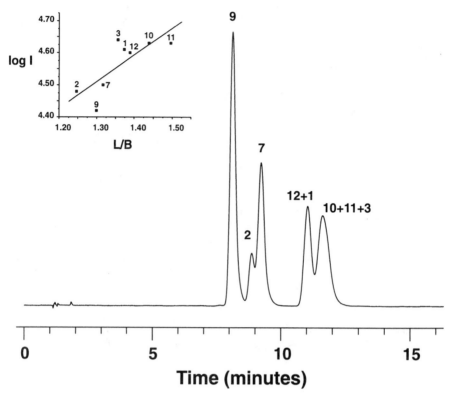

Figure 1.19 Reversed-phase LC separation of methylbenzo[*b*]fluoranthene isomers on polymeric C_{18} phase; position of methyl substitution is indicated above each peak.

pounds (molecular mass 284). Retention index data for these compounds are listed in Table 1.3. Although the relationship between solute shape and retention was not explored in this early work [77], we subsequently determined L/B and other shape descriptors for these PASHs, and these data are provided in Table 1.3.

For the PASHs the correlation of log I with L/B is poor compared with similar correlations for the PAHs, particularly for the four-ring catacondensed PASH. A better correlation exists for the pericondensed and catacondensed five-ring PASHs, although the correlation coefficients are still low ($r = 0.701$ and 0.755, respectively). The low correlation for the PASHs compared to the PAHs may be due to the strong influence of the position of the thiophenic ring in the structure, particularly if the thiophene ring is an external ring (two shared carbons) rather than an internal ring (four shared carbons). The pericondensed five-ring isomers are all planar, whereas many of the catacondensed five-ring isomers are nonplanar, and it appears that nonplanarity does not influence retention for the PASHs as strongly as for the PAHs. These data indicate that molecular shape provides a smaller contribution to retention for PASHs than for PAHs and suggest that other factors, as yet undetermined, are more important to the description of retention relationships for PASHs.

Table 1.3 RETENTION INDICES AND SHAPE PARAMETERS FOR SELECTED POLYCYCLIC AROMATIC SULFUR HETEROCYCLES

Compound	Molecular Mass	Structure[a]	Molecular Dimensions (Å)				Log I	
			Length	Width	Thickness	L/B	Polymeric Phase[b]	Monomeric Phase[c]
Four Rings								
Phenanthro[9,10-b]thiophene	234	CT2	11.76	10.47	4.07	1.12	3.77	3.86
Phenanthro[3,4-b]thiophene	234	CT2	12.10	9.26	4.68	1.31	3.57	3.70
Phenanthro[4,3-b]thiophene	234	CT2	12.09	9.18	4.68	1.32	3.79	3.86
Anthra[1,2-b]thiophene	234	CT2	13.31	8.66	4.07	1.54	3.99	3.93
Phenanthro[2,3-b]thiophene	234	CT2	13.38	8.65	4.06	1.55	3.65	3.70
Phenanthro[3,2-b]thiophene	234	CT2	13.66	8.66	4.07	1.58	3.67	3.67
Anthra[2,1-b]thiophene	234	CT2	13.47	8.50	4.07	1.58	3.78	3.74
Phenanthro[2,1-b]thiophene	234	CT2	13.30	8.04	4.07	1.65	3.80	3.71
Phenanthro[1,2-b]thiophene	234	CT2	13.49	8.08	4.19	1.67	4.08	3.93
Anthra[2,3-b]thiophene	234	CT2	14.07	7.47	4.13	1.88	3.41	2.97
Benzo[b]naphtho[1,2-d]thiophene	234	CT4	12.52	9.23	4.43	1.36	3.79	4.02
Benzo[b]naphtho[2,3-d]thiophene	234	CT4	13.85	8.18	4.06	1.69	4.05	4.05
Benzo[b]naphtho[2,1-d]thiophene	234	CT4	13.65	8.09	4.10	1.69	4.20	4.24
Five Rings								
Pyreno[4,5-b]thiophene	258	PT2	11.39	10.56	4.08	1.08	4.31	4.55
Benzo[1,2]phenaleno[3,4-bc]thiophene	258	PT2	11.83	10.44	4.06	1.13	4.35	4.62
Benzo[1,2]phenaleno[4,3-bc]thiophene	258	PT2	11.82	10.46	4.06	1.13	4.16	4.81
Pyreno[2,2-b]thiophene	258	PT2	13.22	9.29	4.07	1.42	4.35	4.44
Pyreno[1,2-b]thiophene	258	PT2	13.43	9.30	4.06	1.44	4.45	4.62
Benzo[4,5]phenaleno[1,9-bc]thiophene	258	PT3	13.27	9.31	4.07	1.43	4.37	4.48
Benzo[4,5]phenaleno[9,1-bc]thiophene	258	PT3	13.29	9.31	4.10	1.43	4.36	4.51
Triphenyleno[4,5-bcd]thiophene	258	PT4	11.77	10.55	4.08	1.12	4.30	4.61
Benzo[2,3]phenanthro[4,5-bcd]thiophene	258	PT4	13.60	9.30	4.08	1.46	4.65	4.98
Chryseno[4,5-bcd]thiophene	258	PT4	13.75	9.24	4.07	1.49	4.51	4.77

(*Continued*)

Table 1.3 (*Continued*)

Compound	Molecular Mass	Structure[a]	Molecular Dimensions (Å)					Log I	
			Length	Width	Thickness	L/B		Polymeric Phase[b]	Monomeric Phase[c]
Triphenyleno[2,3-b]thiophene	284	CT2	13.57	11.13	4.50	1.22	4.30	4.61
Triphenyleno[2,1-b]thiophene	284	CT2	13.18	10.51	5.03	1.25	4.21	4.53
Triphenyleno[1,2-b]thiophene	284	CT2	13.47	10.49	4.97	1.28	4.40	4.68
Benzo[b]phenanthro[4,3-d]thiophene	284	CT4	12.28	10.17	5.57	1.21	4.18	4.80
Dinaphtho[2,1-b:1',2'-d]thiophene	284	CT4	12.44	10.17	5.33	1.22		
Benzo[b]phenanthro[9,10-d]thiophene	284	CT4	13.62	10.56	4.59	1.29	4.68	5.17
Benzo[b]phenanthro[3,4-d]thiophene	284	CT4	14.19	9.59	4.75	1.48	4.81	5.21
Benzo[b]phenanthro[3,2-d]thiophene	284	CT4	14.94	9.35	4.06	1.60	4.52	4.84
Dinaphtho[1,2-b:1',2'-d]thiophene	284	CT4	14.88	9.25	4.46	1.61	4.95	5.27
Anthra[1,2-b]benzo[d]thiophene	284	CT4	15.76	8.72	4.10	1.81	>5 (1)[d]	5.25
Dinaphtho[2,3-b:2',3'-d]thiophene	284	CT4	16.17	8.89	4.09	1.82	4.84	5.12
Dinaphtho[1,2-b:2',1'-d]thiophene	284	CT4	15.56	8.53	4.11	1.82	>5 (2)[d]	5.51
Benzo[b]phenanthro[2,3-d]thiophene	284	CT4	15.87	8.69	4.08	1.83	4.74	4.91
Dinaphtho[1,2-b:2',3'-d]thiophene	284	CT4	16.04	8.21	4.10	1.95	>5 (4)[d]	5.38
Benzo[b]phenanthro[2,1-d]thiophene	284	CT4	15.86	8.10	4.07	1.96	>5 (3)[d]	5.32

[a] CT2, catacondensed structure containing a thiophene ring with two shared carbons; CT4, catacondensed structure containing a thiophene ring with four shared carbons; PT2, pericondensed structure containing a thiophene ring with two shared carbons; PT3, pericondensed structure containing a thiophene ring with three shared carbons; PT4, pericondensed structure containing a thiophene ring with four shared carbons.

[b] Polymeric C_{18} phase with $\alpha_{TBN/BaP} \approx 0.7$.

[c] Monomeric C_{18} phase with $\alpha_{TBN/BaP} \approx 1.7$.

[d] Log I value greater than 5; number in parentheses indicates elution order of the compounds with log I value greater than 5.

1.5.2 Nonplanarity Effects

The three-dimensional nature of PAH molecules is often overlooked, probably because of the planar structure implied by chemical representations. Some PAH molecules have nonplanar conformations that result from steric hindrance of proximate hydrogen and/or carbon atoms. The effect of solute planarity and linearity on PAH retention and selectivity was investigated using two sets of probe compounds: one set of compounds with similar overall shape, but different planarity, and the second set consisting of polyphenyl arene isomers with linear and nonlinear structures (see Figure 1.20). These two sets of solutes were compared on a monomeric C_{18} phase and on three polymeric C_{18} phases with different phase loadings; the selectivity factors for the planar/nonplanar and linear/nonlinear solutes are summarized in Table 1.4 [31]. In each case, the nonplanar or nonlinear PAH was retained less than the corresponding planar/linear PAH for both monomeric and polymeric C_{18} columns, even though in the case of the planar/nonplanar probes the molecular mass difference would be expected to contribute to the retention of the nonplanar species. For example, phenanthro[3,4-c]phenanthrene (PhPh) has two additional carbon atoms compared with coronene, yet PhPh is less retained than coronene. The difference in shape recognition between monomeric and polymeric C_{18} columns is evident in Table 1.4. Selectivity factors are much larger with the polymeric phases than with monomeric phases, and in one instance (p-quaterphenyl and 1,3,5-triphenylbenzene), the comparison is 46 to 1.5! The selectivity factor increases with the increasing phase loading for the three polymeric phases.

Among planar PAH isomers, solute retention in reversed-phase LC on polymeric C_{18} phases is closely related to L/B, as discussed earlier with respect to long narrow, rodlike molecules being retained longer than square or globular molecules. However, nonplanar solutes elute earlier than expected based only on L/B. As illustrated in Figure 1.4 for the separation of 11 PAH isomers of molecular mass 278, in addition to the trend of increasing retention with increasing L/B, when two PAHs have similar L/B values, the nonplanar PAH typically elutes before the planar PAH. Solute nonplanarity results when portions of a molecule are sterically hindered or strained. For example, for these five-ring isomers, the nonplanarity of dibenzo[c,g]phenanthrene is obviously due to the steric hindrance of the overlapping rings; however several other structures, including benzo[g]chrysene, dibenzo[b,g]phenanthrene, and benzo[c]chrysene, also have varying degrees of nonplanarity. The thickness parameter listed in Table 1.1 provides a quantitative indication of this nonplanarity. Most planar unsubstituted PAHs listed in Table 1.1 are about 3.9 Å thick, whereas these four nonplanar isomers of molecular mass 278 have modeled thicknesses ranging from 5.2 to 6.2 Å. Thus, benzo[c]chrysene ($L/B = 1.53$; nonplanar, with modeled thickness 5.4 Å) elutes before dibenz[a,j]anthracene ($L/B = 1.53$; planar, with modeled thickness 3.9 Å) because of the nonplanar shape. For the four-ring isomers, the elution of benzo[c]phenanthrene ($L/B = 1.28$) before triphenylene ($L/B = 1.12$) can be attributed to its greater nonplanarity based on the thickness parameter (5.0 vs. 4.4 Å, respectively). (Note that even triphenylene is nonplanar.) For the six-ring structures, several of the isomers also

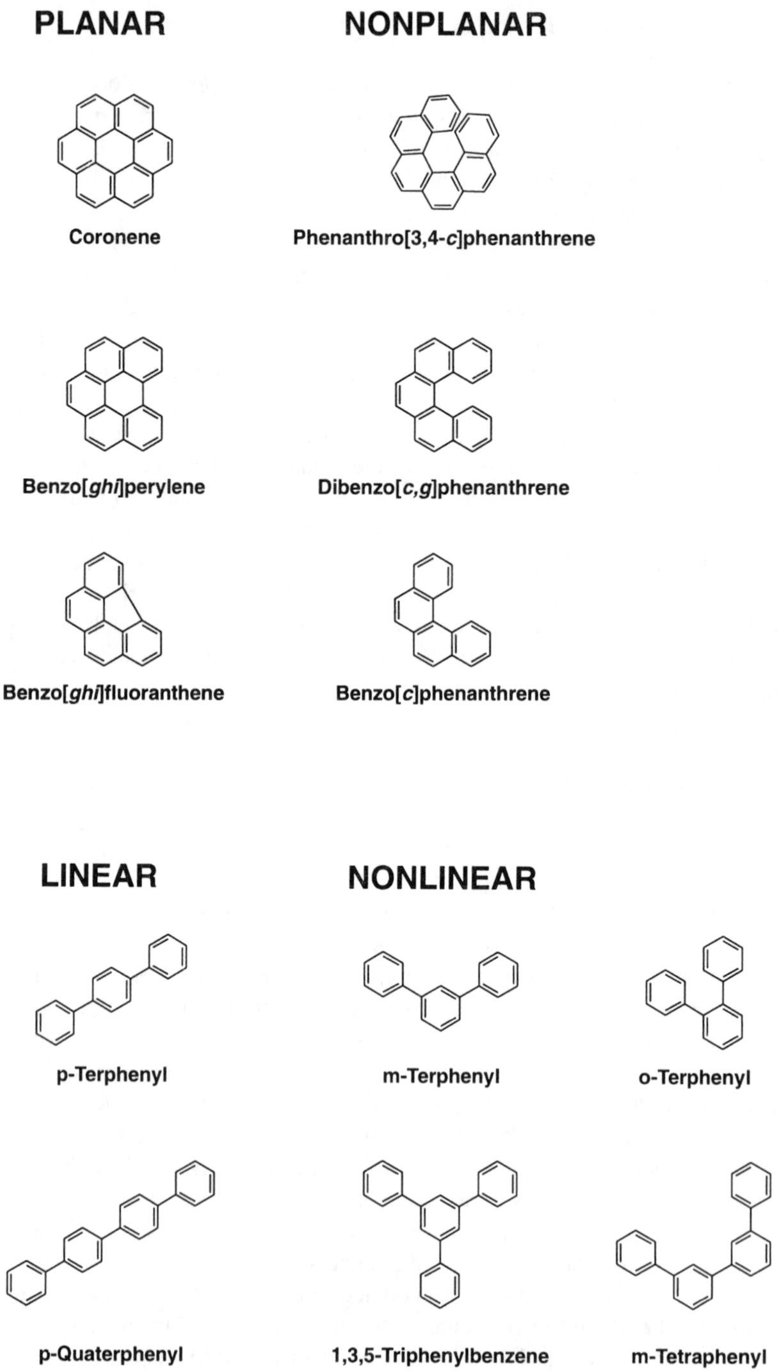

PLANAR

Coronene

Benzo[*ghi*]perylene

Benzo[*ghi*]fluoranthene

NONPLANAR

Phenanthro[3,4-*c*]phenanthrene

Dibenzo[*c,g*]phenanthrene

Benzo[*c*]phenanthrene

LINEAR

p-Terphenyl

p-Quaterphenyl

NONLINEAR

m-Terphenyl

o-Terphenyl

1,3,5-Triphenylbenzene

m-Tetraphenyl

Figure 1.20 Structures of planar/nonplanar and linear/nonlinear solute probes used in the evaluation of shape discrimination of C_{18} columns.

Table 1.4 PLANAR/NONPLANAR SELECTIVITY COEFFICIENTS

Compound	85% Acetonitrile/Water				100% Acetonitrile			
	Monomeric	Polymeric			Monomeric	Polymeric		
		Low	Normal	Heavy		Low	Normal	Heavy
p/m-Terphenyl	1.2	1.3	2.1	3.0	1.2	1.3	2.3	3.5
p/o-Terphenyl	1.3	1.7	2.6	3.8	1.4	1.5	2.9	4.3
p/m-Tetraphenyl	1.4				1.4	4.4	11.7	34.5
p-Quaterphenyl/1,3,5-triphenylbenzene	1.5				1.5	3.0	14.2	46.0
Benzo[ghi]fluoranthene/benzo[c]phenanthrene	1.1	1.2	1.4	1.6	1.2	1.4	1.7	1.8
Benzo[ghi]perylene/dibenzo[c,g]phenanthrene	1.8	2.8	4.5	4.1	2.1	3.3	7.8	7.5
Coronene/phenanthro[3,4-c]phenanthrene	2.7	5.1	9.6		3.4	6.2	12.1	17.1

exhibit nonplanarity (dibenzo[*def,p*]chrysene, benzo[*a*]perylene, dibenzo[*j,l*]fluoranthene, and dibenzo[*b,e*]fluoranthene), although it is not obvious that they elute early because these isomers have the smallest *L/B* ratios for these isomeric groups. Of particular interest is the magnitude of the thickness parameter for phenanthro[3,4-*c*]phenanthrene and dibenzo[*g,p*]chrysene (7.4 and 6.2 Å, respectively), which are the two nonplanar PAHs in SRM 869. This large nonplanarity for dibenzo[*g,p*]chrysene (i.e., TBN) is the basis for the sensitivity of SRM 869 to changes in the stationary phase characteristics.

The methyl-substituted PAHs are excellent model solutes for investigating solute nonplanarity effects on LC retention. Methyl substitution at certain sites on the parent molecule results in a nonplanar structure as indicated by molecular modeling. The effect of nonplanarity on the retention of methyl-substituted PAHs has been studied extensively by Garrigues and co-workers [78] for dimethyl- and ethyl-substituted isomers of phenanthrene and by Wise et al. [22] for the methyl-substituted isomers of chrysene, perylene, and picene. As a measure of the nonplanarity, Garrigues and co-workers calculated the dihedral angle of distortion for dimethyl-phenanthrene isomers, as illustrated in Figure 1.21. For phenanthrene, the angle of distortion is defined as the dihedral angle given by the four carbon atoms (three sides: *d*, *e*, and *f* in Figure 1.21) in the bay region of the PAH molecule. The dihedral angles of distortion for the methyl-substituted PAHs are provided in Table 1.2.

Small values for the dihedral angle of distortion indicate a nearly planar shape, whereas larger values indicate nonplanarity. The dihedral angle molecular descriptor must be defined individually for different parent molecules. For example, chrysene has two bay regions (*def* and *mno* in Figure 1.16), and two dihedral angles can be calculated for various methylchrysene isomers. Similarly, picene has three bay regions (*ghi*, *pqr*, and *tuv*), which results in the calculation of three dihedral angles. Benzo[*c*]phenanthrene has one bay region defined by four sides, and two separate dihedral angles of distortion can be calculated depending on the three sides selected

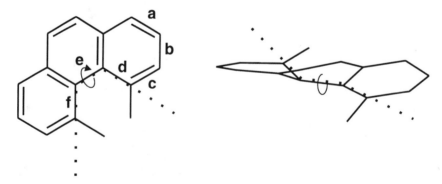

Figure 1.21   The dihedral angle of distortion is defined by the angle between the planes containing the twisted rings under consideration. In this example sides *def* are used to calculate the dihedral angle of distortion.

(see Figure 1.16 and Table 1.2). Consequently, the dihedral angle descriptor is useful for comparing molecular shape only within a single isomer set. Another difficulty exists for this molecular descriptor. The magnitude of the dihedral angle is strongly influenced by the force field used in modeling the molecular structure. This of course is a direct reflection of the molecular modeling software utilized. Dihedral angles determined from molecular structures determined using different force fields will likely differ, sometimes dramatically.

The calculation of the angle of distortion by Garrigues et al. [78] represented the first attempt to quantify a nonplanarity parameter for correlation with LC retention behavior. The LC retention results of these investigators for the dimethylphenanthrenes and ethylphenanthrenes are summarized in Table 1.5. Dihedral angles and the other molecular descriptors listed were determined by our group as described earlier. It should be noted that the dihedral angles we have determined using the MMX force field differ significantly from those reported by Garrigues et al. [78] using a MM2MP2 force field. We have recalculated these structures to obtain angles of distortion and L/B values that are consistent with the other data presented in this chapter. Despite differences between the two methods in the magnitude of the molecular descriptors, particularly the angle of distortion, agreement exists for the overall molecular shape (i.e., whether the structure is planar or nonplanar). As shown in Table 1.5, the dimethylphenanthrene isomers vary significantly in L/B ratios (i.e., a range of 0.60 index unit, which is larger than any other group of methyl-substituted isomers studied). The isomers with methyl groups in the bay region (position 4 or 5) have a strained molecular structure resulting in large angles of distortion, ranging from about 20° to as high as 32° for 4,5-dimethylphenanthrene. Garrigues et al. [78] plotted the retention (log I) against the L/B ratio and observed that the isomers with the large dihedral angles generally eluted earlier than expected based on the L/B ratio. They found a correlation coefficient of 0.67 for the complete set of 25 dimethylphenanthrene isomers. The correlation increased to 0.79 when 3,4- and 4,5-dimethylphenanthrene (the two isomers with the largest angles of distortion) were eliminated and increased to 0.85 when only the 14 planar isomers were plotted.

In Figure 1.22 we have replotted the retention data of Garrigues et al. [78] versus the L/B ratios from Table 1.5 calculated as just described. A correlation coefficient of 0.77 was found for all 25 dimethylphenanthrene isomers, and it increased to 0.85 (the same as with the Garrigues L/B calculations) when only the planar isomers were plotted. This improvement in the correlation for the nonplanar isomers indicates that the incorporation of the three-dimensional shape into the calculation of the L/B value is a useful refinement to the procedure.

The angles of distortion for the methyl-substituted PAHs are included in Table 1.2 as one of the molecular descriptors for those structures with bay regions. For each of the PAHs, substitution of a methyl group in the bay regions results in varying degrees of distortion between the carbon atoms in the aromatic rings. As would be expected, structures with large angles of distortion also have larger thickness parameters (i.e., due to the nonplanarity). For example, 4-methylphenanthrene has an angle of $-19.5°$ and a thickness of 4.7 Å compared to 4.2 Å for the other

Table 1.5 RETENTION INDICES AND SHAPE PARAMETERS FOR DIMETHYL- AND ETHYLPHENANTHRENE ISOMERS

Compound	Dihedral Angle (deg)	Molecular Dimensions (Å)			L/B	Polymeric Phase (log I)[a]
		Length	Width	Thickness		
4,10-Dimethylphenanthrene	21.1	11.56	10.11	4.79	1.14	3.64
4,9-Dimethylphenanthrene	21.2	11.34	9.60	4.90	1.18	3.65
3,10-Dimethylphenanthrene	0.1	11.81	9.85	4.22	1.20	3.71
3,9-Dimethylphenanthrene	0.2	11.82	9.60	4.22	1.23	3.75
4,5-Dimethylphenanthrene	32.2	11.51	9.10	6.22	1.26	3.42
1,9-Dimethylphenanthrene	0	11.88	9.33	4.22	1.27	3.68
9,10-Dimethylphenanthrene	-0.1	11.72	9.19	4.22	1.28	3.64
3,6-Dimethylphenanthrene	0.4	11.74	9.18	4.21	1.28	3.69
3,5-Dimethylphenanthrene	-20.1	11.64	9.06	4.95	1.28	3.67
1,10-Dimethylphenanthrene	-12.4	11.64	9.06	4.92	1.29	3.65
1,3-Dimethylphenanthrene	0.4	11.81	9.02	4.22	1.31	3.72
1,4-Dimethylphenanthrene	-21.8	11.68	9.04	4.83	1.29	3.67
1,5-Dimethylphenanthrene	-21.9	11.96	8.90	4.98	1.34	3.67
3,4-Dimethylphenanthrene	20.8	12.05	8.86	4.80	1.36	3.63
2,9-Dimethylphenanthrene	0.1	12.86	9.28	4.22	1.39	3.84
2,10-Dimethylphenanthrene	0.1	12.86	9.28	4.22	1.39	3.78
1,6-Dimethylphenanthrene	0.1	12.31	8.73	4.22	1.41	3.76
2,5-Dimethylphenanthrene	-19.9	12.82	9.07	4.70	1.41	3.76
2,4-Dimethylphenanthrene	-19.6	12.85	9.07	4.85	1.42	3.75
2,3-Dimethylphenanthrene	-0.1	12.68	8.72	4.21	1.45	3.72
2,6-Dimethylphenanthrene	0.2	12.93	8.71	4.21	1.48	3.83
1,8-Dimethylphenanthrene	-0.2	11.88	8.03	4.22	1.48	3.79
1,2-Dimethylphenanthrene	1.6	12.87	8.03	4.21	1.60	3.80
1,7-Dimethylphenanthrene	0	12.95	8.04	4.22	1.61	3.91
2,7-Dimethylphenanthrene	0	13.98	8.02	4.21	1.74	4.01
9-Ethylphenanthrene	0	11.78	10.49	4.22	1.12	3.65
4-Ethylphenanthrene	21.6	10.85	9.24	5.48	1.18	3.55
3-Ethylphenanthrene	0.2	11.97	9.21	4.21	1.30	3.65
1-Ethylphenanthrene	-0.1	12.98	8.07	4.22	1.61	3.70
2-Ethylphenanthrene	0.3	14.00	8.03	4.21	1.74	3.78

[a] Retention data from Garrigues et al. [78].

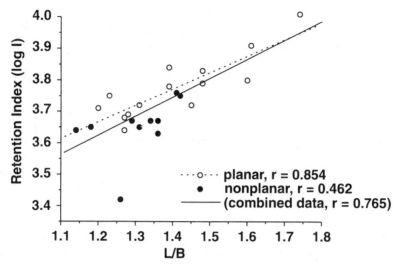

Figure 1.22 Log I plotted as a function of L/B for dimethylphenanthrene isomers. Log I values from Garrigues et al. [78]; L/B values from Table 1.5.

methylphenanthrenes. For the methylbenz[a]anthracenes, the 1-methyl and 12-methyl isomers have angles of $-24.7°$ and $-26.7°$ and thicknesses of 5.0 and 5.2 Å compared to 4.2 Å for the other isomers. 4-Methyl- and 5-methylchrysene have angles of $-23.4°$ and $-24.0°$ in the *def* bay region, each resulting in thicknesses of 5.0 Å, compared to 4.2 Å for the other isomers. For picene, which has three bay regions, distortion of varying amounts occurs in all three regions for 1-methyl-, 6-methyl-, and 13-methylpicene with thicknesses of 5.4–5.6 Å compared to 4.2–5.1 Å for the other isomers.

As discussed earlier for the unsubstituted PAHs, the nonplanar methyl-substituted PAHs have reduced retention compared to that expected on the basis of L/B alone. For example, nonplanarity, as indicated by angle of distortion and thickness, accounts for the early elution of 4-methylphenanthrene ($L/B = 1.29$) compared to 9-methylphenanthrene ($L/B = 1.27$). The same justification can be used for the elution of 12-methylbenz[a]anthracene ($L/B = 1.54$) prior to 7-methylbenz[a]anthracene ($L/B = 1.54$), and the elution of 10-methyl- and 11-methylbenzo[a]pyrene ($L/B = 1.43$ and 1.35, respectively) prior to 4-methylbenzo[a]pyrene ($L/B = 1.34$). For the methylbenzo[c]phenanthrene isomers, which are all nonplanar, the 1-methyl isomer ($L/B = 1.26$) has the greatest nonplanarity (thickness 5.9 Å) and elutes earlier than the 2-methyl isomer ($L/B = 1.20$, thickness 5.2 Å).

The extent of the discrepancy between elution and L/B ratio increases with the degree of polymeric phase loading of the stationary phase. This trend was demonstrated by investigating the retention of all possible monomethyl isomers of chrysene, perylene, and picene on a series of C_{18} phases including both monomeric and polymeric C_{18} phases with varying degrees of polymeric loading based on the

$\alpha_{TBN/BaP}$ value described earlier. The plot of retention of the methylpicene isomers relative to picene versus column selectivity ($\alpha_{TBN/BaP}$) for the different C_{18} phases is shown in Figure 1.23. The reversed-phase LC separation of the methylpicene isomers on three different C_{18} phases is illustrated in Figure 1.24. As shown in Figure 1.23, the selectivity (α) increases as the phase becomes more polymeric (i.e., decreasing $\alpha_{TBN/BaP}$ values). All methyl-substituted isomers were observed to elute after picene with the monomeric C_{18} phases, but with the polymeric phases certain methylpicene isomers elute earlier than picene. For the methylpicene isomers with the largest L/B values (3-methyl and 4-methyl), the α values increase as the phase becomes more polymeric. In contrast, the isomers with the smallest L/B ratios and the greatest nonplanarity (13-methyl, 6-methyl, and 1-methyl with angles of distortion from about 7–31°) elute the earliest, and the selectivity ratio decreases and even becomes less than 1 (i.e., it elutes prior to picene) for all but two of the isomers (3-methyl and 4-methyl) on the phases with the polymeric character.

The selectivity ratios for the three methylperylene isomers relative to perylene are plotted versus stationary phase characteristics in Figure 1.25, and the chromatograms for the separation of the isomers on two different phases are shown in Figure 1.26. Similar trends are observed for the methylperylene isomers; the elution order follows L/B ratios, and the elution of 1-methylperylene prior to perylene on the polymeric phases can be attributed to the nonplanarity of this isomer because the methyl group in the 1-position results in an angle of distortion of 29°. It is interesting to note that for the 2-methyl isomer, which has less distortion from planar, the retention relative to perylene changes very little on the different phases.

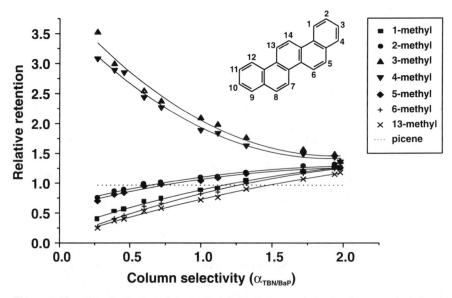

Figure 1.23   Plot of retention of the methylpicene isomers relative to picene versus column selectivity ($\alpha_{TBN/BaP}$) for 11 different commercial C_{18} columns.

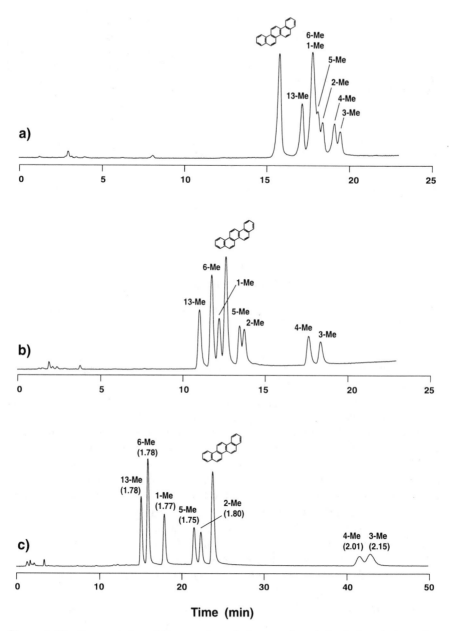

Figure 1.24 Reversed-phase LC separation of picene and methyl-substituted picene isomers with (a) a monomeric C_{18} column, (b) an "intermediate" C_{18} column, and (c) a polymeric C_{18} column; L/B values given in parentheses.

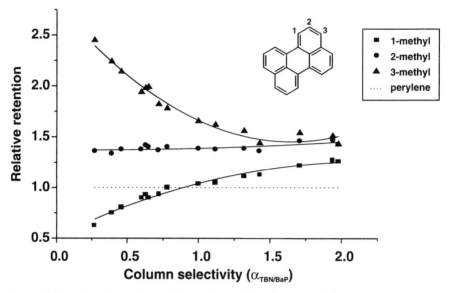

Figure 1.25 Plot of retention of the methylperylene isomers relative to perylene versus column selectivity ($\alpha_{TBN/BaP}$) for 11 different commercial C_{18} columns.

The following trends can be summarized for methyl-PAH isomers: (1) the relative separation of the methyl-substituted PAH isomers increases as the C_{18} phase becomes more polymeric ($\alpha_{TBN/BaP}$ decreases); (2) the elution order of methyl-PAH isomers generally follows increasing L/B ratio on polymeric phases; (3) as the polymeric nature of the phase increases, isomers with some nonplanarity and small L/B values eventually elute prior to the parent PAH; (4) as the polymeric nature of the phase increases, the retention of the isomers with large L/B values and planar structures increases relative to the parent PAH; and (5) on monomeric C_{18} phases the methyl isomers elute after the parent PAH and tend to elute as a group with little separation of the individual isomers. Thus, nonplanarity is a factor influencing retention on polymeric C_{18} phases more than on monomeric C_{18} phases, and heavily loaded polymeric phases showed a greater effect than lightly loaded phases. It is of particular interest that the absolute retention of many nonplanar methyl PAH isomers was less than that of the unsubstituted parent PAH. This behavior is remarkable because based on hydrophobic retention theory and other theories of solute retention, the addition of alkyl groups to a solute molecule is expected to increase retention.

1.5.3 Comparison of Shape Selectivity in LC and GC

As mentioned already, the development of the L/B shape parameter and its correlation with LC retention were based on observations of similar shape-selective behavior for liquid crystalline stationary phases in GC. Thus, a comparison of the retention behavior of various PAH isomer groups both in reversed-phase LC on polymeric C_{18} phases and in GC on a liquid crystalline stationary phase would help

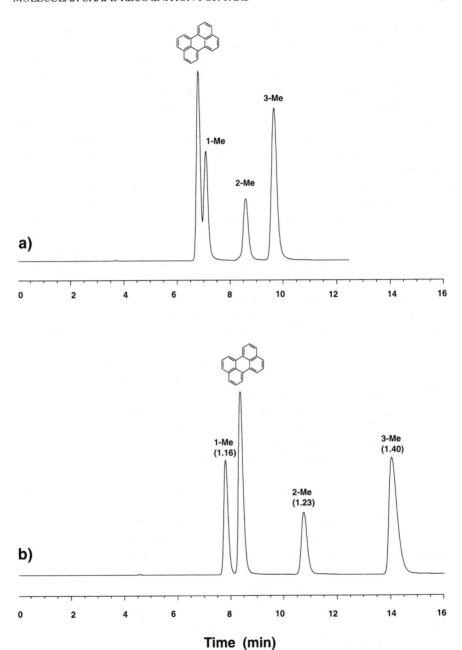

Figure 1.26 Reversed-phase LC separation of perylene and methyl-substituted perylene isomers with (a) an "intermediate" C_{18} column ($\alpha_{TBN/BaP} = 1.12$) and (b) a polymeric C_{18} column ($\alpha_{TBN/BaP} = 0.60$); L/B values given in parentheses.

to determine whether similar retention mechanisms and/or the physical nature of the stationary phases are responsible for the unique shape recognition characteristics. In the original study describing the relationship of L/B and LC retention, a limited amount of data from the literature for GC separations on liquid crystalline phases was briefly compared with the LC data [10]. A later investigation provided a more detailed comparison of PAH retention in reversed-phase LC on polymeric C_{18} phases and in GC on a smectic liquid crystalline phase [28]. Two groups of unsubstituted PAHs (four- and five-ring catacondensed isomers) and three groups of methyl-substituted PAHs (phenanthrene, chrysene, and benz[a]anthracene) were compared in this study.

Figure 1.27 plots retention versus L/B ratio for both GC and LC data for the five-ring catacondensed isomers, methylchrysene isomers, and methylbenz[a]anthracene isomers, respectively. For the five-ring catacondensed isomers, the relative elution orders for LC and GC are similar, with the same anomalies in elution order relative to L/B ratios in both systems; that is, dibenz[a,c]anthracene and benzo[a]naphthacene both have significantly longer retention than predicted based on L/B value. For the methylchrysenes, excellent correlation is observed between the LC and GC elution orders. In GC even the 6-methyl and 5-methyl isomers are separated and elute in the order predicted by L/B. For the methylbenz[a]anthracenes, even though there are more deviations from the elution order predicted by L/B ratios, the GC and LC systems exhibit similar behavior. For example, in both systems, the 5-methyl isomer elutes significantly later than would be predicted by its L/B value, and the 11-methyl isomer elutes earlier than expected. However, as with the groups of isomers discussed earlier, the significant observation is the similarity of the elution behavior of both chromatographic systems.

Sander et al. [76] reported GC separation of methylbenzo[a]pyrene isomers, methylbenzo[b]fluoranthene isomers, and molecular mass 252 isomers on a liquid crystalline stationary phase. The following elution order was observed for the 12 methylbenzo[a]pyrene isomers: 12-methyl, 5-methyl, 11-methyl, 9-methyl, 10-methyl, 7-methyl, 4-methyl, 1-methyl, 6-methyl, 2-methyl, 8-methyl, 3-methyl. This elution order is similar to that obtained in LC with the 9-methyl and 4-methyl isomers eluting earlier in GC relative to the other isomers and as expected based on the L/B values, and the 6-methyl and the 3-methyl isomers eluting later in GC. For the methylbenzo[b]fluoranthene isomers, the elution behavior is again similar, with the isomers clustering in two groups in both GC and LC (see Figure 1.19). For the 12 molecular mass 252 isomers, the GC and LC elution orders have remarkable similarities when these isomers are grouped as in Table 1.1. For both the P6 and the PF5 isomer groups of molecular mass 252, the elution orders for both the LC and GC are identical. For the PF4 isomers, the elution order of benz[l]acephenanthrylene and benz[e]aceanthrylene is reversed.

1.5.4 LC Retention Models for PAHs

1.5.4.1 Slot Model

As an attempt to consolidate all the foregoing observations of the effect of shape (L/B ratio) and nonplanarity on LC retention of PAHs, we developed an empirical

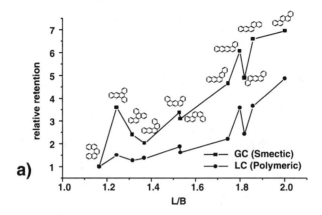

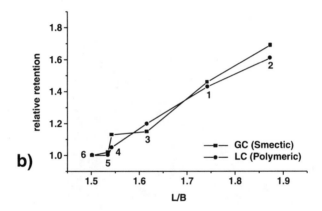

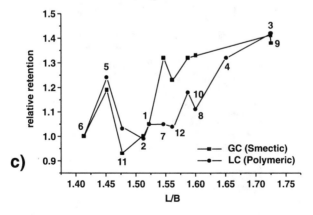

Figure 1.27 Plots of relative LC and GC retention versus *L/B* ratios for (a) five-ring catacondensed isomers, (b) methylchrysene isomers, and (c) methylbenz[*a*]anthracene isomers; numbers in (b) and (c) refer to position of methyl substitution.

retention model termed the "slot model" [31]. In this model, the stationary phase is viewed schematically as an environment consisting of slots into which the PAH solute molecules can penetrate (see Figure 1.28). These slots can be thought of as the space between the alkyl chains in C_{18} phases. For a slot of finite height, width, and depth, solute interaction (penetration) will vary depending on the solute shape characteristics. Planar molecules will fit readily into some slots, whereas nonplanar molecules may be excluded from the phase, resulting in decreased retention. Among planar isomer sets, molecules with large L/B ratios have the potential for the greatest interaction with the phase because long, narrow solutes will fit into a larger fraction of the available slots. In contrast, square molecules (i.e., $L/B \approx 1$) may be excluded from a fraction of the slots on the basis of solute width, resulting in weak interaction and reduced retention.

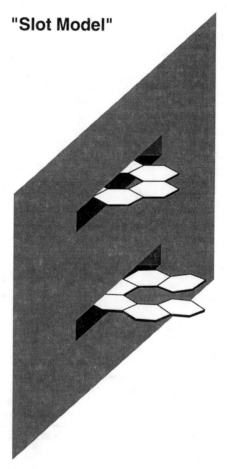

Figure 1.28   Schematic representation of solute interaction with the bonded stationary phase. Slots represent spaces between bonded alkyl chains.

The observed retention behavior of nonplanar PAHs also can be explained in terms of the slot model. The nonplanar structure imposed by steric hindrance can be viewed as reducing the interaction with the stationary phase (compared with the unsubstituted PAH solute), because the bulkier structure requires more space (thicker slots) to fit between the alkyl chains. If this concept is extended further, the differences in shape selectivity that are observed between monomeric and polymeric C_{18} phases (and among polymeric phases with varying loadings) can also be represented by the slot model. We have shown that polymeric C_{18} phases are thicker than monomeric C_{18} phases, and more dense [59]. In terms of the slot model representation, polymeric phases have thinner, narrower (and deeper) slots (i.e., less space between alkyl chains) that monomeric phases. Because the alkyl chains are more disordered with monomeric phases, the slots are poorly defined, and less discrimination on the basis of shape is predicted. Because phase thickness corresponds to slot depth, increasing alkyl chain length produces greater shape recognition. Higher density phases produce narrower slots thereby increasing the shape selectivity.

1.5.4.2 Other LC Retention Models as Applied to PAHs

A number of theories have been developed to explain retention in reversed-phase LC. Several of these theories were discussed recently by Sander and Wise [36] in terms of the current understanding of both the alkyl bonded stationary phase morphology and the solute characteristics. Of particular interest to this discussion are the comprehensive models developed by Martire and co-workers [79–82] based on statistical thermodynamics. These models are not limited to LC, but include applications to GC and supercritical fluid chromatography. It is intended to describe the retention of rigid "block like" molecules such as PAHs on ordered stationary phases. Examples of ordered phases include liquid crystalline stationary phases and surfaces covalently modified with long chain normal hydrocarbons (i.e., LC stationary phases).

The basis of Martire's model is a lattice description of an anisotropic (ordered) stationary phase and an isotropic mobile phase. Partitioning of solute molecules between these phases depends on repulsive and attractive contributions of each species as well as the molecular structure of the stationary phase. Martire and Yan rigorously derived relationships between the solute distribution coefficient K and state variables (temperature and density) and molecular parameters (interaction energies and molecular dimensions) [80–82]. These expressions can be simplified as the following equation:

$$\ln(3K) = Q_1 A_{\min} + Q_2 A_{\text{eff}} + Q_3 V_{\text{vdw}} \tag{1.2}$$

where $A_{\min}$ is the minimum cross-sectional area of the solute, A_{eff} is the effective contact area of the solute, and V_{vdw} is the van der Waals volume of the solute; Q_1, Q_2, and Q_3 are constants that in principle can be determined from state variables and properties of the solute molecule. The term $A_{\min}$ is defined by a rectangle expanded about the solute molecule in mutually orthogonal directions such that when the area of this rectangle is multiplied by the molecular thickness, the van der Waals volume is obtained (see Figure 1.29).

Equation 1.2 can be simplified further for isomeric solutes. Among PAH isomers, V_{vdw} and A_{eff} differ only slightly, so equation 1.3 can be written:

$$\ln(K) = Q_1 A_{min} + \text{constant} \tag{1.3}$$

This relation suggests that among isomer sets, $\ln k'$ should be linearly related to A_{min}. Such a plot for the 11 five-ring catacondensed PAH isomers (molecular mass 278) was made by Yan and Martire [80] from published GC data on a liquid crystalline phase [28]. A reasonably good fit of the data was achieved (correlation coefficient 0.98), although larger deviations from this relation were observed for other data sets. Similar relations have been reported for supercritical fluid chromatography retention data using a smectic liquid crystalline column [82,83]. The data indicate a "preferential orientation of PAH solutes such that the plane of the minimum cross sectional area is perpendicular to the extended bonded phase ligands" [83]. Thus, rodlike molecules should be retained in preference to more cubical molecules. The lattice–fluid retention model in its present form is based on complete alignment of the bonded phase chains, and neglects swelling effects of the mobile phase.

1.5.4.3 Comparison of Lattice Model and L/B Ratio

The more rigorous lattice model of Yan and Martire is consistent with the empirical slot model discussed in connection with Figure 1.28, and the minimum area parameter these authors describe (A_{min}) can be viewed as a mathematical representation of the required slot dimensions for a given solute. To evaluate whether the A_{min} parameter is a better predictor of retention for PAHs, the LC retention data for two sets of isomers, the 11 five-ring catacondensed PAHs of molecular mass 278 and the 9 six-ring pericondensed (six-membered rings only) PAHs of molecular mass 302,

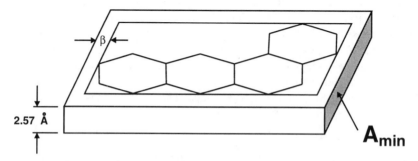

Figure 1.29 A_{min} as determined by Yan and Martire [80]; A_{min} is defined by the end of a box drawn to enclose the structure of a molecule such that the volume of the box is equal to the van der Waals volume of the molecule. The thickness is set at 2.57 Å for planar molecules. For nonplanar molecules the value determined by x-ray crystallographic data is utilized.

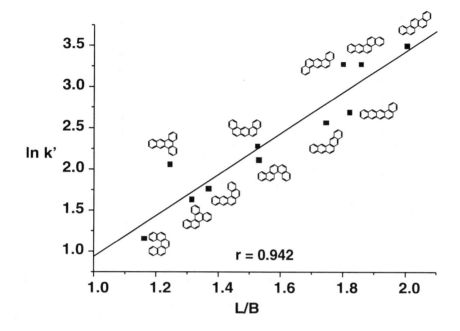

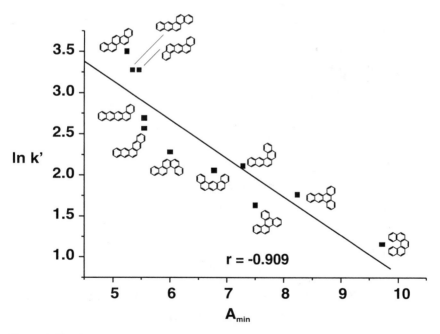

Figure 1.30   Correlation between the natural log of the relative LC retention and A_{min} for five-ring catacondensed isomers of molecular mass 278 (A_{min} data from references 80 and 81).

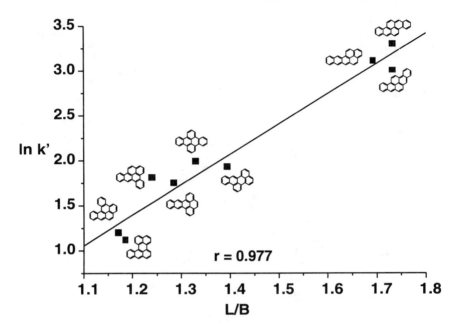

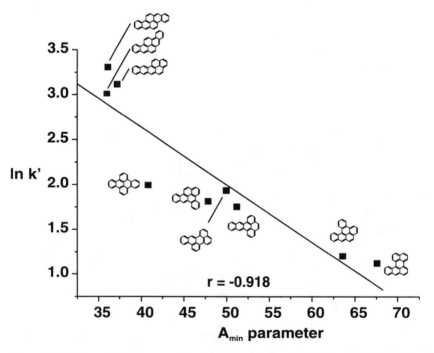

Figure 1.31   Correlation between the natural log of the relative LC retention and A_{min} for six-ring pericondensed isomers of molecular mass 302 ("A_{min} parameter" calculated as described in text).

were plotted against the L/B ratio and A_{min}. These plots are illustrated in Figures 1.30 and 1.31 for the 278 and 302 molecular mass isomers, respectively. The A_{min} values in Figure 1.30 are scaled as reported by Yan and Martire [80], whereas the "A_{min} parameter" plotted in Figure 1.31 was calculated from sides yz for the box enclosing the molecular model (i.e., the width and thickness parameters in Table 1.1; see Figure 1.13b). A linear relation exists between these two descriptors ($r = 0.975$ for a comparison of A_{min} and A_{min} parameter for the 278 molecular mass isomers). For both isomer sets a better correlation was achieved using L/B ratio than A_{min} (0.942 vs. -0.909 for the 278 molecular mass isomers and 0.977 vs. -0.918 for the 302 molecular mass isomers). These differences are small, and both L/B and A_{min} molecular descriptors provide a good correlation with LC retention measurements. It should be noted, however, that the L/B is an empirical descriptor, whereas A_{min} has its basis in a statistical mechanical model of the stationary phase. The consistency of these descriptors lends credence to the empirical slot model discussed earlier.

References

1. J.A. Schmit, R.A. Henry, R.C. Williams, and J.F. Dieckman, *J. Chromatogr. Sci.* **9**, 645 (1971).

2. S.A. Wise, in A. Bjørseth, Ed., *Handbook for Polycyclic Aromatic Hydrocarbons*. Dekker, New York, 1983, p. 183.

3. S.A. Wise, in A. Bjørseth and T. Ramdahl, Eds., *Handbook for Polycyclic Aromatic Hydrocarbons*, Vol. II. Dekker, New York, 1985, p. 113.

4. S.A. Wise, L.C. Sander, and W.E. May, *J. Chromatogr.* **642**, 329 (1993).

5. J.C. Fetzer, in T. Vo-Dinh, Ed., *Chemical Analysis of Polycyclic Aromatic Compounds*. Wiley, New York, 1993, p. 59.

6. K.D. Bartle, M.L. Lee, and S.A. Wise, *Chem. Soc. Rev.* **10**, 113 (1981).

7. S.A. Wise, W.J. Bonnett, F.R. Guenther, and W.E. May, in A. Bjørseth and A.J. Dennis, Eds., *Polynuclear Aromatic Hydrocarbons: Chemistry and Biological Effects*. Battelle Press, Columbus, OH, 1980, p. 791.

8. S.A. Wise, S.N. Chesler, H.S. Hertz, L.R. Hilpert, and W.E. May, *Anal. Chem.* **49**, 2306 (1977).

9. M. Popl, V. Dolansky, and J. Mostecky, *J. Chromatogr.* **117**, 117 (1976).

10. S.A. Wise, W.J. Bonnett, F.R. Guenther, and W.E. May, *J. Chromatogr. Sci.* **19**, 457 (1981).

11. K.L. Ogan and E.D. Katz, *J. Chromatogr.* **188**, 115 (1980).

12. E.D. Katz and K.L. Ogan, *J. Liquid Chromatogr.* **3**, 1151 (1980).

13. A. Colmsjo and J.C. MacDonald, *Chromatographia*. **13**, 350 (1980).

14. R. Amos, *J. Chromatogr.* **204**, 469 (1981).

15. R.A. Hites and Simonsick, Jr., *Calculated Molecular Properties of Polycyclic Aromatic Hydrocarbons*. Elsevier, Amsterdam, 1987.

16. K. Jinno, S. Shimura, N. Tanaka, K. Kimata, J.C. Fetzer, and W.R. Biggs, *Chromatographia*, **27**, 285 (1989).

17. J.C. Fetzer, *Adv. Chem. Ser.* **217**, 309 (1988).

18. J.C. Fetzer and W.R. Biggs, *J. Chromatogr.* **322**, 275 (1985).

19. K. Jinno and K. Kawasaki, *Chromatographia*, **17**, 445 (1983).

20. R.H. Rohrbaugh and P.C. Jurs, *Anal. Chem.* **59**, 1048 (1987).

21. H. Budzinski, M. Radke, P. Garrigues, S.A. Wise, J. Bellocq, and H. Willsch, *J. Chromatogr.* **627**, 227 (1992).

22. S.A. Wise, L.C. Sander, R. Lapouyade, and P. Garrigues, *J. Chromatogr.* **514**, 111 (1990).

23. L.C. Sander and S.A. Wise, *Anal. Chem.* **67**, 3284 (1995).

24. K.L. Williams, L.C. Sander, S.H. Page, and S.A. Wise, *J. High Resolution Chromatogr.* **18**, 477 (1995).

25. L.C. Sander, R.M. Parris, S.A. Wise, and P. Garrigues, *Anal. Chem.* **63**, 2589 (1991).

26. L.C. Sander and S.A. Wise, *Anal. Chem.* **61**, 1749 (1989).

27. L.C. Sander and S.A. Wise, *J. High Resolution Chromatogr., Chromatogr. Commun.* **11**, 383 (1988).

28. S.A. Wise, L.C. Sander, H. Chang, K.E. Markides, and M.L. Lee, *Chromatographia*, **25**, 473 (1988).

29. L.C. Sander and S.A. Wise, *Anal. Chem.* **59**, 2309 (1987).

30. S.A. Wise and W.E. May, *Anal. Chem.* **55**, 1479 (1983).

31. S.A. Wise and L.C. Sander, *J. High Resolution Chromatogr., Chromatogr. Commun.* **8**, 248 (1985).

32. L.C. Sander and S.A. Wise, *Anal. Chem.* **56**, 504 (1984).

33. L.C. Sander and S.A. Wise, *J. Chromatogr.* **316**, 163 (1984).

34. S.A. Wise, L.C. Sander, and W.E. May, *J. Liquid Chromatogr.* **6**, 2709 (1983).

35. L.C. Sander and S.A. Wise, in M.W. Cooke and A.J. Dennis, Eds., *Polynuclear Aromatic Hydrocarbons: Eighth International Symposium on Mechanism, Method and Metabolism.* Battelle Press, Columbus, OH, 1983, p. 1133.

36. L.D. Sander and S.A. Wise, in N. Tanaka and K.K. Unger, Eds., *Stationary Phases for Reversed-Phase Liquid Chromatography.* Elsevier Science Publishers, Amsterdam, 1996.

37. L.C. Sander and S.A. Wise, in R.M. Smith, Ed., *Retention and Selectivity Studies in HPLC.* Elsevier Science Publishers, Amsterdam, 1994, p. 337.

38. L.C. Sander and S.A. Wise, *J. Chromatogr.* **656**, 335 (1993).

39. L.C. Sander and S.A. Wise, *LC GC*, **8**, 378 (1990).

40. L.C. Sander and S.A. Wise, *CRC Crit. Rev. Anal. Chem.* **18**, 299 (1987).

41. L.C. Sander and S.A. Wise, in J.C. Giddings, E. Grushka, J. Cazes, and P.R. Brown, Eds., *Advances in Chromatography.* Dekker, New York, 1986, p. 139.

42. S.A. Wise, L.C. Sander, and W.E. May, in D.E. Leyden, Ed., *Silanes Surfaces and Interfaces.* Gordon & Breach, New York, 1986, p. 349.

43. *Fed. Regist.* **44**, 233 (1979).

44. S.A. Wise, B.A. Benner, H. Liu, G.D. Byrd, and A. Colmsjo, *Anal. Chem.* **60**, 630 (1988).

45. L.C. Sander and S.A. Wise, *Certificate of Analysis, Standard Reference Materials Program.* National Institute of Standards and Technology, Gaithersburg, MD, 1990.

46. L.C. Sander, *J. Chromatogr. Sci.* **26**, 380 (1988).

47. D. Baschke, M. Jendziezyck, D. Youngs, and M. Henry. *Twelfth International Symposium of Polynuclear Aromatic Hydrocarbons* (1989) (abstract).

48. M. Henry, *J. Chromatogr.* **544**, 413 (1991).

49. *Perspectives in Chromatography,* Vydac, Hesperia, CA, 1995 (pamphlet, issue 1-PP).

50. B. Buszewski, Z. Suprynowicz, P. Staszczuk, K. Albert, B. Pfleiderer, and E. Bayer, *J. Chromatogr.* **499**, 305 (1990).

51. B. Buszewski, *J. High Resolution Chromatogr., Chromatogr. Commun.* **13**, 410 (1990).

52. K.D. Lork and K.K. Unger, *Chromatographia,* **26**, 115 (1989).

53. K.B. Sentell and J.G. Dorsey, *Anal. Chem.* **61**, 930 (1989).

54. K.B. Sentell and J.G. Dorsey, *J. Chromatogr.* **461**, 193 (1989).

55. J. Yamaguchi, T. Hanai, and H. Cai, *J. Chromatogr.* **441**, 183 (1988).

56. K.B. Sentell, K.W. Barnes, and J.G. Dorsey, *J. Chromatogr.* **455**, 95 (1988).

57. K. Szabo, N. Le Ha, P. Schneider, P. Zeltner, and E. Kovats, *Helv. Chim. Acta,* **67**, 2128 (1984).

58. K.A. Dill, *J. Phys. Chem.* **91**, 1980 (1987).

59. L.C. Sander, C.J. Glinka, and S.A. Wise, *Anal. Chem.* **62**, 1099 (1990).

60. C.J. Glinka, L.C. Sander, S.A. Wise, and N.F. Berk, *Mater. Res. Soc. Symp. Proc.* **166**, 415 (1990).

61. H. Hemetsberger, W. Maasfeld, and H. Ricken, *Chromatographia,* **9**, 303 (1976).

62. P. Spacek, M. Kubin, S. Vozka, and B. Porsch, *J. Liquid Chromatogr.* **3**, 1465 (1980).

63. N. Tanaka, K. Sakagami, and M. Araki, *Chem. Lett.* 587 (1980).

64. G.E. Berendsen and L. de Galan, *J. Chromatogr.* **196**, 21 (1980).

65. C.H. Lochmüller and D.R. Wilder, *J. Chromatogr. Sci.* **17**, 574 (1979).

66. L.C. Sander, J.B. Callis, and L.R. Field, *Anal. Chem.* **55**, 1068 (1983).

67. E. Kovats, *Helv. Chim. Acta,* **41**, 1915 (1958).

68. M. Popl, V. Dolansky, and J. Mostecky, *J. Chromatogr.* **91**, 649 (1974).

69. M.L. Lee, D.L. Vassilaros, C.M. White, and M. Novotny, *Anal. Chem.* **51**, 768 (1979).

70. S.A. Wise, B.A. Benner, G.D. Byrd, S.N. Chesler, R.E. Rebbert, and M.M. Schantz, *Anal. Chem.* **60**, 887 (1988).

71. D.L. Vassilaros, R.C. Kong, D.W. Later, and M.L. Lee, *J. Chromatogr.* **252**, 1 (1982).

72. G.M. Janini, G.M. Muschik, J.A. Schroer, and W.L. Zielinski, *Anal. Chem.* **48**, 1879 (1976).

73. A. Radecki, H. Lamparczyk, and R. Kaliszan, *Chromatographia,* **12**, 595 (1979).

74. H.J. Issaq, G.M. Janini, B. Poehland, R. Shipe, and G.M. Muschik, *Chromatographia,* **14**, 655 (1981).

75. L.C. Sander and S.A. Wise, *Polycyclic Aromatic Hydrocarbon Structure Index.* NIST Special Publications, Washington, DC, in press.

76. L.C. Sander, M. Schneider, C. Woolley, and S.A. Wise, *J. Microcolumn Sep.* **6,** 115 (1994).

77. S.A. Wise, R.M. Campbell, W.E. May, M.L. Lee, and R.N. Castle, in M. Cooke and A.J. Dennis, Eds., *Polynuclear Aromatic Hydrocarbons, Seventh International Symposium.* Battelle Press, Columbus, OH, 1983, p. 1247.

78. P. Garrigues, M. Radke, O. Druez, H. Willsch, and J. Bellocq, *J. Chromatogr.* **473,** 207 (1989).

79. D.E. Martire, *J. Chromatogr.* **406,** 27 (1987).

80. C. Yan and D.E. Martire, *J. Phys. Chem.* **96,** 3489 (1992).

81. C. Yan and D.E. Martire, *Anal. Chem.* **64,** 1246 (1992).

82. C. Yan and D.E. Martire, *J. Phys. Chem.* **96,** 7510 (1992).

83. C. Yan and D.E. Martire, *J. Phys. Chem.* **96,** 3505 (1992).

2

Molecular Planarity Recognition for Polycyclic Aromatic Hydrocarbons in Liquid Chromatography

Kiyokatsu Jinno

2.1 Introduction

The analysis of polycyclic aromatic hydrocarbons (PAHs) is of great importance in the environmental and biological fields because of increasing concern over the mutagenicity and carcinogenicity of many of these compounds. Since the level of biological activity (mutagenicity or carcinogenicity) of a particular PAHs is very much dependent on structure, similar isomers can range from being highly active to being totally inactive [1]. The quantitation of the PAHs and their present levels in various materials, therefore, could be used to estimate the bioactivity of these substances. Many instrumental techniques have been used for PAH analysis, but many of the more common ones have limited utility in applications entailing the recognition of isomeric PAHs. Most PAHs are not volatile, hence are not readily amenable to analysis by gas chromatography (GC). The huge number of isomers for any particular carbon number limits the use of mass spectrometry (MS) because isomers have identical molecular weights and yield similar fragmentation patterns. Reversed-phase liquid chromatography (LC) has therefore become one of the preferred methods for the separation and analysis of PAHs.

Stationary phases of many different types are available in LC, but the most widely used one is octadecylsilica (ODS). The first step in understanding the molecular planarity recognition mechanism in an LC separation process is to understand the details of the separation properties of such ODS stationary phases; the information obtained by examining ODS phases will then give insights that will be useful in developing different stationary phases with improved selectivity for molecular planarity recognition. This chapter describes such approaches and uses

65

illustrations of real liquid chromatographic data for PAHs to introduce many newly synthesized stationary phases that offer promise of better shape selectivity.

2.2 ODS Stationary Phases

Retention in the LC analysis of many PAHs is generally controlled by the size and two-dimensional shape of the PAH. However, the elution behavior of larger PAHs with various ODS stationary phases is not explained by these properties, especially in nonaqueous reversed-phase separations, where anomalous elution behavior that is not controlled simply by size and shape has apparently been seen. Nonplanar solutes that are larger than small, planar solutes are eluted very early on particular chemically bonded stationary phases. To understand such observations, the ODS phases must be evaluated systematically.

Generally ODS phases can be divided into two different types: monomeric and polymeric, on the basis of the bonding chemistry used for their preparation. Monomeric phases are synthesized by monochlorosilane as the starting material, and trichlorosilane is the starting material for the polymeric type, as shown in Figure 2.1. The difference between these two ODS phase types is defined by the C_{18} alkyl chains on the silica surface: the monomeric phase has isolated single C_{18} chains but the possibility of networked chains exists for the other type (hence the designation "polymeric").

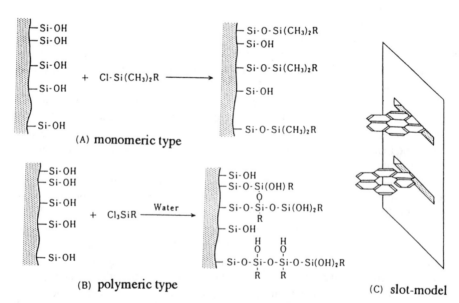

Figure 2.1 Methods for the preparation of two types of ODS stationary phase and the slot model for the structure of the polymeric-type ODS phase.

Are there any differences between the two ODS phases that are significant in terms of the shape and size recognition ability for PAHs?

Wise and Sander have examined this topic intensively [2–5] (some of their work is described in Chapter 1 of this book), and they have found that polymeric ODS has a stronger tendency to recognize the molecular planarity of PAHs than with the monomeric phase. These authors have concluded that the structure of the polymeric ODS is more rigid and slitlike (they developed a "slot model" to describe it) than that of the monomeric ODS. Thus the polymeric ODS cannot retain or interact well with nonplanar PAHs because of its slitlike structure, while planar solutes can be retained more strongly, because they more easily enter the slits. The big question raised by Wise's proposed slot model is this: What is the actual situation of the bonded phases in polymeric ODS compared to monomeric ODS? It was to find the most reasonable explanation of the slot model proposed by Wise and Sander that the following experiments and evaluation were performed.

2.2.1 Planarity Recognition of PAHs with Various ODS Phases

Various ODS phases were synthesized in the laboratory, using the same base silicas for all phases to prevent differences of the base silicas from influencing the chromatographic properties. The silica used had a nominal diameter of 5 μm and a pore size of 110 Å. In the synthetic procedure employed, the silica gel was first dried at 150°C for 2 hours, then dried under reduced pressure (30 mmHg) for an additional 2 hours. The derivatization reactions were performed with 10 g of silica gel and 8.5 μmol/m² equivalents of silane compound in pyridine and toluene. Trichlorooctadecylsilane was used for the preparation of trifunctional phases, dichloromethyloctadecylsilane for difunctional phases, and monochlrodimethyloctadecylsilane for monofunctional phases, respectively. Each mixture of silica gel and a silane was refluxed for 5 hours at 110°C in toluene. After cooling, the packings were dispersed in chloroform, filtered, and washed several times with chloroform. The final packings were dried at 80°C for 8 hours. End capping was performed using 8.5 μmol/m² equivalents of trimethylchlorosilane and hexamethyldisilazane under refluxing pyridine and toluene for 5 hours. Half-loadings were produced using 2–2.5 μmol/m² equivalents of each silane.

The synthesized packings are categorized by functionality (tri-, di-, and mono-) of the silane, with or without end capping, and the maximum loading or half-loading of the material as in Table 2.1. Some commercially available ODS phases were selected for reference, as shown in Table 2.2.

Figure 2.2 presents schematically the structures of the four PAHs that comprised the test probe selected in this evaluation. Coronene and dibenzo[cd,lm]perylene are planar, and tribenzo[a,cd,lm]perylene and tetrabenzo[a,cd,j,lm]perylene are nonplanar; the molecular size increases from coronene to tetrabenzo[a,cd,j,lm]perylene. Figure 2.3 clearly indicates the nonplanarity of tetrabenzo[a,cd,j,lm]perylene.

For the LC evaluation, the phases listed in Table 2.1 were packed in 0.53 mm i.d. fused-silica capillaries, 15–30 cm long, and the mobile phase was a 50:50 mixture

Table 2.1   LABORATORY-PREPARED ODS PHASES USED IN THIS STUDY

Abbreviation	Functionality	Degree of Loading	Carbon Content (%)	End Capping
TM	Tri-	Maximum	18.20	No
TME	Tri-	Maximum	19.07	Yes
TH	Tri-	Half	8.50	No
THE	Tri	Half	11.18	Yes
DM	Di-	Maximum	18.48	No
DME	Di-	Maximum	18.88	Yes
MM	Mono-	Maximum	17.79	No
MME	Mono-	Maximum	18.40	Yes
MH	Mono-	Half	8.20	No
MHE	Mono-	Half	10.46	Yes

of dichloromethane and methanol with a flow rate of 4 μL/min. The retention of those four PAHs with various ODS phases was measured and the retention factor (k') values were used for the evaluation. Those values are summarized in Table 2.3, where the relative retention to coronene is also calculated for each experimental condition.

Table 2.2   COMMERCIALLY AVAILABLE ODS PHASES EVALUATED IN THIS WORK

Product Name	Manufacturer	Particle Diameter (μm)	Pore Size (Å)
Vydac 201 TPB5	Separations Group (Hesperia, CA, USA)	5	300
Vydac 218 TPB5	Separations Group (Hesperia, CA, USA)	5	300
Develosil ODS-T5	Nomura Chemicals (Seto, Japan)	5	110
Develosil ODS-K5	Nomura Chemicals (Seto, Japan)	5	110
Develosil ODS-A5	Nomura Chemicals (Seto, Japan)	5	110
TSK ODS-120T	Tosoh (Tokyo, Japan)	5	120
RP-18	Merck (Darmstadt, Germany)	5	100
Nucleosil 100-10	Macherey-Nagel (Duren, Germany)	10	100
FinSIL	JASCO (Tokyo, Japan)	10	120

Figure 2.2 Structures of peropyrene-type PAHs used in this work: (A) coronene, (B) dibenzo[*cd,lm*]perylene, (C) tribenzo[*a,cd,lm*]perylene, and (D) tetrabenzo[*a,cd,j,lm*] perylene.

The data in Table 2.3 indicate the different chromatographic behaviors for planar and nonplanar solutes with respect to ODS functionality, highlighting the impact of ODS functionality on the separation process. Plots of relative retention and the structure of the PAHs clearly show trends (Figure 2.4). For ODS packings prepared using a trifunctional silane, for example, high-loading packings (TME and TM) show exclusive planarity recognition, such that nonplanar solutes eluted faster than planar solutes, even though the latter molecules are smaller. For example, with the TM packing, tetrabenzo[*a,cd,j,lm*]perylene eluted faster than coronene. On the TME phase, the retention times of the two compounds were almost identical. The elution order of PAHs on TM and TME depends on the degree of nonplanarity of the solute. The planar solutes coronene and dibenzo[*cd,lm*]perylene eluted in relation to their size. However, with the TH packing, the trend is completely different from the behavior with the TM and TME packings. With TH the elution order depends only on the size of the solutes. The THE packing seems to be intermediate between the TM and TME packings, and the TH phase. It is apparent that the low-loading packing is fundamentally different from the high-loading packing, even though both are generated from the same silane functionality.

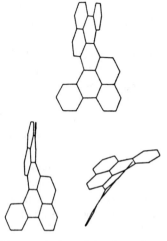

Figure 2.3 Computer graphics, rendering of the three-dimensional structure of tetrabenzo[*a,cd,j,lm*]perylene.

Table 2.3 RETENTION FACTORS AND RELATIVE RETENTIONS OF FOUR LARGE
PAHs ON VARIOUS STATIONARY PHASES

Stationary Phase	Coronene	Dibenzo[cd,lm]-perylene	Tribenzo[a,cd,lm]-peryrene	Tetrabenzo[a,cd,j,lm]-perylene
		Retention Factors (relative retention)[a]		
TM	0.885	1.521 (1.719)	0.890 (1.006)	0.726 (0.821)
TME	0.871	1.439 (1.653)	1.000 (1.149)	0.852 (0.978)
TH	0.356	0.347 (1.051)	0.389 (1.094)	0.389 (1.095)
THE	0.322	0.444 (1.380)	0.404 (1.257)	0.398 (1.237)
DM	0.725	0.935 (1.290)	0.905 (1.248)	0.847 (1.168)
DME	0.543	0.633 (1.164)	0.632 (1.163)	0.618 (1.137)
MM	0.542	0.667 (1.231)	0.686 (1.267)	0.830 (1.533)
MME	0.529	0.596 (1.126)	0.625 (1.182)	0.859 (1.624)
MH	0.256	0.274 (1.070)	0.301 (1.173)	0.354 (1.381)
MHE	0.296	0.332 (1.121)	0.358 (1.211)	0.402 (1.360)

[a] Relative retention to coronene with 50:50 dichloromethane/methanol as the mobile phase.

With ODS packings prepared using a difunctional silane, the trend in planarity discrimination is not as obvious as with the trifunctional ODS phases, but it is still observed with both the DM and DME phases. This behavior is very similar to that of THE packing.

With ODS phases prepared by monofunctional silane, the situation changes drastically. The elution order of the PAHs is now dependent on molecule size alone, being independent of compound planarity. No difference is found between end capping and non–end capping. Low-loading packings, with and without end capping, also gave weaker retention in this case. But unlike the case of the trifunctional packings, the elution behavior was not different from that of the high-loading packings.

Clearer discrimination between packings prepared using various silane functionalities was found by statistical cluster analysis. The cross-correlation coefficients of each data set were calculated and are summarized in Table 2.4. Using these data, cluster analysis was performed. The dendrogram obtained is shown in Figure 2.5. The results indicate that three large categories clearly exist. One category is for monofunctional and low-loaded trifunctional packings without end capping. Another category is for difunctional and low-loaded trifunctional packings with end cap-

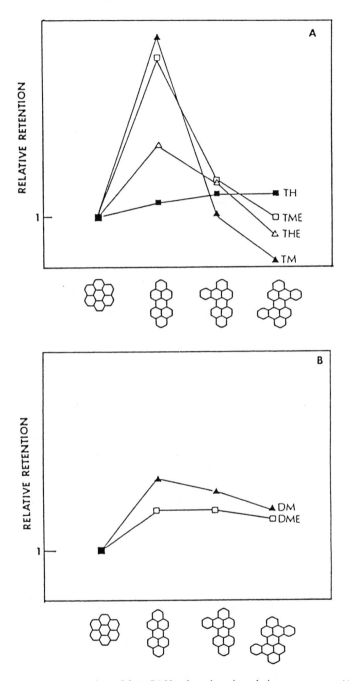

Figure 2.4 Relative retention of four PAHs plotted against their structures on (A) trifunctional, (B) difunctional, and (C) monofunctional phases. [5] The phases are identified in Table 2.1. Reproduced from K. Jinno et al. in *Chromatographia* **27**, 285 (1989) by permission of Vieweg Publishing, Wiesbaden, Germany.

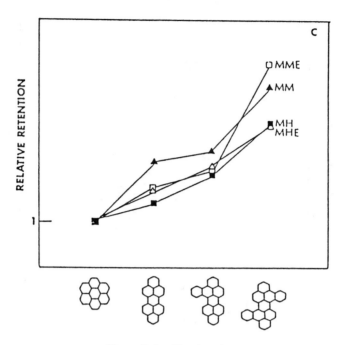

Figure 2.4 (*Continued*)

ping, and the last category is high-loaded packings prepared using a trifunctional silane. From the results of the cluster analysis, the elution behaviors observed in the data discussed earlier can be interpreted as arising from bonded phase structure differences generated by the various derivatization schemes.

As an extension of the evaluation for the different functional ODS phases synthesized, several commercially available ODS phases were also evaluated using four PAHs as the test probes. Figure 2.6 graphs the retention behavior of the four PAHs with those stationary phases, and from these results it is easy to state that Vydac 218 and 201 are trifunctional, as are TSK and Develosil A5 and T5. Develosil K5 is monofunctional; Nucleosil may be difunctional. Only the RP phase is difficult to judge—it may be a low-loaded trifunctional or difunctional phase.

2.2.2 CP-MAS NMR Spectroscopic Measurements of Various ODS Phases

The discussion in Section 2.2.1 about the differences in characteristics of functional ODS phases is mainly based on evidence from chromatographic retention studies. Recently, however, the approach of characterizing the chromatographic bonded phases with various spectroscopic methods has become extremely popular. Besides infrared, fluorescence, and photoacoustic spectroscopy, nuclear magnetic resonance (NMR) spectroscopy is a technique that has attracted considerable research interest.

Table 2.4 CORRELATION MATRIX OF 10 LABORATORY-PREPARED STATIONARY PHASES

	TM	TME	TH	THE	DM	DME	MM	MME	MH	MHE
TM	—									
TME	0.980	—								
TH	0.227	0.030	—							
THE	0.598	0.741	0.638	—						
DM	0.544	0.701	0.686	0.974	—					
DME	0.335	0.515	0.842	0.946	0.968	—				
MM	0.262	0.106	0.867	0.552	0.487	0.679	—			
MME	0.454	0.334	0.727	0.302	0.213	0.434	0.956	—		
MH	0.517	0.374	0.827	0.316	0.271	0.501	0.962	0.980	—	
MHE	0.387	0.222	0.909	0.477	0.446	0.654	0.985	0.947	0.982	—

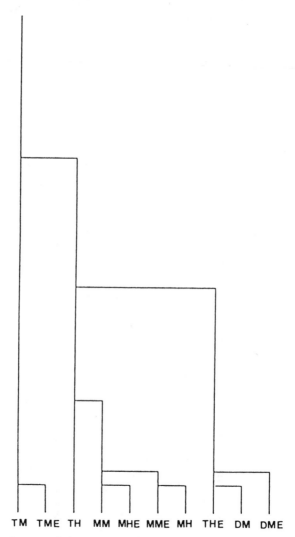

TM TME TH   MM MHE MME MH THE DM DME

Figure 2.5   Dendrogram of cluster analysis for the retention of four PAHs. The phases are identified in Table 2.1. [5] Reproduced from K. Jinno et al. in *Chromatographia* **27**, 285 (1989) by permission of Vieweg Publishing, Wiesbaden, Germany.

The emergence of high resolution NMR techniques for solids has made it possible to bring the great power of these methods to the study of molecular structure and dynamics for the investigation of retention mechanisms in reversed-phase LC. This progress is mainly due to the advent and development of solid state NMR with cross-polarization (CP) and magic-angle-spinning (MAS) techniques. CP-MAS is the technique of combining MAS, which removes the line-broadening effect of chemical shift anisotropy, with CP, which is a double-resonance technique for the transfer of magnetization from an abundant spin set (usually protons) to a dilute spin set, thereby circumventing what can otherwise be a debilitating time bottleneck due to inefficient spin–lattice relaxation. From this basic principle of CP, the approach is effective only for relatively immobile systems, thereby providing a way of distinguishing such systems on a surface from the highly mobile species that would give rise to efficient spin–lattice relaxation. Therefore, CP-MAS should provide new insights for the study of structures of the stationary phase used in reversed-phase LC, typically alkyl bonded phases based on silica gels such as ODS phases.

A number of research groups have been committed to the application of high resolution, solid state NMR techniques for studying retention interactions in LC [6–20]. As a result of their efforts, it is clear that [^{13}C]CP-MAS spectra can be highly useful in characterizing alkyl-bonded stationary phases.

As discussed in the preceding section, typical ODS phases are categorized into three functional phases (mono-, di-, and tri-), and those different functionalities induce different molecular planarity recognition ability for PAHs. To explain why this should be so, we look to CP-MAS, because this very promising technique makes available new views of phase structures bonded on silica gel. The results of attempts to identify ODS functionalities are very promising for the elucidation of the configuration of the organosilyl moiety in ODS, that is, for determining the extent to which the chains can be described by such popularly discussed models as "bristle brush," "haystack," and "blanket" [11,19], as pictured in Figure 2.7, and the extent to which structure and dynamics are influenced by various chromatographic environments and/or interactions of solvent or solute species with the silica surface or the alkyl chains.

The CP-MAS spectra obtained for the various ODS materials synthesized are shown in Figures 2.8, 2.9, and 2.10. The spectral assignments of each peak position are as follows:

$$-O-\underset{\underset{CH_3}{|}}{\overset{\overset{CH_3}{|}}{Si}}-CH_2-CH_2-(CH_2)_{12}-CH_2-CH_2-CH_3$$

C-4 to

C-1′   C-1   C-2   C-15   C-16   C-17   C-18

These assignments had been made earlier on the basis of well-established empirical guidelines for liquid state ^{13}C NMR [18–20]. Eight distinct peaks are observed—six of them assignable to specific carbon positions and two corresponding to overlap of two or more resonance, depending on the characteristics of the stationary phases. The signal at 2 ppm can be assigned to CH_3 of the end-capped trimethyl groups.

In Figure 2.8, the trifunctional ODS gave very few signals compared to those found in the spectra of the difunctional and the monofunctional ODS phases. The NMR spectrum of the trifunctional polymeric phase, which shows no signals f C-1 and C-2 and only a very weak signal for C-3, appears to indicate the v

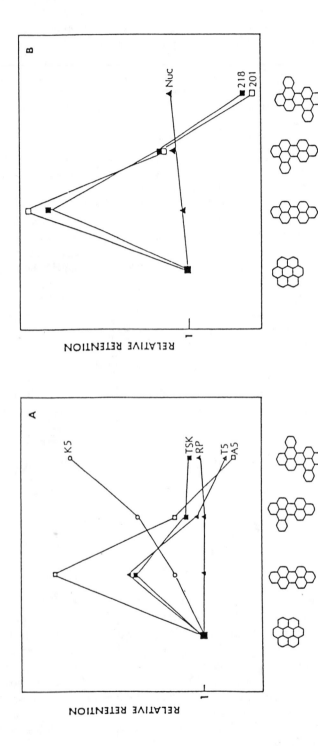

Figure 2.6 Relative retention of four PAHs plotted against their structures on two sets of commercially available phases [5] (identified in Table 2.2). Reproduced from K. Jinno et al. in *Chromatographia* **27**, 285 (1989) by permission of Vieweg Publishing, Wiesbaden, Germany.

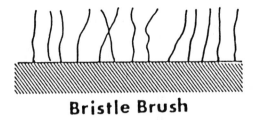

Bristle Brush

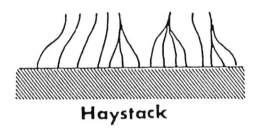

Haystack

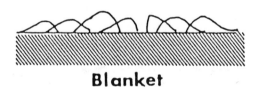

Blanket

Figure 2.7 Suggested models for octadecyl chains of a derivatized silica gel. [21] Reproduced from K. Jinno in *J. Chromatogr. Sci.* **27**, 729 (1989) by permission of Preston Publications, Inc., Niles, IL, USA.

restricted situation for the alkyl bonded moiety. The absence or near-absence of signals for those carbons, which is due to the inability of the bottom part of the alkyl bonded phase to move freely, means that the trifunctional ODS could have a "bristle brush" or "haystack" structure, as shown in Figure 2.7. The monofunctional ODS, however, can move more freely, and the signal intensities of C-1, C-2, and C-3 are stronger than those of C-16, C-17, and C-18. These results indicate that the monofunctional phase may have the "blanket" structure and the top terminal part of the alkyl bonded phase is bent to the surface of the base materials shown in Figure 2.7. It seems that the difunctional polymeric ODS gives the intermediate pattern of the NMR spectrum between tri- and monofunctional ODS phases, and rather close to that of the monomeric phase. These signal pattern differences clearly explain the dissimilar surface structures among the various functionalities of the ODS bonded phases. Figure 2.9 shows the effect of end capping on the NMR spectral changes.

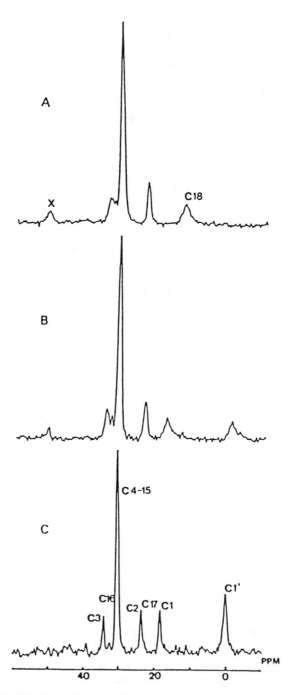

Figure 2.8 ^{13}C CP-MAS NMR spectra of three ODS silicas, showing chemical shift assignments: (A) TM, (B) DM, and (C) MM. [21] Reproduced from K. Jinno in *J. Chromatogr. Sci.* **27**, 729 (1989) by permission of Preston Publications, Inc., Niles, IL, USA.

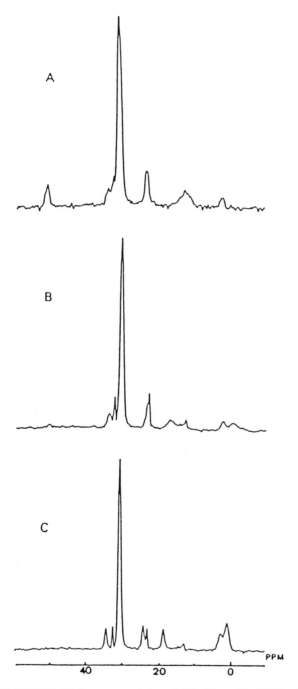

Figure 2.9 ^{13}C CP-MAS NMR spectra of three ODS silicas: (A) TME, (B) DME, and (C) MME. [21] Reproduced from K. Jinno in *J. Chromatogr. Sci.* **27**, 729 (1989) by permission of Preston Publications, Inc., Niles, IL, USA.

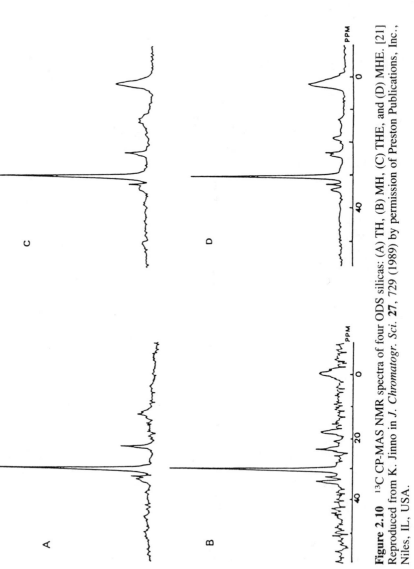

Figure 2.10 ^{13}C CP-MAS NMR spectra of four ODS silicas: (A) TH, (B) MH, (C) THE, and (D) MHE. [21] Reproduced from K. Jinno in *J. Chromatogr. Sci.* **27**, 729 (1989) by permission of Preston Publications, Inc.., Niles, IL, USA.

Of course, the distinct difference in the spectra of non-end-capped and end-capped materials is the signal at 2 ppm of the methyl groups of the end capping trimethyl groups bonded on the surface. For all the phases, the end capping process induces the more freely movable situation of the top part of the alkyl phase, because the NMR spectra indicate that the signal intensities of C-18, C-17, and C-16 are enhanced by end capping. The enhancement is very distinctive for the monomeric and difunctional polymeric phases, suggesting that the surface structures of those phases can change as a result of end capping from the "blanket"-like form to the other two forms. The space between each pair of bonded alkyl chains is decreased by the end capping process, and this makes the surface structure more ordered. In the trifunctional phase the ordered phase originally exists before the end capping process, and afterward there is little change. This is why the signal enhancements for carbon atoms at the tops of chains are not very large in those trifunctional phases. Even for the low-loading trifunctional and the monofunctional phases, the situation is very similar. The NMR spectrum of the low-loading trifunctional phase indicates that C-3 can move more easily than that of its high-loading counterpart, but still the movement of C-1 and C-2 is restricted. However, the signal peak widths of C-4 to C-15 of the low-loading trifunctional phase are narrower than those of the high-loading trifunctional phase, and this suggests that the long carbon chains of this material are more freely movable than those of the high-loading phases. Therefore, the structure of the material is more likely to be of the "haystack" type.

The NMR features are summarized in Table 2.5, and it is clearly demonstrated that CP-MAS ^{13}C NMR spectra can be a powerful tool for the depiction of basic surface structures of various ODS phases.

The characteristics of the CP-MAS spectra of ODS phases can be useful in the identification of the functionality of commercial ODS phases. To confirm this

Table 2.5 FEATURES OF ^{13}C CP-MAS NMR SPECTRA OF VARIOUS ODS PHASES [21]

	Assignments of Carbons[a]								
ODS Phase	C-18	C-17	C-2	C-4 to C-15	C-16	C-1	C-3	C-1′	50 ppm[b]
TME	B	B	B	B	B	—	WB	O	O
TM	B	B	WB	B	B	—	WB	—	O
THE	B	S	B	S	S	—	B	B	—
TH	B	S	B	S	S	—	B	—	—
MME	O	S	S	S	S	O	S	D	—
MM	—	W	S	S	W	O	S	O	—
MHE	W	S	S	S	S	O	S	B	—
MH	—	W	S	S	S	O	S	O	—
DME	W	S	S	B	S	W	B	D	—
DM	—	B	WB	B	B	O	B	B	O

[a] B, broad signal; S, sharp signal, W, weak signal; O, signal; WB, weak broad signal; D, doublet signal; —, no signal.

[b] This signal is tentatively assigned to $—OCH_3$.

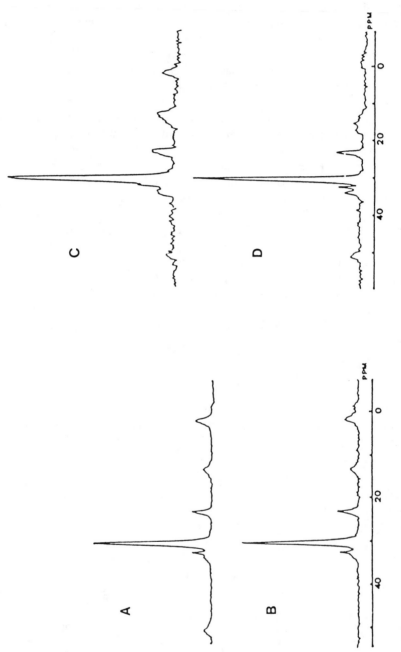

Figure 2.11 ^{13}C CP-MAS NMR spectra of (A) Nucleosil ODS, (B) FineSIL ODS, (C) TSK ODS, and (D) RP-18. [21] Reproduced from K. Jinno in *J. Chromatogr. Sci.* **27**, 729 (1989) by permission of Preston Publications, Inc., Niles, IL, USA.

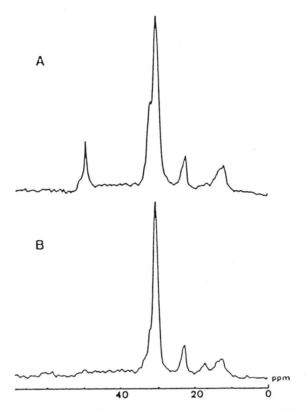

Figure 2.12 ^{13}C CP-MAS NMR spectra of (A) Vydac 201 and (B) Vydac 218. [21] Reproduced from K. Jinno in *J. Chromatogr. Sci.* **27**, 729 (1989) by permission of Preston Publications, Inc., Niles, IL, USA.

possibility, several commercial ODS phases were examined by NMR spectroscopy. The NMR spectra obtained are found in Figures 2.11 to 2.14. Several features of the measured NMR spectra were extracted and summarized in Table 2.6. Comparing those features with those of the synthesized ODS phases in Table 2.5, one can identify the functionality of the ODS phases as shown in Table 2.7, which gives the functionality determined by the CP-MAS method. This information can be easily compared to the retention behavior method described in the preceding section, and Table 2.7 presents such a comparison. Thus it is clear that both methods can recognize the functionality of ODS phases very easily and accurately.

2.2.3 Suspension NMR Spectroscopic Measurements of ODS Phases

In liquid chromatography the mobile phase exists between the stationary phase and the sample solutes. Therefore, it is reasonable to ask why CP-MAS measurements of the stationary phases fail to produce spectra that show any evidence of bonded

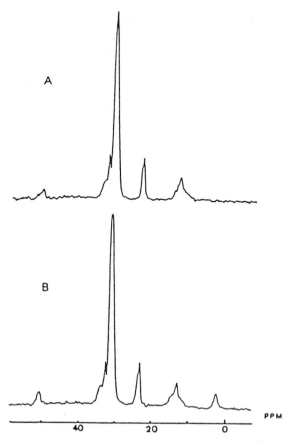

Figure 2.13 ^{13}C CP-MAS NMR spectra of (A) Develosil ODS-A5 and (B) ODS-T5. [21] Reproduced from K. Jinno in *J. Chromatogr. Sci.* **27**, 729 (1989) by permission of Preston Publications, Inc., Niles, IL, USA.

phase structures under LC conditions. To clarify to this matter, one must examine the NMR spectroscopic measurements under conditions similar to those of the LC process; suspension NMR measurements are required. Therefore suspension NMR spectra of each ODS phase have been obtained (Figures 2.15 and 2.16). Figure 2.15 shows the differences among high-loaded trifunctional, difunctional, and mono-functional phases. The main resonances, found at 13, 22, and 30 ppm, can be assigned to the methyl group at the end of the C_{18} chains, the methylene groups at the β-position from the ends of ODS chains and the base silicon atoms, and the bulk methylene groups, respectively. Apparently, the freedom of the bulk methylene groups in octadecyl chains is different among bonded phases prepared from the three different silane functionalities. McNally and Rogers [22] and Yonker et al. [23] proposed using the width at half-height of the 30 ppm peak (the bulk methylene groups) as a measure of the liquidlike nature of the bonded alkyl chains. Wider peaks result when there is less freedom of movement due to various interactions,

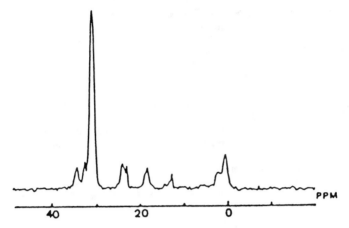

Figure 2.14 ^{13}C CP-MAS NMR spectrum of Develosil ODS-K. [21] Reproduced from K. Jinno in *J. Chromatogr. Sci.* **27**, 729 (1989) by permission of Preston Publications, Inc., Niles, IL, USA.

such as the bonded phase chain with the surface, other bonded phase chains, and unreacted silanols. As seen in Figure 2.15, the half-width of the 30 ppm peak for ODS packings generated from trifunctional silane is the largest, and that of difunctional ODS packings is larger than that of monofunctional ODS phases. The order means that the bonded chains of trifunctional ODS have the least freedom of molecular motion and monofunctional ODS have the most among the three types of packings.

Table 2.6 FEATURES OF ^{13}C NMR SPECTRA OF VARIOUS COMMERCIAL ODS PHASES [21]

Product Name				Assignments of Carbons[a]					
	C-18	C-17	C-2	C-4 to C-15	C-16	C-1	C-3	C-1'	50 ppm[b]
Vydac 201	B	B	WB	GB	B	—	WB	—	—
Vydac 218	B	B	WB	B	B	—	WB	O	O
ODS-K	O	S	S	S	S	O	S	D	c
ODS-A5	B	B	WB	B	S	—	WB	—	O
ODS-T5	B	B	WB	B	S	—	WB	O	O
RP-18	W	B	B	B	S	W	B	W	O
TSK-120T	B	B	WB	B	B	—	B	O	W
Nucleosil	B	B	WB	S	S	—	WB	O	O
FinSIL	B	B	WB	S	B	—	WB	O	—

[a] Abbreviations as in Table 2.5 plus GB, great broad signal.

[b] This signal is tentatively assigned to —OCH$_3$.

[c] Spectrum not measured at this region.

Source: Reproduced from K. Jinno in *J. Chromatogr. Sci.* **27**, 729 (1989) by permission of Preston Publications, Inc., Niles, IL, USA.

Table 2.7   STATIONARY PHASE FUNCTIONALITY IDENTIFIED BY NMR AND RETENTION BEHAVIOR METHODS

Product Name	NMR Method	Retention Behavior Method
Vydac 201	Tri-, high polymeric	Tri-, maximum load
Vydac 218	Tri-, medium polymeric, end capping	Tri-, maximum load, end capping
ODS-T5	Tri-, medium polymeric, end capping	Tri-, maximum load, end capping
ODS-A5	Tri-, medium polymeric	Tri-, maximum load
ODS-K	Mono-, end capping	Mono-, maximum load, end capping
TSK-120T	Tri-, medium polymeric, end capping	Tri-, maximum load, end capping
Nucleosil	Tri-, low polymeric, end capping	Tri-, low load
FineSIL	Tri-, low polymeric, end capping	
RP-18	Di-, low polymeric	Di- or tri-, low load

As shown in Figure 2.16, the order of this molecular freedom is not changed by the end capping process. Very interesting NMR spectra (Figure 2.17) were obtained for half-loaded materials. The peak widths at half-height for both these half-loaded packings are much smaller than those of their high-loaded counterparts. This means low-loaded materials have greater freedom of molecular motion than their high-loaded counterparts. Thus the suspension NMR studies permit the conclusion that the freedom of molecular motion of ODS chains in high-loaded packings decreases with the functionality of the silanes used for preparation in the following order: mono, di, tri. For the case of low-loaded phases, the trend still remains. Even the bulk methylene groups of the low-loaded trifunctional phase have more freedom than the high-loaded difunctional phases. The order of the molecular freedom of the alkyl chains is:

MH > TH > MM, MME, MHE > DM, DME, THE > TM, TME

This ordering is totally consistent with CP-MAS NMR spectral information and also can explain the retention behavior of PAHs with various ODS phases, if we assume that the planarity recognition capability of the bonded phases produced from trifunctional silanes is due to the limited freedom of motion possessed by the alkyl chains. Difunctional phases have freedom intermediate between mono- and trifunctional phases and, therefore, the planarity recognition capability is also intermediate. The highly mobile methylene chains of monofunctional ODS phases and low-loaded mono- and trifunctional ODS phases have low planarity recognition capability because of their nonrigid structures. Rigid structures can recognize planar solutes well, and nonplanar solutes are excluded from the stationary phase chains. As the "slot model" proposed by Wise and Sander [2–5] suggests, polymeric phases have rigid structure and monomeric phases do not. The ^{13}C CP-MAS and suspension spectral studies performed here clearly indicate these structural differences among the bonded phases prepared from siloxanes having different functionalities.

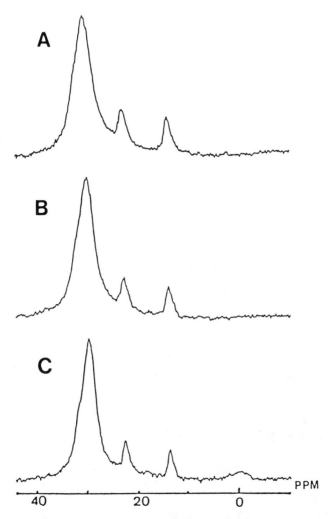

Figure 2.15 Suspension ^{13}C NMR spectra for (A) TM, (B) DM, and (C) MM ODS phases (identified in Table 2.1). [5] Reproduced from K. Jinno et al. in *Chromatographia* **27**, 285 (1989) by permission of Vieweg Publishing, Wiesbaden, Germany.

2.2.4 More Detailed Study on Polymeric ODS Phases for PAH Planarity Recognition

The literature details a number of attempts to understand the function of the stationary phase in reversed-phase LC by means of spectroscopic techniques such as NMR [6,9,11,14,15,24–27], FTIR [10,28], and thermal analysis [29]. However, those reports concerned mainly characteristics of the starting silane materials and stationary phases other than ODS chains, and sometimes no distinction was made between

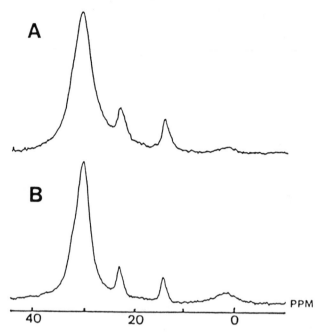

Figure 2.16 Suspension ^{13}C NMR spectra for (A) TME and (B) MME phases (identified in Table 2.1). [5] Reproduced from K. Jinno et al. in *Chromatographia* **27**, 285 (1989) by permission of Vieweg Publishing, Wiesbaden, Germany.

monomeric and polymeric ODS phases. Those publications also failed to relate the phase data to chromatographic retention data obtained with the stationary phases. It is reasonable to suppose that our understanding of reversed-phase retention mechanisms will be enhanced by the interpretation of chromatographic retention data related to the characteristics of stationary phases examined in the light of supplemental spectroscopic information. Therefore, in this section as an extension of the preceding material, we provide more details concerning the spectroscopic interpretation of planarity recognition for PAHs in reversed-phase LC with a polymeric ODS stationary phase (Vydac 201).

The chromatographic elution order of the four PAHs introduced in Figure 2.2 with different mobile phase compositions is illustrated in Figure 2.18, where the mobile phase compositions range between 20 and 50% dichloromethane in methanol. The typical behavior shown in Figure 2.18 can be interpreted to mean that the elution order of PAHs depends mainly on their size and planarity. In this evaluation it has been found that tetrabenzo[*a,cd,j,lm*]perylene (nine rings) is always eluted earlier than dibenzo[*cd,lm*]perylene (seven rings). At higher dichloromethane concentrations the solute was eluted even earlier than coronene, the smallest molecule used here. This is due to the nonplanarity of tetrabenzo[*a,cd,j,lm*]perylene as shown in Figure 2.3. The contribution of dichloromethane to this exclusion-like retention behavior of the nonplanar PAHs may be due to the modification of the structure of

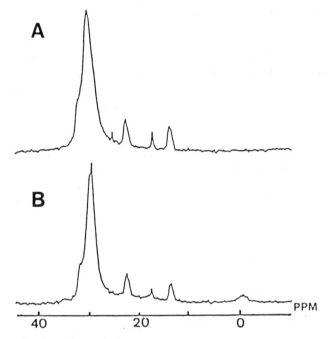

Figure 2.17 Suspension ^{13}C NMR spectra for (A) TH and (B) MH phases (identified in Table 2.1). [5] Reproduced from K. Jinno et al. in *Chromatographia* **27**, 285 (1989) by permission of Vieweg Publishing, Wiesbaden, Germany.

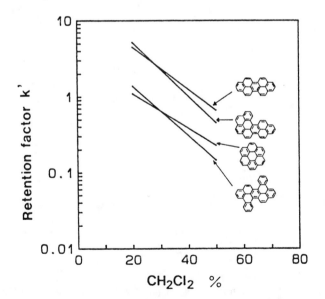

Figure 2.18 Plots of log k' of four peropyrene-type PAHs versus dichloromethane concentration in methanol as the mobile phase. [23] Reproduced from K. Jinno et al. in *J. Chromatogr.* **461**, 209 (1989) by permission of Elsevier Science-NL, Amsterdam, The Netherlands.

the polymeric ODS, the solute planarity, or both, which occurs with changes in the mobile phase composition.

To determine which factor is dominant in LC, NMR measurements were performed. Figure 2.19 shows ^{13}C NMR spectra of the polymeric ODS suspended in different compositions of deuteromethanol/deuterodichloromethane. It has been proposed by McNally and Roger [22] that the width at half-height of the bulk CH_2 peaks at 33 ppm can be used as a measure of the liquidlike nature of the bonded alkyl chains. Wider peaks are a result of less freedom of movement due to interactions of the bonded phase chain with the surface, including other bonded phase chains and unreacted silanols. As seen in Figure 2.19, only small changes occur with increasing dichloromethane concentration. It appears that as the solvent is changed from pure methanol to pure dichloromethane, the peak width and intensity increase for the peak at 33 ppm as well as for the peaks at 24 ppm (assigned to β-CH_2) and 15 ppm (assigned to the terminal methyl groups). These small changes

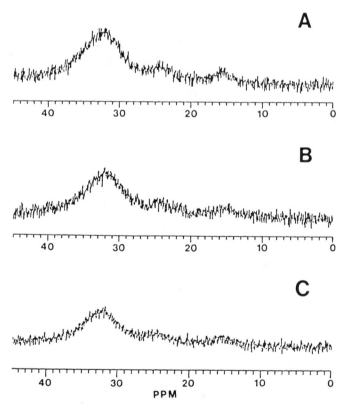

Figure 2.19 ^{13}C NMR spectra of Vydac 201 TPB5 in various solvents: (A) 100% [^{2}H$_2$]dichloromethane, (B) [^{2}H$_2$]dichloromethane/[^{2}H$_4$]methanol (50:50), and (C) 100% [^{2}H$_4$]methanol. [23] Reproduced from K. Jinno et al. in *J. Chromatogr.* **461**, 209 (1989) by permission of Elsevier Science-NL, Amsterdam, The Netherlands.

might indicate a solvent effect on this polymeric ODS. The work of Shah et al. [19] indicates that in 100% dichloromethane, the difunctional polymeric ODS chains move relatively freely and interact with the solvent. Although Figure 2.19 does not show this clearly, it is reasonable to consider that as the percentage of methanol increases, the molecular motion of the bonded alkyl chain becomes more restricted, but not as much as Shah et al. [19] observed for a difunctional polymeric phase.

Since solute structure changes seem to be more reasonable in the sense of solution chemistry, ^{1}H NMR spectra were collected for two representative solutes (coronene and tetrabenzo[a,cd,j,lm]perylene) at three deuterodichloromethane concentrations (20, 50, and 100% in deuteromethanol). The results appear in Figure 2.20. The signals from tetrabenzoperylene at 9.2 ppm were drastically shifted to higher magnetic field when the dichloromethane concentration in the mobile phase was increased from 20% to 50%. No difference in the spectra was observed between 50 and 100% dichloromethane. The signals at 9.2 ppm with 20% dichloromethane concentration showed four lines, but with 50 and 100% only two lines remained and two lines were shifted to a higher field. The signals at 9.2 ppm can be assigned to protons in the A region of its structure, and the shift of the signals to 9.05 ppm suggests that the environment of those protons has been changed to be similar to those in the B region. In contrast, the signal of coronene at 8.95 ppm did not show any shift with changing solvent composition. Tetrabenzoperylene may become more nonplanar with increasing dichloromethane concentration. Evidence in support of this possibility is provided by UV spectrometric studies conducted by Fetzer and Biggs [30]. This planarity change is the main cause of the anomalous chromatographic retention behavior. The structure change of the stationary phase seems to be a second factor contributing to the phenomenon. Because the signal change in NMR measurements in the molecular planarity study is more drastic than that observed for the stationary phase when the mobile phase compositions are changed, changes in solute planarity dominate elution behavior.

To study in more detail the contribution of the stationary phase to the retention behavior of PAHs, one needs to examine the temperature dependence of the retention. As a general rule, lower temperatures are preferable for planarity recognition.

Retention data for the four PAHs were obtained at various temperatures, and the resulting van't Hoff plots for coronene and tetrabenzoperylene are shown in Figure 2.21. Although if the retention mechanism were constant, the plots should be linear for a normal chromatographic process, two linear relationships were found in this case and the critical temperature is about 45°C. Another perspective on retention behavior can be obtained by plotting the retention ratio of coronene and tetrabenzoperylene versus temperature, which indicates the temperature dependence of the planarity recognition capability of the stationary phase. In Figure 2.22 it appears that increasing temperature decreases recognition ability. The critical point found again seems to be between 40 and 50°C. Thus two important pieces of evidence obtained from the examination of the temperature dependence of the retention of planar and nonplanar PAHs with the polymeric ODS phase indicate the existence of some critical temperature around 40–50°C. *What is this temperature?*

Thermal analysis is the best way to accomplish the more detailed examination

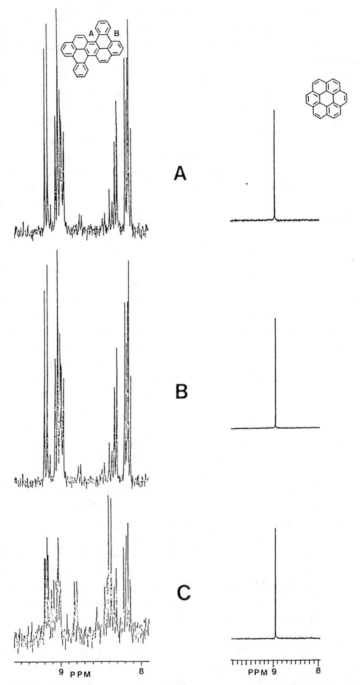

Figure 2.20 ^{1}H NMR spectra of tetrabenzo[a,cd,j,lm]perylene and coronene in various solvents: (A) and (B) as in Figure 2.19; (C) [^{2}H$_2$]dichloromethane/[^{2}H$_4$]methanol (20:80). [23] Reproduced from K. Jinno et al. in *J. Chromatogr.* **461**, 209 (1989) by permission of Elsevier Science-NL, Amsterdam, The Netherlands.

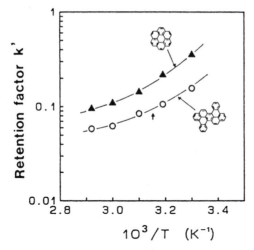

Figure 2.21 van't Hoff plots for coronene and tetrabenzo[*a,cd,j,lm*]perylene with methanol/chloroform (50:50) as the mobile phase; arrow indicates the critical point. [23] Reproduced from K. Jinno et al. in *J. Chromatogr.* **461**, 209 (1989) by permission of Elsevier Science-NL, Amsterdam, The Netherlands.

necessary to determine this critical temperature. Figure 2.23 indicates the differential scanning calorimetry (DSC) chart of this polymeric ODS phase. The data clearly reveal a drastic change around 45°C above the melting point of octadecane (29–30°C). Typical monomeric ODS phases have a weak transition at 35°C, as determined by similar DSC measurements. The DSC results indicate that a transition of the polymeric ODS occurred at about 45°C and may be caused by a phase

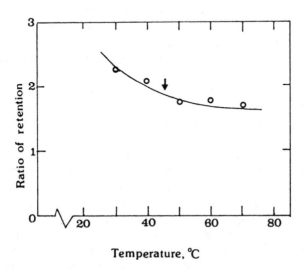

Figure 2.22 Retention ratio of coronene and tetrabenzo[*a,cd,j,lm*]perylene versus temperature; arrow indicates the critical point. [23] Reproduced from K. Jinno et al. in *J. Chromatogr.* **461**, 209 (1989) by permission of Elsevier Science-NL, Amsterdam, The Netherlands.

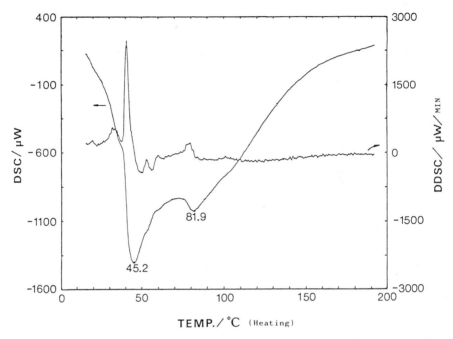

Figure 2.23 DSC chart of Vydac 201 TPB5; DDSC, differential of DSC curve. [23] Reproduced from K. Jinno et al. in *J. Chromatogr.* **461**, 209 (1989) by permission of Elsevier Science-NL, Amsterdam, The Netherlands.

transition from a solidlike to a liquidlike structure. An important discovery is the very good agreement of such transition temperatures with those found in measurements of the temperature dependence of retention in van't Hoff plots and of planarity recognition ability. Therefore one can say that something happened at 45°C for the polymeric ODS phase. To ensure accuracy in discussing this transition temperature, several spectroscopic measurements were used.

First FTIR spectra of the stationary phase itself were collected by means of the diffuse reflectance spectroscopic technique (DRIFT). The spectra obtained (Figure 2.24) showed an obvious change with temperature, a shift of 2 cm^{-1} in the band at 2850 cm^{-1}. Figure 2.25C plots the center of gravity at the band at 2850 cm^{-1} versus temperature. This plot can be considered to be sigmoidal, centered near 45°C. Then we tried to determine the full width of the peak. Figure 2.25A shows the full width at half-height versus temperature, an unusual result, because bandwidth usually increases with temperature if the absorption is due to a single structural functionality. It is reasonable to rationalize this result in terms of two highly overlapped bands, separated by less than the full width at half-height (FWHH). The band at higher wavenumber would correspond to the more liquidlike region and that at lower wavenumber to the more solidlike region, as reported for alkanes by Casal et al. [31]. If this model actually describes what is happening in practice, one might expect the width at, for example, 75% height to be more likely to show temperature effects. And indeed, as Figure 2.25B indicates, a distinct break was found between

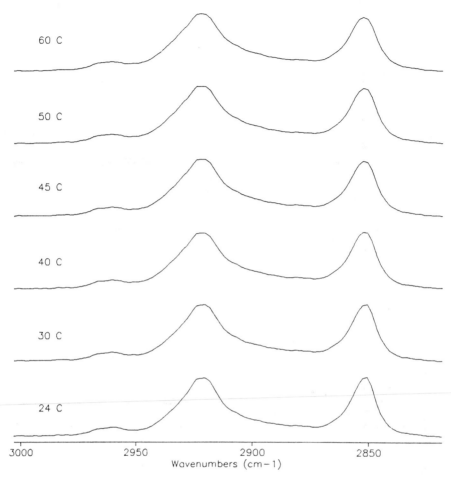

Figure 2.24 Diffuse reflectance infrared Fourier transform spectra of Vydac 201 TPB5 at various temperatures. [23] Reproduced from K. Jinno et al. in *J. Chromatogr.* **461**, 209 (1989) by permission of Elsevier Science-NL, Amsterdam, The Netherlands.

30 and 45°C. Therefore, it is concluded that IR spectrometric information appears to point to a transitional structure change with increasing temperature up to 50°C. This corresponds well with the thermal transition at 45°C observed by DSC, and one may conclude that at 45°C a transition occurs in the CH_2 chains of octadecyl groups.

NMR spectroscopic measurements provide a clearer interpretation of the polymeric ODS than FTIR measurements by DRIFT. Figure 2.26 presents the CP-MAS spectra of the phase at different temperatures. The most drastic change is seen at 30 ppm for the bulk CH_2 signal. At room temperature, the peak has two maxima, which merge with increasing temperature; above 45°C, the maxima at 32 ppm disappears and becomes a shoulder. This tendency is much clearer if the peak deconvolution procedure has been applied to three different spectra at room tem-

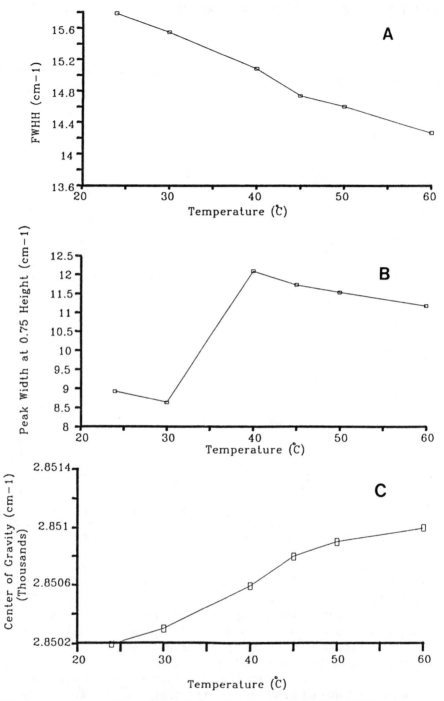

Figure 2.25   Interpretation of FTIR spectra of Vydac 201 TPB5: (A) FWHH versus temperature, (B) peak width at 0.75 height versus temperature, and (C) center of gravity versus temperature. [23] Reproduced from K. Jinno et al. in *J. Chromatogr.* **461**, 209 (1989) by permission of Elsevier Science-NL, Amsterdam, The Netherlands.

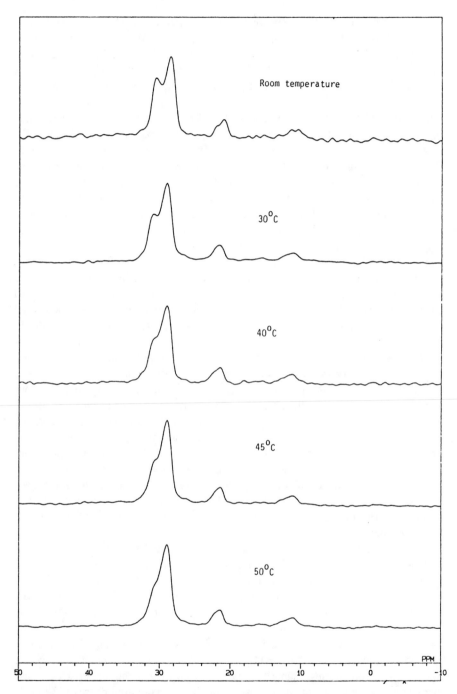

Figure 2.26   CP-MAS ^{13}C NMR spectra of Vydac 201 TPB5 at various temperatures. [23] Reproduced from K. Jinno et al. in *J. Chromatogr.* **461**, 209 (1989) by permission of Elsevier Science-NL, Amsterdam, The Netherlands.

perature, 40°C, and 50°C, as shown in Figure 2.27. The peak at 32 ppm shifted toward 30.8 ppm with increasing temperature. This situation is reminiscent of the case of monomeric and low-loaded polymeric ODS phases. The NMR spectrum at 50°C is very similar to that of Vydac 218 reported by Shah et al. [19]. Vydac 218 is a low-loaded polymeric ODS, and the distance between each alkyl chain on the surface may be larger than that of Vydac 201. From the NMR spectra, which indicate the existence of different types of CH_2 chain in the bulk, one can assume that the polymeric alkyl chain at room temperature exhibits the ordered phase suggested by Sander et al. [2–5] or "fold"-like structure [11,19], which becomes disordered with increasing column temperature, adopting various conformations and becoming freely mobile. Therefore, the NMR signal at 30 ppm becomes unified and similar to those of monomeric and low-loaded polymeric phases.

Similar discussion will be required to study the solute conformation change with temperature in terms of the dependence of the retention behavior with temperature change on the stationary phase conformation change with temperature. Figure 2.28 shows the NMR spectra of tetrabenzo[a,cd,j,lm]perylene dissolved in 50:50 deuterochloroform/deuteromethanol at three temperatures. The change in mobile phase composition produced a drastic shift of the signals around 9.2 ppm. As discussed earlier, the solvents having higher concentrations of deuterodichloromethane shifted two lines at 9.2 ppm, which showed four lines at lower dichloromethane concentrations. A similar tendency, but less pronounced than in the case of deuterodichloromethane, was seen in deuterochloroform: note the two split signals in the spectrum of Figure 2.28C. By increasing the chloroform concentration, a pair of signals at 9.2 ppm could be shifted to higher magnetic field. Figure 2.28B shows the four sharp lines observed when the chloroform concentration was changed from 20% to 50% in methanol. A change in temperature also gave a directionally similar, but smaller, signal shift at 9.2 ppm. This behavior is identical to that observed upon increasing the concentration of dichloromethane or chloroform in the mobile phase. As seen in the preceding section, increasing the dichloromethane concentration in the solvent induced distortion of the molecular structure of tetrabenzoperylene. An increase in column temperature also induced such distortion. At higher temperatures the nonplanarity of the solute is increased, but to a smaller extent than when the solvent is changed.

The foregoing discussions are evidence that changing the mobile phase composition causes a drastic change in solute conformation and a small change in the polymeric ODS phase conformation to a more freely mobile state. From the CP-MAS spectra it is seen that an increase in temperature causes a drastic change in the dry stationary phase from solidlike to liquidlike, and solution NMR spectroscopic data indicate only a small change in the solute conformation. In conclusion, in the absence of actual LC NMR data, it is possible to say that the solidlike polymeric ODS phase tends to retain planar PAHs more strongly than nonplanar PAHs, and the liquidlike structure has less planarity recognition ability than the former state. These characteristics are very similar to those of the liquid crystalline phases [32–34].

The data in this section suggest that the ordering of the polymeric ODS phase in the solidlike state is sufficient to permit recognition of the planarity of molecules

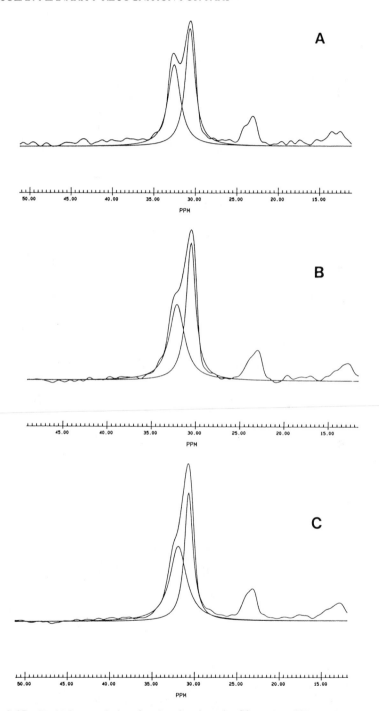

Figure 2.27 Peak deconvolution for the signal at the 30 ppm at (A) room temperature, (B) 40°C, and (C) 50°C. [23] Reproduced from K. Jinno et al. in *J. Chromatogr.* **461**, 209 (1989) by permission of Elsevier Science-NL, Amsterdam, The Netherlands.

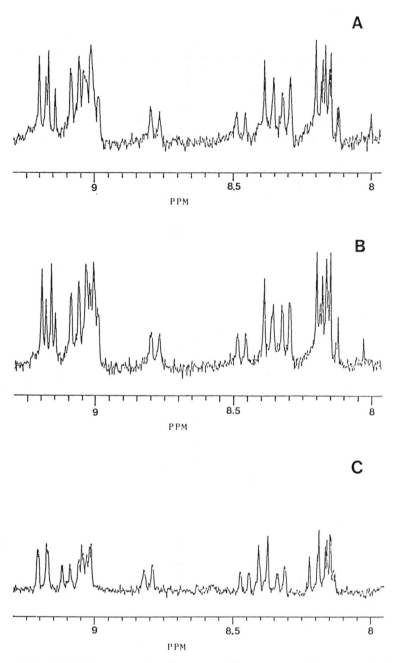

Figure 2.28 ¹H NMR spectra of tetrabenzo[*a,cd,j,lm*]perylene at different mobile phase compositions and temperatures: (A) [²H₂]dichloromethane/[²H₄]methanol (50:50) at 0°C, (B) as (A) but at 27°C, and (C) as (B) but the solvent ratio is 20:80. [23] Reproduced from K. Jinno et al. in *J. Chromatogr.* **461**, 209 (1989) by permission of Elsevier Science-NL, Amsterdam, The Netherlands.

like those of the liquid crystalline phases discussed in the next section. Thus the great influence of steric factors would appear to make the slot model appropriate for the polymeric ODS phase. The Dill interphase model [35] predicts that as alkyl surface densities increase, the corresponding configurational constraints are also increased, creating a more rigid and ordered chain packing structure. This anisotropy of the bonded chains gives rise to additional shape selectivity among solute molecules, since the molecules that can most effectively align with the chains are those that are most effectively retained. In this model, the driving force for retention is the creation of a solute-sized cavity in the stationary phase. As the bonding density and consequently chain ordering are increased, the free energy required for cavity formation also increases. Greater free energy is required to insert solute substructures that are parallel to the silica bonded chain interface; therefore selectivity for linear and planar molecules will increase with alkyl bonding density, as predicted by this theory and as shown by the foregoing discussions. However, more precise and informative NMR studies are required to clarify these matters at the molecular level.

2.3 Liquid Crystal Bonded Phases for PAH Planarity Recognition

We have seen that molecular planarity recognition can be induced for PAHs by means of a rigid and very ordered stationary phase conformation bonded to the silica gel surface. To obtain much higher selectivity for molecular planarity recognition than normally is provided by polymeric ODS phases, very highly ordered and rigid bonded phases are preferred. Liquid crystalline bonded phases are among the most appropriate phases ever thought of by chemists [30–34].

Liquid crystals as stationary phases were introduced in gas chromatography (GC), and extensive applications have been carried out for difficult separation problems [36–40], taking advantage of the strong shape selectivity of liquid crystal stationary phases with respect to both enantiomeric and diastereomeric compounds. Since liquid crystalline compounds have low volatility, they were coated on a solid support or a capillary wall for GC in early experimental studies [39,40]. These columns have high stability, but there is no chemical bonding of the stationary phases to the surfaces. In contrast to the many publications on GC, there have been few reports about such phases in LC [41,42]. To achieve long column lifetime, in LC, however, the liquid crystals must be chemically bonded to the solid support via a chemical reaction, and this reaction, in turn, is influenced by the solubility under normal mobile phase solvent conditions [43].

Recent studies have shown that some liquid crystals can be bonded to a polysiloxane backbone and still retain liquid crystalline properties [33]. These bonded phases possess high selectivity toward structural isomers, particularly for PAHs, in supercritical fluid chromatography [44] as well as in GC [45–48].

Here we introduce two liquid crystal bonded phases that are chemically linked to a particulate silica surface (Figure 2.29) [49] and in the sections that follow, we will

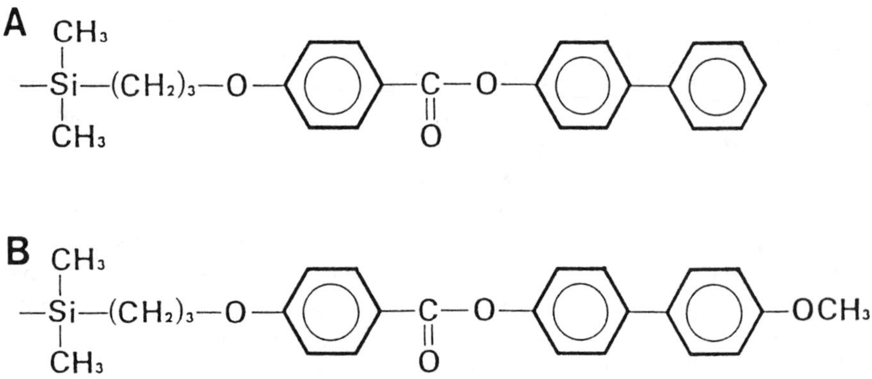

A

CH₃
|
—Si—(CH₂)₃—O—⬡—C(=O)—O—⬡—⬡
|
CH₃

B

CH₃
|
—Si—(CH₂)₃—O—⬡—C(=O)—O—⬡—⬡—OCH₃
|
CH₃

Figure 2.29 Chemical structures of liquid crystal bonded phases: (A) phase 1 and (B) phase 2 [49]. Reproduced from Y. Saito et al. in *Chromatographia* **38**, 295 (1994) by permission of Vieweg Publishing, Wiesbaden, Germany.

observe the retention behavior of PAHs at various column temperatures and mobile phase compositions in reversed-phase LC.

2.3.1 Bonded Phase Synthesis

The silica support used for liquid crystal phases is Nucleosil 300-10. The bonded structures 4-{[4-(allyloxy)benzoyl]oxy}biphenyl bonded phase (liquid crystal bonded phase 1) and 4-{[4-(allyloxy)benzoyl]oxy}-4'-methoxybiphenyl bonded phase (liquid crystal bonded phase 2), shown in Figure 2.29, are discussed in Sections 2.3.1.1 and 2.3.1.2.

2.3.1.1 Synthesis of Liquid Crystal Phase 1: 4-{[4-(Allyloxy)benzoyl]oxy}biphenyl

To a two-necked, 250 mL, round-bottomed flask equipped with a 50 cm reflux condenser, a Teflon septum with a nitrogen line, a magnetic stirring bar, and an oil bath, 4.93 g of 4-phenylphenol was added and dissolved in 50 mL of dry pyridine under nitrogen. Next, 6.13 g of 4-(allyoxy)benzoyl chloride was slowly added to the resulting solution with stirring. After the mixture had been stirred for 3 hours at 25°C, 200 mL of acidic (20% HCl) solution was added. The resulting precipitate was collected by vacuum filtration and washed with 200 mL of cold deionized water. The crude product was recrystallized from acetone/ethanol (1:1), yielding 8.45 g of phase 1 (88%, mp 138–140°C).

2.3.1.2 Synthesis of Liquid Crystal Phase 2: 4-{[4-(Allyloxy)benzoyl]oxy}-4'-methoxybiphenyl

First 400 mL of 10% aqueous NaOH was cooled to 0°C and placed in a three-necked, 1000 mL, round-bottomed flask equipped with a self-equalizing addition funnel, a 50 cm reflux condenser, a PTFE septum with a nitrogen line, a magnetic stirring bar and an ice bath. 4,4'-Dihydroxybiphenyl (75.50 g) was added to the flask with vigorous stirring, and then dimethylsulfate (48.64 g) was added and maintained at 0°C with vigorous stirring over one hour. When the addition was complete, the resulting precipitate was filtered by vacuum filtration with a sintered glass filter and redissolved in 400 mL of 10% aqueous NaOH. The solution was heated briefly to boiling and was allowed to cool, whereupon the resulting solid was filtered again. The solid was then placed in 200 mL of water, heated to boiling, and filtered while hot. The filtrate was heated to 70°C and acidified with 20% HCl to obtain crude 4-hydroxy-4'-methoxybiphenyl as a precipitate. The product was re-crystallized twice from ethanol.

The recrystallized material (3.62 g), dissolved in 100 mL of dry pyridine and under a nitrogen blanket, was placed in a three-necked, 250 mL, round-bottomed flash equipped with a 100 mL self-equalizing addition funnel, a 50 cm reflux condenser, a magnetic stirrer, and an oil bath. 4-(Allyoxy)benzoyl chloride (4.19 g), dissolved in 20 mL of dry pyridine, was added with stirring over one hour. The reaction mixture was stirred for an additional 3 hours at 25°C, and then heated to 60°C and stirred for 2 hours more. The reaction mixture was cooled to 25°C and 400 mL of water was added. The solution was then acidified with 20% HCl. The precipitate was filtered by a vacuum filtration before washing with 400 mL of sodium bicarbonate solution and 400 mL of deionized water. The crude product was recrystallized twice from acetone yielding 4.27 g of phase 2 (84%) with the following melting points: crystalline to nematic, 147°C; nematic to isotropic liquid, 249°C.

2.3.1.3 Synthesis of Silane Reagents and Silica Bonding

Either compound (i.e., liquid crystal phase 1 or 2, 11.8 mmol), dissolved in 20 mL of dry toluene, was placed in a two-necked, 100 mL, round-bottomed flash equipped with a 50 cm reflux condenser, a PTFE septum with a nitrogen line, a magnetic stirrer, and an oil bath. Dimethylchlorosilane (11.8 mmol) was added to the reaction flask while stirring and purging with nitrogen. After 5 minutes, 8 mg of hexachloroplatinic acid was added. The reaction mixture was heated to 65°C and stirred for 10 days under nitrogen. Then 20 mL of freshly distilled toluene was added, followed by 6.5 g of silica and 0.5 mL of dry pyridine. The reaction mixture was stirred for another 10 days at 40°C under nitrogen. The solid was then filtered and washed with 60 mL of toluene followed by 60 mL of ethanol. The washing procedure was repeated six times.

2.3.2 Basic Selectivity of Liquid Crystal Bonded Phases for PAHs

Fundamental assessments of molecular size and shape selectivity for PAHs in both liquid crystal bonded phases have been carried out, and these retention data versus F

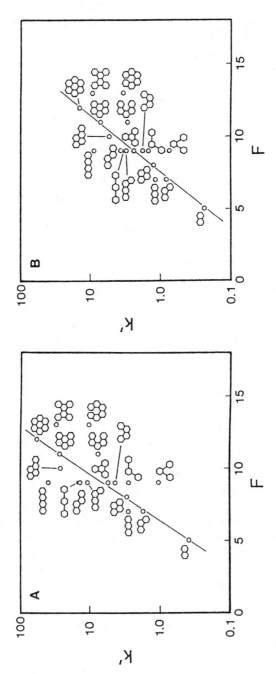

Figure 2.30 Relationship between log k' and F number with liquid crystal bonded phases; mobile phase, methanol/water (80:20); column temperature, 20°C: (A) phase 1 and (B) phase 2 [49]. Reproduced from Y. Saito et al. in *Chromatographia* **38**, 295 (1994) by permission of Vieweg Publishing, Wiesbaden, Germany.

number are plotted in Figure 2.30. The F number is a molecular size descriptor proposed by Hurtubise et al. [50] as follows:

F = (number of double bonds) + (number of primary and
secondary carbons) $-$ 0.5 $\times$ (number of nonaromatic rings)

A high linear correlation between log k' and F was found in aqueous reversed-phase LC [51–53]. Both phases are observed to have similar selectivity toward PAHs, although liquid crystal phase 1 is higher in molecular shape recognition capability than phase 2. Thus, nonplanar PAHs such as *o*-terphenyl, dibenzo[*c,g*]phenanthrene, and PhPh elute faster than planar ones of similar size, and relatively rodlike solutes such as naphthacene and chrysene are retained more strongly than more compact planar PAHs.

Retention data for seven PAHs with the liquid crystalline phases are summarized in Table 2.8. In the liquid crystal phases, a methoxy-substituted bonded phase (phase 2) results in less retention for PAHs than a nonsubstituted one (phase 1), although the surface coverage value of the former is the larger.

2.3.3 Effect of Mobile Phase Composition on Planarity and Shape Recognition Capability

To investigate the planarity recognition power of liquid crystal phases, the separation factors of triphenylene and *o*-terphenyl with the two stationary phases were evaluated (see Table 2.9). The separation factor of triphenylene and *o*-terphenyl has been proposed by Tanaka et al. [54] and Jinno, Wise, et al. [55–58] as a good indicator of the planarity recognition capability of stationary phases in reversed-phase LC. In general, monomeric ODS phases have a value of about 1–2, and polymeric ones give a value of about 2–3 in aqueous mobile phase compositions. It has been observed that these values are larger when pure methanol is used as the

Table 2.8 RETENTION DATA FOR SEVEN
PAHs WITH LIQUID CRYSTAL
PHASES 1 AND 2[a]

PAH	Retention Factor k'	
	Phase 1	Phase 2
Naphthalene	0.096	0.082
Anthracene	0.553	0.252
Pyrene	0.617	0.297
Triphenylene	0.894	0.429
Chrysene	1.96	0.562
Benzo[*a*]pyrene	3.36	0.938
Coronene	6.63	1.95

[a] Mobile phase, methanol; column temperature, 20°C.

Source: Ref. 49. Reproduced from Y. Saito et al. in *Chromatographia* **38**, 295 (1994) by permission of Vieweg Publishing, Wiesbaden, Germany.

Table 2.9   RETENTION FACTORS AND SEPARATION FACTORS OF
o-TERPHENYL AND TRIPHENYLENE WITH LIQUID CRYSTAL
PHASES 1 AND 2[a]

	Retention Factor, k'		Separation Factor, α
Phase 1	0.181	0.894	4.94
Phase 2	0.114	0.429	3.76

[a]   Mobile phse, methanol; column temperature, 20°C

Source: Ref. 49. Reproduced from Y. Saito et al. in *Chromatographia* **38**, 295 (1994) by permission of Vieweg Publishing, Wiesbaden, Germany.

mobile phase solvent [57]. As found in Table 2.9, the liquid crystal bonded phases are comparable in planarity recognition capability to the commercially available polymeric ODS phase, and the separation factor decreases linearly with increasing water concentration in the mobile phase, as demonstrated in Figure 2.31.

Another planar/nonplanar selectivity descriptor of stationary phases is the retention factor ratio among benzo[*a*]pyrene (BaP; $F = 10$), TBN ($F = 13$), and PhPh ($F = 13$), which is recommended by Sander and Wise [5]. Table 2.10 shows these factors for the liquid crystal phases. The data indicate the strong planarity recognition power of the liquid crystalline bonded phases. The selectivity is controlled by

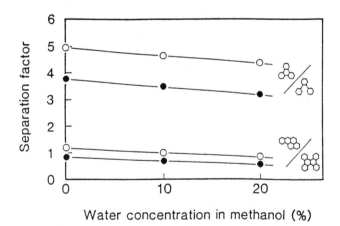

Figure 2.31   Separation factors for triphenylene/o-terphenyl and BaP/TBN versus mobile phase composition with liquid crystal bonded phases 1 (○) and 2 (●); column temperature, 20°C [49]. Reproduced from Y. Saito et al. in *Chromatographia* **38**, 295 (1994) by permission of Vieweg Publishing, Wiesbaden, Germany.

Table 2.10   SEPARATION FACTORS BETWEEN PhPh, BaP, AND TBN WITH LIQUID
                  CRYSTAL PHASES 1 AND 2[a]

	Separation Factor, α		
Phase 1	1.19	2.76	2.31
Phase 2	0.80	1.82	2.27

[a] Mobile phase methanol; column temperature, 20°C.

Source: Ref. 49. Reproduced from Y. Saito et al. in *Chromatographia* **38**, 295 (1994) by permission of Vieweg
Publishing, Wiesbaden, Germany.

two factors, F number and planarity. More planar solutes are retained longer than
nonplanar ones for solutes that are similar in size and F number.

Further chromatographic measurements with three sets of solute pairs, having the
distortion differences of nonplanar solutes, were carried out to investigate the planarity recognition power of these bonded phases. The results are summarized in
Table 2.11. All the upper solutes in Table 2.11 (benzo[*ghi*]fluoranthene, benzo[*ghi*]perylene, and coronene) are planar, while the lower ones (benzo[*c*]phenanthrene, dibenzo[*c,g*]phenanthrene, and PhPh) are nonplanar. The nonplanarity increases with the number of the aromatic rings in the respective structure. Table 2.11

Table 2.11   SEPARATION FACTORS FOR
                  BENZO[*ghi*]FLUORANTHENE/BENZO[*c*]PHENANTHRENE,
                  BENZO[*ghi*]PERYLENE/DIBENZO[*c, g*]PHENANTHRENE, AND
                  CORONENE/PhPh WITH LIQUID CRYSTAL PHASES 1 AND 2[a]

	Separation Factor, α		
Phase 1	1.25	3.58	5.45
Phase 2	1.25	2.47	3.79

[a] Mobile phase methanol; column temperature, 20°C.

Source: Ref. 49. Reproduced from Y. Saito et al. in *Chromatographia* **38**, 295 (1994) by permission of Vieweg
Publishing, Wiesbaden, Germany.

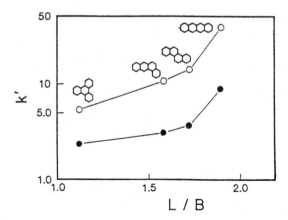

Figure 2.32 Relationship between log k' and L/B value for planar four-ring PAHs with liquid crystal bonded phases 1 (○) and 2 (●); mobile phase, methanol; column temperature, 20°C [49]. Reproduced from Y. Saito et al. in *Chromatographia* **38**, 295 (1994) by permission of Vieweg Publishing, Wiesbaden, Germany.

indicates not only excellent planarity recognition capabilities of the liquid crystal bonded phases as discussed earlier, but also a clear trend in which the difference among planar/nonplanar selectivities becomes more evident with increasing distortion of the nonplanar PAHs.

Figure 2.32 plots log k' values for planar four-ring PAHs with liquid crystal phases versus length-to-breadth ratio (L/B; Table 2.12) [61,62]. L/B represents the ratio of maximized length to breadth of the rectangle enclosing the solute molecule and, in general, linear correlation has been reported between L/B value and log k' of geometrical isomers of PAHs in reversed-phase LC [57]. It is clearly observed in Figure 2.32 that with liquid crystal bonded phases, relatively rodlike PAHs were retained longer than squarelike ones.

Similar trends were seen in the separation factors among *o*-terphenyl/*m*-terphenyl/*p*-terphenyl and naphthacene/triphenylene/pyrene, plotted in Figures 2.33 and 2.34, respectively. The general behavior found with both liquid crystal phases is that retention increases with increasing L/B ratio of solutes, and these separation

Table 2.12 LENGTH-TO-BREADTH RATIO (L/B) VALUES FOR PLANAR FOUR-RING PAHs

PAH	L/B
Triphenylene	1.12
Benz[*a*]anthracene	1.58
Chrysene	1.72
Naphthacene	1.89

Source: Ref. 49. Reproduced from Y. Saito et al. in *Chromatographia* **38**, 295 (1994) by permission of Vieweg Publishing, Wiesbaden, Germany.

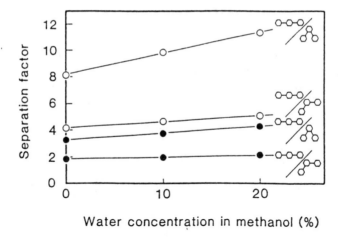

Figure 2.33 Separation factors for *p*-terphenyl/*o*-terphenyl and *p*-terphenyl/*m*-terphenyl versus mobile phase composition with liquid crystal bonded phases 1 (○) and 2 (●); column temperature, 20°C [49]. Reproduced from Y. Saito et al. in *Chromatographia* **38**, 295 (1994) by permission of Vieweg Publishing, Wiesbaden, Germany.

factors also increase linearly with the water concentration in the mobile phase, although phase 1 possesses much higher shape recognition power than phase 2.

Figure 2.35, which plots the relative retentions for orthofused four-ring PAH isomers to pyrene versus water content in the mobile phase, also shows the linear dependence of PAH retention on mobile phase composition and the selectivity difference between phase 1 and phase 2.

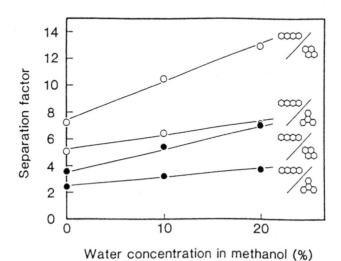

Figure 2.34 Separation factors for naphthacene/pyrene and naphthacene/triphenylene versus mobile phase composition with liquid crystal bonded phases 1 (○) and 2 (●); column temperature, 20°C [49]. Reproduced from Y. Saito et al. in *Chromatographia* **38**, 295 (1994) by permission of Vieweg Publishing, Wiesbaden, Germany.

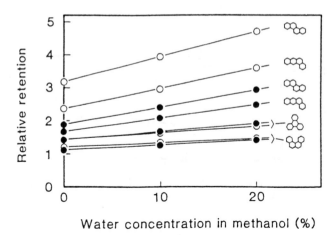

Figure 2.35 Relative retention to pyrene versus mobile phase composition with liquid crystal bonded phases 1 (○) and 2 (●); column temperature, 20°C [49]. Reproduced from Y. Saito et al. in *Chromatographia* **38**, 295 (1994) by permission of Vieweg Publishing, Wiesbaden, Germany.

Normally, the bonded phase ordering on the support surfaces significantly affects the shape selectivity toward PAH isomers and highly ordered bonded phases have strong shape recognition power for PAHs, as discussed in the preceding sections. Thus, the observations reported here clearly indicate that these liquid crystal bonded phases are highly ordered. This result is mainly due to the central part of the bonded phase structure, and it should be noted that the methoxy group at the top of the liquid crystal bonded phase may affect to a different extent the ordering on the silica support.

2.3.4 Effect of Temperature on Planarity Recognition with Liquid Crystal Phases

To investigate the temperature dependence of the retention factors of PAHs with the liquid crystal bonded phases, van't Hoff plots (log k' vs. $1/T$) were evaluated for all the solutes studied here.

The fundamental relationship between temperature and retention factor is:

$$2.303(\log k') = -\frac{\Delta H}{RT} + \frac{\Delta S}{R} + (\ln \phi) \tag{2.1}$$

where k' is the solute retention factor, ΔH and ΔS are the enthalpy and the entropy of solute transfer from the mobile phase to the stationary phase, respectively, R is the gas constant, T is the absolute temperature, and ϕ is the volume phase ratio of the stationary phase to the mobile phase. Thus, equation 2.1 is a linear relationship between log k' and $1/T$.

Apfel et al. [46] and Markides et al. [47] proposed that nonlinear van't Hoff plots indicate a phase transition with polysiloxane-based liquid crystal bonded phases in GC.

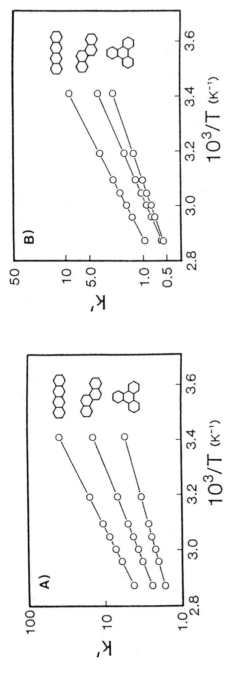

Figure 2.36 van't Hoff plots for triphenylene, chrysene, and naphthacene, mobile phase, methanol/water (80:20): (A) liquid crystal bonded phase 1; (B) liquid crystal bonded phase 2 [49]. Reproduced from Y. Saito et al. in *Chromatographia* **38**, 295 (1994) by permission of Vieweg Publishing, Wiesbaden, Germany.

Figure 2.36 shows the van't Hoff plots of the orthocondensed four-ring PAHs triphenylene, chrysene, and naphthacene with liquid crystal phases 1 and 2. Linear van't Hoff plots were also obtained for all solutes investigated with two liquid crystal phases in this temperature range.

Although higher temperatures should be investigated, our results clearly indicate the absence of a phase transition of the liquid crystal type of bonded silica phases over the temperature range of 20–75°C.

Figure 2.37, which plots the separation factors among naphthacene, triphenylene, and pyrene with liquid crystal phases against the column temperature, illustrates that decrease of retention is more pronounced for more "rodlike" molecules when the column temperature is elevated. Similar decreases in the molecular shape recognition capability of liquid crystal phases are also observed in Figures 2.38 and 2.39, which show the temperature dependence on planar/nonplanar selectivity. Thus it is implied that with increasing temperature, the orderliness and conformation of the liquid crystal phase must be changed and the recognition power of molecule length must be decreased. And indeed, the data of Wise and Sander [5] suggest that increasing temperature with polymeric-type ODS phases induces a similar selectivity change.

Figure 2.40 illustrates typical chromatograms with the liquid crystal bonded phase indicating good separation for the PAH mixture. Chromatograms from a typical polymeric ODS (Vydac 201 TPB-5; Separations Group, Hesperia, CA) and a monomeric ODS (Develosil ODS-5; Nomura Chemical, Seto, Japan) are also shown in Figure 2.40. In the case of the monomeric ODS phase, poor separation was observed for this mixture. The mixture also could not be separated for each PAH with the Vydac phase, although as expected, the polymeric ODS phase provides significantly better separation than the monomeric ODS phase. With the liquid crystal bonded phase, however, all PAHs were separated to distinguishable peaks.

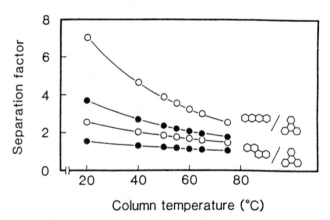

Figure 2.37 Separation factors for naphthacene/triphenylene and chrysene/triphenylene versus column temperature with liquid crystal bonded phases 1 (○) and 2 (●); mobile phase, methanol/water (80:20) [49]. Reproduced from Y. Saito et al. in *Chromatographia* **38**, 295 (1994) by permission of Vieweg Publishing, Wiesbaden, Germany.

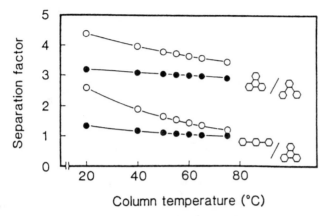

Figure 2.38 Separation factors for triphenylene/o-terphenyl and p-terphenyl/triphenylene versus column temperature with liquid crystal bonded phases 1 (○) and 2 (●); mobile phase, methanol/water (80:20) [49]. Reproduced from Y. Saito et al. in *Chromatographia* **38**, 295 (1994) by permission of Vieweg Publishing, Wiesbaden, Germany.

As shown in Table 2.13, the best separation of benz[*a*]anthracene and chrysene was obtained with the liquid crystal phase. This good separation has not been accomplished by using other commercial stationary phases in reversed-phase LC. Furthermore, it is also clear from Table 2.13 that the liquid crystal phase 1 is a more powerful separator of phenanthrene and anthracene than the typical polymeric ODS. This is because liquid crystal phase 1 has stronger shape recognition capability for "rodlike" PAHs than other bonded phases.

Since the "rodlike" shape selectivity of the liquid crystal phases significantly decreases with increasing column temperature, it can be said that when using the

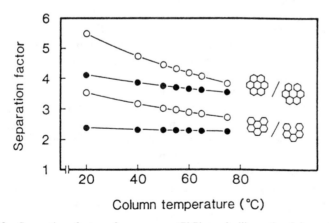

Figure 2.39 Separation factors for coronene/PhPh and dibenzo[*c,g*]phenanthrene/benzo[*ghi*]perylene versus column temperature with liquid crystal bonded phases 1 (○) and 2 (●); mobile phase, methanol/water (80:20) [49]. Reproduced from Y. Saito et al. in *Chromatographia* **38**, 295 (1994) by permission of Vieweg Publishing, Wiesbaden, Germany.

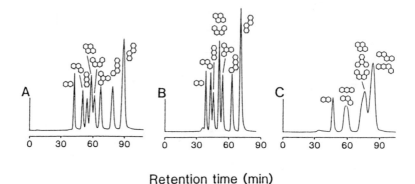

Retention time (min)

Figure 2.40 Typical chromatograms for PAH mixtures with (A) liquid crystal bonded phase 1, (B) Vydac 201 TPB-5, and (C) Develosil ODS-5; mobile phase, methanol; column temperature, 20°C [49]. Reproduced from Y. Saito et al. in *Chromatographia* **38**, 295 (1994) by permission of Vieweg Publishing, Wiesbaden, Germany.

liquid crystal phases, changing both mobile phase composition and column temperature will easily lead to improved separation conditions for PAHs.

2.4 Multilegged Bonded Phases for PAH Planarity Recognition Ability

It has been shown that the achievement of improved molecular planarity recognition capability for PAHs depends on the rigidity and the orderliness of bonded phase structures on a silica gel surface. Thus the converse—namely, that selectivity for molecular planarity recognition will be differentially distributed—would be expected. That is, the stationary phase moiety is conceived of as being located horizontally rather than vertically, like polymeric ODS and liquid crystal bonded phases, where the bonded phases do not have any slitlike structures to promote retention during molecular interactions between the phase and the solutes. The schematic concept is drawn in Figure 2.41 for typical multilegged phases. The novel phases described in this section have multiple legs to permit Si—O—Si bonding to the silica gel silanol groups for constructing the chemically bonded stationary phases. The phases use "bidentate" or "tridentate" silanes containing one reactive atom on each two or three silicon atoms, which in this case are connected through a phenyl ring. Discussions herein indicate that stationary phases that can offer high selectivity for particular solute groups can be designed on the basis of the consideration of molecular–molecular interactions between solutes and stationary phase ligands.

2.4.1 Synthesis of New Multilegged Phases

New stationary phases were synthesized using reactions developed in our laboratory [58]. The base silica had a particle size of 5 μm, a 120 Å pore diameter, and a

Table 2.13 RETENTION FACTORS AND SEPARATION FACTORS FOR THE SEPARATION OF ANTHRACENE/PHENANTHRENE AND CHRYSENE/BENZ[a]ANTHRACENE WITH LIQUID CRYSTAL PHASE 1 AND VYDAC 201 TPB-5

	Retention Factor, k'				Separation Factor, α	
Liquid crystal phase 1	0.373	0.553	1.46	1.96	1.48	1.34
Vydac 201 TPB-5	0.213	0.287	0.787	1.02	1.34	1.29

Source: Ref. 49. Reproduced from Y. Saito et al. in *Chromatographia* **38**, 295 (1994) by permission of Vieweg Publishing, Wiesbaden, Germany.

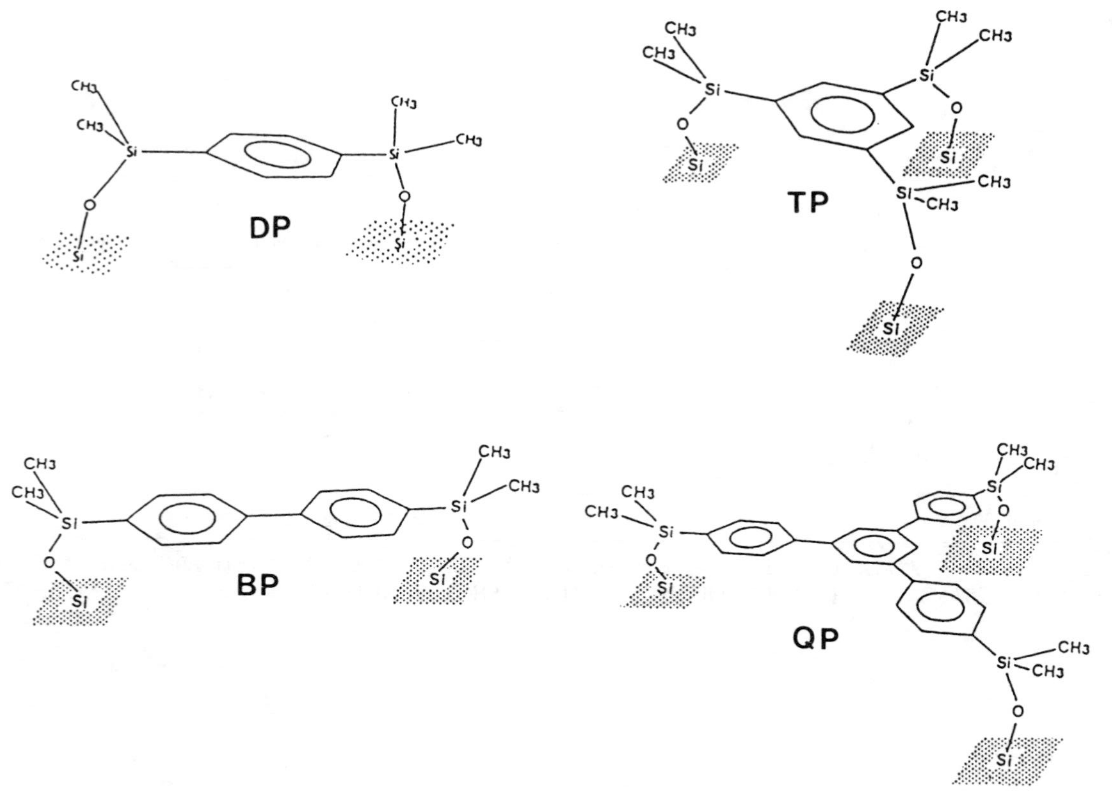

Figure 2.41 Chemical structures of multilegged bonded phases [58]. Reproduced from K. Jinno et al. in *J. Chromatogr.* **461**, 209 (1989) by permission of Elsevier Science-NL, Amsterdam, The Netherlands.

270 m^2 surface area. The new materials are p-bis(dimethylphenyl) (DP), 1,3,5-tris(dimethylphenyl) (TP), p-bis(dimethylbiphenyl) (BP), and 1,1′,1″-tris(dimethyl quaternary phenyl) (QP) bonded silicas.

The basic difference between these phases and the phases generally available commercially is the structure on the silica surface. As shown in Figure 2.41, DP, TP, BP, and QP are unique in that their respective structures cover the silica surface horizontally, whereas typical LC stationary phases are attached vertically to the silica surface by siloxane bonding. Since the bidentate or tridentate reagents can form two or three bonds to the silica surface for each molecule of silane and these multiple bonds produce a multimembered ring with the silica structure, they should lead to increased stability of the attached ligand. In considering the multilegged silanes, it is important to determine whether each silicon atom from the silane has reacted with the surface silanol groups for a true multidentate attachment, or whether the silane is connected to the surface by only one Si—O—Si bond. To confirm the structural differences among the phases, a typical phenyl bonded phase such as the reference (P), and DP and TP, both solution and solid state was subjected to NMR measurements.

The ^{29}Si solution state NMR spectra of the authentic methoxysilanes shown in Figure 2.42 indicate that these compounds have almost identical structures with regard to the silicon atoms, since a signal of approximately 10 ppm can be assigned to the silicon atom desired. The signal at −4 ppm is due to 1,1′,3,3′-tetramethylsiloxane as the internal standard.

The similar chemical shifts in these spectra suggest that the three silanes have the same silicon structure because the silicon atoms are connected to two methyl, one phenyl, and one methoxy group as the active site, although the trisphenylsilane has three silicon atoms, the bisphenylsilane has two, and the phenylsilane has only one.

Figure 2.43 illustrates the ^{13}C solution state NMR spectra for our three silanes. Depending on the location of carbon atoms, the signals around 130–150 ppm are different from each other. The spectra clearly indicate that TP and DP have two positionally different carbons, whereas P has four chemically different carbons, although the signal intensities do not accurately indicate the number of carbons. Figure 2.44 summarizes some ^{13}C CP-MAS solid state NMR spectra for these bonded silica phases. Even though the signal peaks are much broader than those of the authentic silanes, the signals are in good agreement with those that appeared in the solution state spectra. The peak numbers in Figures 2.43 and 2.44 demonstrate this consistency.

The ^{29}Si CP-MAS NMR spectra shown in Figure 2.45 offer different information about the synthesized phases. The signals appearing at approximately 2 ppm for P, DP, and TP can be assigned to silicon atoms connected to the phenyl ring and the silica surface by siloxane bonds, and the two signals around −100 ppm are caused by SiOH and SiO— bondings in the substrate silica. The spectrum for P suggests that it has only one such connecting silicon atom, whereas the spectra of TP and DP indicate the presence of another type of silicon atom in their structure because of the signals at about 16 ppm. Since Kirkland et al. [59] reported similar ^{29}Si NMR spectra for bidentate alkyl bonded phases, it seems that these spectra are not unusu-

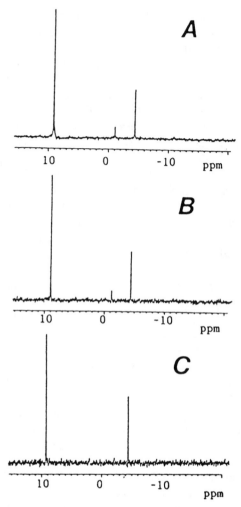

Figure 2.42 ^{29}Si solution state NMR spectra of the authentic silanes: (A) 1,3,5-tris(di-methylmethoxysilyl)benzene, (B) p-bis(dimethylmethoxysilyl)benzene, and (C) dimethyl-methoxysilylbenzene [58]. Reproduced from K. Jinno et al. in J. Chromatogr. **517**, 193 (1990) by permission of Elsevier Science-NL, Amsterdam, The Netherlands.

al. The presence of the two low-field peaks could be explained by one of the following assumptions: (1) dimerization of TP and DP occurred in the synthetic reaction, or (2) different bond angles and conformations of the Si—O—Si bonds led to positionally different silicon atoms. However, it is apparent from the spectroscopic data that the synthesized bonded phases are the desired materials.

For comparison, monomeric (Develosil ODS-5) and polymeric (Vydac 201 TPB-5 ODS) ODS phases, were also evaluated for chromatographic separation performance.

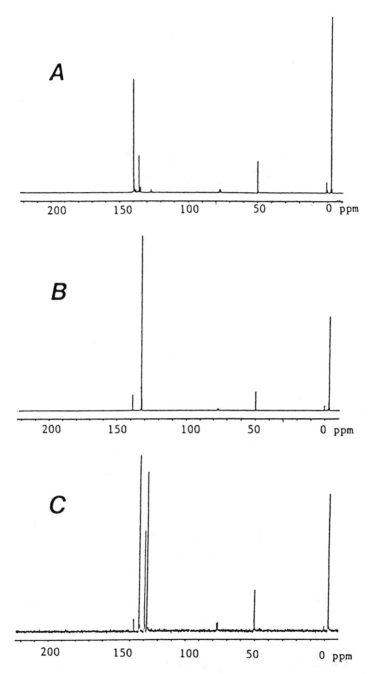

Figure 2.43 ^{13}C solution state NMR spectra of the authentic silanes: (A) 1,3,5-tris(di-methylmethoxysilyl)benzene, (B) *p*-bis(dimethylmethoxysilyl)benzene, and (C) dimethyl-methoxysilylbenzene [58]. Reproduced from K. Jinno et al. in *J. Chromatogr.* **517**, 193 (1990) by permission of Elsevier Science-NL, Amsterdam, The Netherlands.

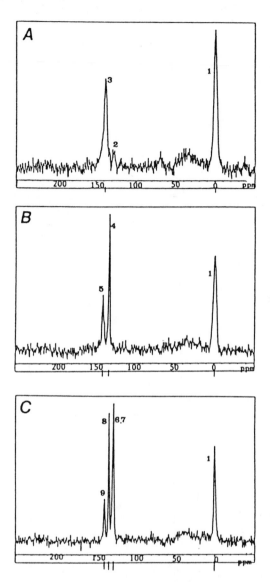

Figure 2.44 ^{13}C CP-MAS solid state NMR spectra of the chemically modified silicas: (a) TP, (B) BP, and (C) P [58]. Reproduced from K. Jinno et al. in *J. Chromatogr.* **517**, 193 (1990) by permission of Elsevier Science-NL, Amsterdam, The Netherlands.

2.4.2 Planarity Recognition for PAHs

We use two different methods of evaluating the PAHs planarity recognition ability of the phases. One useful method is to measure the selectivity for the three PAHs, proposed by Sander and Wise [5]: benzo[*a*]pyrene (BaP), tetrabenzonaphthalene (TBN), and phenanthrophenanthrene (PhPh). BaP is a planar molecule, whereas

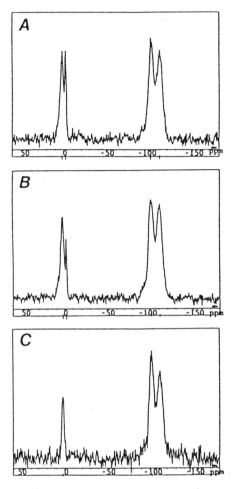

Figure 2.45 ^{29}Si CP-MAS solid state NMR spectra of the chemically modified silicas: (a) TP, (B) BP, and (C) P [58]. Reproduced from K. Jinno et al. in *J. Chromatogr.* **517**, 193 (1990) by permission of Elsevier Science-NL, Amsterdam, The Netherlands.

TBN and PhPh are nonplanar. The degree of planarity decreases in the following order: BaP > TBN > PhPh. An alternative approach is to evaluate the selectivities for *o*-terphenyl and triphenylene as discussed earlier, as well as coronene (Cor) and PhPh, and dibenzo[*bc,ef*]coronene (DBCO) and tetrabenzo[*a,cd,j,lm*]perylene (TEBP).

Figure 2.46 shows the plots of relative retention of BaP and PhPh compared to TBN. With the polymeric phase, as expected, the nonplanar solutes were eluted faster than the planar solute, even though the latter is smaller. The elution order depends on the degree of planarity of the solutes. This result is agreement with the discussion of the preceding sections. On the other hand, BaP was retained longer than PhPh and was eluted faster than TBN on the monomeric ODS, Develosil. This

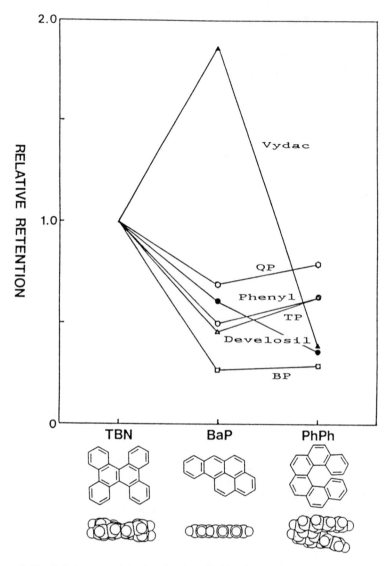

Figure 2.46 Relative retention plots for three PAHs with six phases: mobile phase, methanol/water (90:10); flow rate, 2 μL/min; solute, tetrabenzonaphthalene (TBN), benzo[a]pyrene (BaP), and phenanthro[3,4-c]phenanthrene (PhPh) [58]. Reproduced from K. Jinno et al. in *J. Chromatogr.* **517**, 193 (1990) by permission of Elsevier Science-NL, Amsterdam, The Netherlands.

suggests a slight planarity recognition capability for the monomeric ODS, but not as high as the polymeric ODS. In contrast, both the BP and QP phases showed different behavior from those on two ODS phases. However, the elution order was the same on BP and QP phases as on TP and P phases. It seems that there is no significant difference in column selectivity among BP, QP, TP, and P using those PAHs.

Tables 2.14 through 2.16 summarize the retention data for the three pairs of Figure 2.46 in the alternative approach. The size of PAH pairs increases with the table number; the solute on the left is planar and that in the right nonplanar. Typical chromatograms for o-terphenyl and triphenylene with BP and TP are presented in Figure 2.47. It is clear that the TP phase has specific selectivity for this pair of the solutes. The reason for the larger retention of o-terphenyl than of triphenylene with TP phase can be explained in terms of molecular–molecular interactions between the two solutes and TP phase as dipicted in Figure 2.48, where a computer molecular graphic technique was applied to this problem. Figure 2.48A shows that the methyl groups favor the planar solute by hydrophobic interaction and maintain the distance between the solute and TP to reduce the possibility of π-π interactions, where triphenylene can interact with the phenyl ring of the TP phase. However, as shown in Figure 2.48B, o-terphenyl has a suitable steric structure that can interact well with the phenyl ring of the TP phase. o-Terphenyl has a high possibility of retention with the TP phase by two mechanisms: hydrophobic interaction of the methyl groups with two phenyl rings of the solute, and π-π interaction of the phenyl ring of the phase with the central phenyl ring. This means that in addition to the hydrophobic interaction between the methyl groups and the solute, more pronounced π-π interactions induce a longer retention of o-terphenyl with the TP phase than those with other phases. The retention factors and the separation factors of Cor and PhPh using a methanol mobile phase are summarized in Table 2.15. This pair is larger than the pair discussed earlier, and it is evident that only the QP phase shows planarity recognition capability among other phases, while even the BP phase cannot recognize the difference in planarity of those solutes.

Table 2.14 RETENTION FACTORS k' AND SEPARATION FACTORS α FOR TRIPHENYLENE (Tri) AND o-TERPHENYL (Ote) WITH VARIOUS BONDED PHASES[a]

	k'		
Stationary phase	Tri	Ote	α k'_{Tri}/k'_{Ote}
P	11.0	13.7	0.81
TP	14.8	22.9	0.65
BP	149	94.7	1.57
QP	1.72	1.47	1.17
Develosil	115	69.3	1.66
Vydac	38.5	15.4	2.50
DP	16.58	17.33	0.96

[a] Mobile phase, methanol/water (60 : 40).

Table 2.15   RETENTION FACTORS k' AND SEPARATION FACTORS α FOR CORONENE (Cor) AND PHENANTHRO[3,4-c]PHENANTHRENE (PhPh) WITH VARIOUS BONDED PHASES[a]

| Stationary phase | k' | | α |
	Cor	PhPh	k'_{Cor}/k'_{PhPh}
P	0.28	0.22	0.81
TP	0.30	0.24	0.65
BP	4.31	2.08	1.57
QP	0.13	0.06	1.17
Develosil	3.56	0.95	1.66
Vydac	5.07	0.36	2.50

[a] Mobile phase methanol.

From the preceding data, it appears that the planarity recognition capability of multilegged phases depends on the size of the solutes. The nonplanarity recognition capability was observed with TP for small PAHs, while the planarity recognition capability was seen with QP for large PAHs. The BP phase was intermediate between TP and QP, and this is consistent with the size of the phenyl moiety of the

Table 2.16   RETENTION FACTORS k' AND SEPARATION FACTORS α FOR DIBENZO[bc, ef]CORONENE (DBCO) AND TETRABENZO[a, cd, j, lm]PERYLENE (TEBP) WITH VARIOUS BONDED PHASES[a]

| Stationary phase | k' | | α |
	DBCO	TEBP	k'_{DBCO}/k'_{TEBP}
P	0.38	0.40	0.95
TP	0.15	0.17	0.87
BP	2.84	2.87	0.99
QP	0.20	0.09	2.21
Develosil	4.85	2.22	2.19
Vydac	22.0	1.03	21.4

[a] Mobile phase, methanol/dichloromethane (70:30).

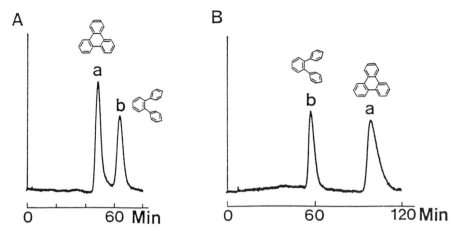

Figure 2.47 Chromatograms for the separation of *o*-terphenyl (a) and triphenylene (b) with (A) TP and (B) BP phases: flow rate, 2 μL/min; UV detection at 254 nm; mobile phases, methanol/water 70:30 for (A) and 80:20 for (B). Reproduced from K. Jinno et al. in *J. Chromatogr.* **517**, 193 (1990) by permission of Elsevier Science-NL, Amsterdam, The Netherlands.

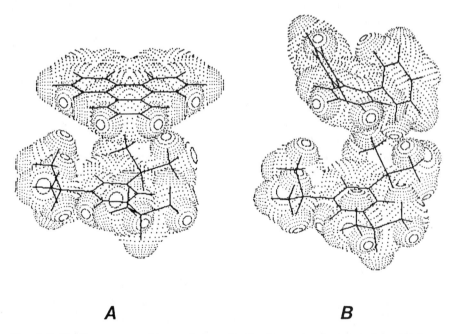

Figure 2.48 Computer graphics rendering of molecular–molecular interactions: (A) interaction between triphenylene and the TP phase and (B) interaction between *o*-terphenyl and the TP phase [58]. Reproduced from K. Jinno et al. in *J. Chromatogr.* **517**, 193 (1990) by permission of Elsevier Science-NL, Amsterdam, The Netherlands.

multilegged phases. Therefore, it is concluded that the size and shape of the solute molecule can be recognized by differences in the space made by cavity-like structures formed by methyl groups and phenyl rings. When the experimentally obtained data are interpreted by using the concept of molecular–molecular interaction, we have a clear explanation of the retention mechanism found in the separation system. This kind of theoretical approach can open up the possibility of designing new stationary phases suitable for particular problems such as isomer separations or chiral separations.

2.5 Compound Stationary Phases

The criteria for stationary phases in the liquid chromatographic separation of PAHs, as discussed in earlier sections, include:

high selectivity for molecular size and shape
high chemical stability during use
availability of graduated particle sizes for scaling up the separation

The first criterion is of most concern in this chapter, and it can be met by several phases, such as polymeric ODS, liquid crystal bonded phases, and multilegged bonded phases. If, however, the other two requirements are strong concerns, one needs to find other, more chemically stable stationary phases having specific separation ability for PAHs; then LC analysis would be a much more powerful analytical method for PAHs.

2.5.1 Dicoronylene Stationary Phase

To ensure high selectivity, one can choose a pure large PAH as the stationary phase for LC separation of PAHs. A stable compound typically chosen is dicoronylene ($C_{48}H_{20}$: benzo[1,2,3bc,4,5,6$b'c'$]dicoronene), whose structure is shown in Figure 2.49. Since this compound has a very planar structure, an enhanced specificity for planarity recognition of PAHs should be expected, assuming that the "slitlike" structure of polymeric ODS affords such recognition. Because dicoronylene requires only small amount of packing and the amount available is not large enough for conventional size column packing, it can be used as the stationary phase in microcolumn separation techniques.

Dicoronylene was synthesized and purified according to Lempka et al. [60]. The particle diameter distribution was measured as shown in Figure 2.50; the sizes are generally in the range of usable chromatographic particles. The mean diameter of the material is about 11 μm. The compound is only sparingly soluble in chlorobenzene and insoluble in methanol and dichloromethane, which could be used as the mobile phase solvents.

Figure 2.51 shows a typical separation of PAHs with a dicoronylene-packed column. The mobile phase is 60:40 methanol/dichloromethane, and the flow rate is

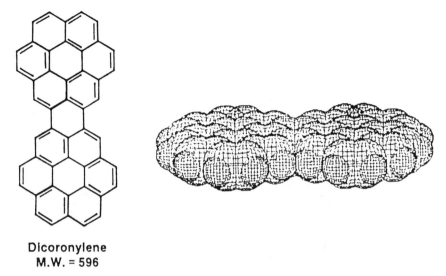

Dicoronylene
M.W. = 596

Figure 2.49   Structure of dicoronylene [61,62]. Reproduced from K. Jinno et al. in *Polycyclic. Aromat. Comp.* **1**, 151 (1990) by permission of Gordon and Breach Publishers, Lausanne, Switzerland.

2 μL/min. Even though the efficiency is not high, and the material was simply packed into the fused-silica capillary column without special treatment, a separation was achieved. This successful separation shows that one can indeed use an insoluble, large PAH as the stationary phase for the separation of PAHs. Two results support the usefulness of this phase as reversed-phase packing for mobile phase systems: (1) the dichloromethane-rich mobile phase eluted the compounds faster

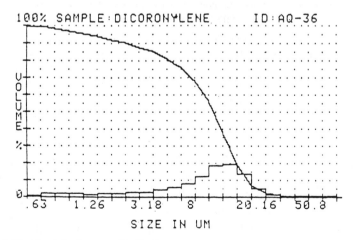

Figure 2.50   Particle diameter distribution of dicoronylene [61]. Reproduced from K. Jinno et al. in *J. High Resolu. Chromatogr. Chromatogr. Commun.* **11**, 673 (1988) by permission of Huethig Verlag, Heidelberg, Germany.

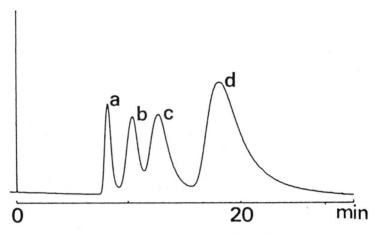

Figure 2.51 Chromatogram of PAH separations with a dicoronylene-packed column, showing peaks for benzene (a), phenanthrene (b), pyrene (c), and chrysene (d): mobile phase, 60:40 methanol/dichloroethane; flow rate, 2 μL/min; UV detection at 254 nm [61]. Reproduced from K. Jinno et al. in *J. High Resolu. Chromatogr. Chromatogr. Commun.* **11**, 673 (1988) by permission of Huethig Verlag, Heidelberg, Germany.

than the weaker mobile phase, and (2) the basic elution order followed the size of the planar PAH molecules.

Reference ODS phases were also evaluated. Experiments were performed with mobile phases having two different compositions of methanol and dichloromethane. Tables 2.17 and 2.18 summarize the data for methanol/dichloromethane in 50:50 and for 20:80 compositions, respectively. Although the data for ODS phases are small retention factors, it is considered better to adjust mobile phase concentration for comparison of retention behaviors with dicoronylene, two polymeric ODS phases, and one monomeric ODS phase. In general, PAH retention in reversed-phase LC with a typical monomeric ODS correlates well with F number and the L/B parameter. Compact molecules give a number close to $L/B = 1$, while long rectangular molecules give large L/B values. If the retention of a PAH is controlled by typical reversed-phase elution behavior under these mobile phase conditions, the logarithm of the retention factor log k' must show a linear relationship with F and L/B. The calculated correlation factors for four stationary phases are shown in Table 2.19. The correlations were poor for the polymeric ODS phases and dicoronylene, although the correlation with the monomeric ODS phase at low dichloromethane concentration is very high. Thus there is similarity between the polymeric ODS and dicoronylene phases, rather than between the polymeric and monomeric ODS phases. This is true as well if one calculates the retention ratio of triphenylene and *o*-terphenyl. The value for dicoronylene with the mobile phase of 50:50 methanol/dichloromethane is calculated as about 7.33, while for Vydac 201, Vydac 218, and Intersil we have 3.40, 2.67, and 1.75, respectively. Therefore dicoronylene significantly exceeds the polymeric ODS phases in planarity recognition ability.

The foregoing discussion indicates that there are some similarities among the polymeric phases and dicoronylene in planarity recognition power. The similarity

Table 2.17   RETENTION DATA FOR PAHs WITH 50:50
METHANOL/DICHLOROMETHANE AS THE MOBILE PHASE

Solute	F	L/B	Retention Factor k' for Stationary Phases			
			Dicoronylene	Vydac 201	Vydac 218	Intersil
Naphthalene	5	1.23	0.05	0.04	0.07	0.12
Biphenyl	6	1.74	0.13	0.02	0.07	0.10
Phenanthrene	7	1.45	0.25	0.06	0.14	0.14
Anthracene	7	1.58	0.48	0.08	0.15	0.16
Pyrene	8	1.26	0.60	0.10	0.16	0.22
o-Terphenyl	9	1.11	0.06	0.05	0.06	0.12
p-Terphenyl	9	2.34	0.33	0.08	0.04	0.14
Triphenylene	9	1.12	0.44	0.17	0.16	0.21
Triphenylmethane	9	1.03	0.01	0.02	0.03	0.17
Chrysene	9	1.65	1.33	0.18	0.18	0.22
Perylene	10	1.27	a	0.27	0.33	0.25
Coronene	12	1.01	a	0.69	0.42	0.42
Dibenzo[cd,lm] Perylene	13	1.70	a	1.48	1.10	0.79
Tribenzo[a,cd,lm] Perylene	15	1.37	a	0.81	0.68	0.83
Tetrabenzo[a,cd,j,lm] Perylene	17	1.22	a	0.37	0.42	0.89

a Huge retention observed and difficult to determing.

Table 2.18   RETENTION DATA FOR PAHs WITH 20:80
METHANOL/DICHLOROMETHANE AS THE MOBILE PHASE

Solute	Retention Factor, k' for Stationary Phases			
	Dicoronylene	Vydac 201	Vydac 218	Intersil
Naphthalene	0.05	0.006	0.06	0.07
Biphenyl	0.05	0.003	0.01	0.07
Phenanthrene	0.13	0.02	0.08	0.10
Anthracene	0.11	0.02	0.07	0.15
Pyrene	0.15	0.03	0.10	0.12
o-Terphenyl	0.03	0.006	0.02	0.12
p-Terphenyl	0.10	0.01	0.03	0.16
Triphenylene	0.22	0.04	0.12	0.10
Triphenylmethane	0.002	0.006	0.02	0.10
Chrysene	0.26	0.06	0.14	0.15
Perylene	0.42	0.07	0.19	0.17
Coronene	2.82	0.17	0.23	0.24
Dibenzo[cd,lm]perylene	5.89	0.57	0.38	0.28
Tribenzo[a,cd,lm]perylene	1.70	0.15	0.31	0.28
Tetrabenzo[a,cd,j,lm]perylene	1.35	0.10	0.26	0.27

Table 2.19   CORRELATION COEFFICIENTS FOR THE
RELATIONSHIPS BETWEEN LOG k' AND
F AND L/B:

$$\log k' = PF + Q(L/B) + R,$$
where P, Q, and R are regression coefficients

	Correlation Coefficient, r	
Stationary Phase	$50:50^a$	$20:80^a$
Dicoronylene	0.470	0.692
Vydac 201	0.784	0.746
Vydac 218	0.697	0.696
Intersil	0.931	0.915

^{a} Mobile phase, composition of methanol and dichloromethane.

can be better seen in the correlation matrix of Table 2.20, which summarizes cross-correlations among data sets for four different stationary phases. The similarity between Vydac 201 and dicoronylene is high for the 50:50 mobile phase and even higher at 20:80. Among all four stationary phases except Intersil, the similarity largely increases as the dichloromethane concentration in the mobile phase increases. One notable observation from Table 2.20 is that the behavior of dicoronylene is closer to that of Vydac 201 than to that of Vydac 218. Inasmuch as Vydac 201 is a more heavily loaded and a more highly polymerized ODS phase than Vydac 218, the greater similarity of Vydac 201 and dicoronylene suggests that the latter phase does have a very strong planarity recognition capability because of its planar and rigid molecular structure. The superiority of dicoronylene's planarity

Table 2.20   CORRELATION MATRIX FOR FOUR STATIONARY PHASES

	Mobile Phase, 50:50 Methanol/Dichloromethane			
	Dicoronylene	Vydac 201	Vydac 218	Intersil
Dicoronylene	—			
Vydac 201	0.839	—		
Vydac 218	0.755	0.980	—	
Intersil	0.733	0.794	0.848	—
	Mobile Phase, 20:80 Methanol/Dichloromethane			
	Dicoronylene	Vydac 201	Vydac 218	Intersil
Dicoronylene	—			
Vydac 201	0.979	—		
Vydac 218	0.835	0.823	—	
Intersil	0.762	0.705	0.898	—

recognition ability over that of Vydac 201 is also thought to be due to the π-π interaction between the stationary phase, and PAHs, which does not exist in the system with Vydac 201. And in fact, the very planar coronene violates the linear relationship between the retention of planar PAHs and F number by showing extraordinarily long retention. This enhanced retention is probably attributable to a strong shape recognition contribution potentially existing between the coronene solute and what is basically the two-coronene structure of the dicoronylene phase [61,62]. This in turn indicates that on polymeric ODS phases the planar–planar interaction exists, and the so-called slot-model seems to offer a reasonable explanation for PAH retention behavior in reversed-phase LC.

2.5.2 C_{60} Stationary Phase

One of the most exciting topics in chemistry is the discovery of all-carbon cluster compounds such as C_{60}, the so-called fullerenes. The compounds contain only carbon atoms, and they have a basic electron-donating character because of their large number of C=C double bonds and because the molecules are aromatic rings. When fullerenes are used as the stationary phase in LC, investigators can take advantage of the electron donor–acceptor complexation mechanism that governs the separation in charge-transfer LC, and the molecular size and shape recognition mechanism. Two approaches are possible for using fullerenes as the stationary phase in LC. One is to obtain a stable chromatographic stationary phase by covalently bonding the fullerenes to a support or particle surface like silica gel. Stalling and others tried this approach first in 1993 [63,64]. Another possibility is to use the fullerenes in their pure form as a stationary phase packing material, paralleling the use of dicoronylene as the stationary phase, as described in the preceding section. We begin with the second approach [65], considering the former approach in the Section 2.6 [66].

Buckminsterfullerene (C_{60}) is generally produced by the method of Kratchmer, Lamb, and Huffman [67,68] and can be isolated by adsorption chromatography. Here C_{60} was obtained by the procedure we described in 1992 [65].

The isolated C_{60} was analyzed for purity using two spectroscopic methods, DRIFT and ^{13}C NMR. The DRIFT spectrum of C_{60} is shown in Figure 2.52, where the absorption bands agree well with a published spectrum [69] shown as an inset in the same figure. Figure 2.53 is the ^{13}C NMR spectrum of the material. The sample in deuterobenzene was measured, and the sharp, single signal clearly indicates that the material is pure C_{60}. After this purity check, the material was crushed to very fine particles (size distribution, 2–30 μm) and packed into the capillary column.

2.5.2.1 Retention Behavior of PAHs

Figure 2.54 presents typical separations of four PAHs. The elution order depends on the size of the PAH molecules, and increasing water concentration in the mobile phase induces an increase in retention, which seems to be controlled by a reversed-phase mechanism. Retention data of several PAHs with two different mobile phase

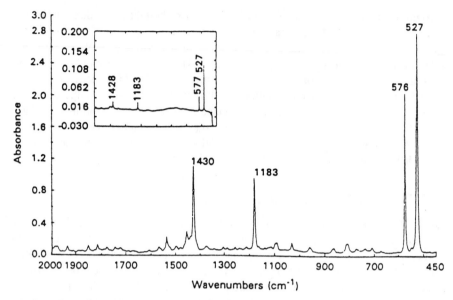

Figure 2.52 DRIFT spectrum of isolated C_{60} [66]. Reproduced from K. Jinno et al. in *J. Microcolumn Sep.* **5**, 517 (1993) by permission of John Wiley and Sons, New York, NY, USA.

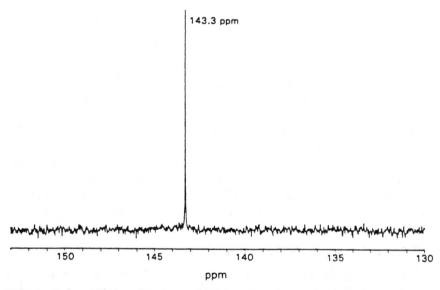

Figure 2.53 ^{13}C NMR spectrum of isolated C_{60} in perdeuterobenzene [66]. Reproduced from K. Jinno et al. in *J. Microcolumn Sep.* **5**, 517 (1993) by permission of John Wiley and Sons, New York, NY, USA.

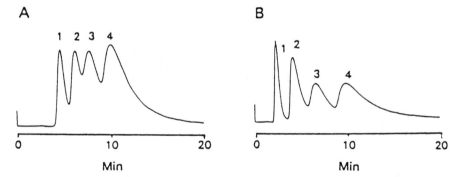

Figure 2.54 Typical LC separations of standard PAHs on the C_{60} stationary phase using a 150 mm $\times$ 0.53 mm i.d. column, showing peaks for naphthalene (1), pyrene (2), chrysene (3), and benzo[a]pyrene (4): flow rate, 2 μL/min; UV detection at 254 nm; mobile phase, methanol/water 95:5 for (A) and 85:15 for (B) [66]. Reproduced from K. Jinno et al. in *J. Microcolumn Sep.* **5**, 517 (1993) by permission of John Wiley and Sons, New York, NY, USA.

systems are summarized in Table 2.21. As found in Figure 2.54 and Table 2.21, the retention of PAHs increases with increasing water concentration and decreasing dichloromethane concentration in the mobile phase. These results apparently indicate that the C_{60} operates as the stationary phase for reversed-phase separations. The relationship between log k' and water concentration in the mobile phase is illustrated in Figure 2.55, and the high linearity supports such a role for C_{60}.

Figure 2.56 plots retention factors of various PAHs against F number. As found for a typical monomeric ODS phase, the relationship between log k' and F is basically linear with the C_{60} phase. The distinct feature of the C_{60} phase different from a typical ODS phase is that long, rodlike molecules (e.g., anthracene, naphthacene, p-terphenyl) show lower retention than their counterparts (e.g., phenanthrene, chrysene, triphenylene, and m-terphenyl). With typical ODS phases the latter are retained longer than the former. For example, naphthacene has the longest retention of several four-ring PAHs, and phenanthrene elutes faster than anthracene [52]. It has been reported that the elution order with a typical ODS for four-ring PAH isomers is triphenylene, benz[a]anthracene, chrysene, and naphthacene [52], although with the C_{60} phase the order is naphthacene, benz[a]anthracene, chrysene, and triphenylene.

Figure 2.57 shows the relationships of the relative retention of large PAHs to coronene and their structures. Two column types are shown for comparison: polymeric ODS (Vydac) has a very strong planarity recognition capability, and monomeric ODS (Develosil) has a lower planarity recognition ability for large PAHs. The retention with the C_{60} shows very strong dependence on molecular size and no recognition of molecular planarity.

The tendencies revealed in Figures 2.56 and 2.57 can be explained by noting that the C_{60} molecule, which is configured like a soccer ball, recognizes the molecular shape and size, of PAHs, whereas its essentially round structure does not fit well with the molecules having a rodlike and/or small nonplanar shape. Large nonplanar

Table 2.21   RETENTION FOR PAHs WITH C_{60} STATIONARY PHASE

Compounds	F Number	Retention Factor, k' for Two Methanol/dichloromethane Mobile Phases	
		90:10	100:0
Naphthalene	5	0.099	0.116
Biphenyl	6	0.133	0.181
Phenanthrene	7	0.361	0.544
Anthracene	7	0.286	0.433
Pyrene	8	0.690	0.930
o-Terphenyl	9	0.230	0.298
m-Terphenyl	9	0.713	1.21
p-Terphenyl	9	0.598	0.868
Triphenylene	9	1.71	2.46
Chrysene	9	1.31	1.78
Naphthacene	9	0.80	1.04
Benz[a]anthracene	9	1.09	1.29
Benzo[c]phenanthrene	9	1.30	1.68
Perylene	10	2.62	3.78
Benzo[a]pyrene	10	2.25	3.27
Benzo[e]pyrene	10	2.32	3.47
Coronene	10	5.72	8.21

Source: Ref. 66. Reproduced from K. Jinno et al. in *J. Microcolumn Sep.* **5**, 517 (1993) by permission of John Wiley and Sons, New York, NY, USA.

PAHs, however, can cover some parts of the shape of the C_{60} molecule, and this induces longer retention than other smaller nonplanar and rodlike molecules.

2.5.2.2 Retention Behavior of Alkylbenzenes

The retention behavior of alkylbenzenes on the C_{60} phase has been investigated. The first solute group we discuss is polymethyl-substituted benzenes. The results are summarized in Figure 2.58A, a plot of log k' versus number of methyl groups in the structures. For comparison, Figure 2.58B shows the data with typical monomeric ODS phase. Generally, increasing the number of methyl substituents induces an increase in retention for both phases. The derivatives having methyl groups at the 1,3,5-positions are on the linear line (dashed line in Figure 2.58A) with the C_{60} phase, while all derivatives are on the same line with the ODS phase. The main reason for this behavior is the contribution of the bulkiness of the molecules just described, compared to other derivatives. For example, 1,3,5-trimethylbenzene is bulkier and more extended in shape than 1,2,4- and 1,2,3-trimethylbenzene. 1,2,3-Trimethylbenzene is the most compact molecule among the three isomers. For di- and tetrasubstituted compounds, the elution orders differ only slightly for the two phases. 1,4-Dimethylbenzene and 1,2,3,4-tetramethylbenzene have the smallest retention among the isomers with the C_{60} phase. This behavior is very similar to that found in the study on the retention of PAHs described in 2.5.2.1. Therefore it can be

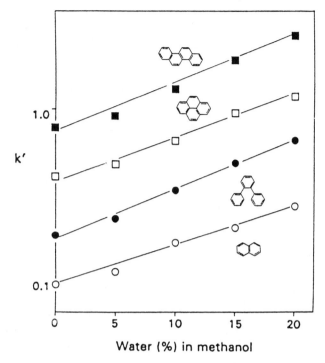

Figure 2.55   Relationship between log k' of standard PAHs and water (vol %) in the mobile phase. Conditions are the same as in Figure 2.54 [66]. Reproduced from K. Jinno et al. in *J. Microcolumn Sep.* **5**, 517 (1993) by permission of John Wiley and Sons, New York, NY, USA.

concluded that molecular shape and bulkiness and total compound size are the keys to estimating retention with C_{60} as the stationary phase.

To learn about the retention characteristics of the C_{60} phase for alkylbenzenes in more detail, normal alkylbenzenes were examined. The relationship between chain length and log k' is shown in Figure 2.59. Most of the alkylbenzenes are on the linear plot, but the retention of *n*-butylbenzene deviates positively from this line. The measurements were performed several times, eliminating the possibility of data obtained by coincidence. Other isomeric alkylbenzenes were also examined under the same chromatographic conditions and the results are summarized in Figure 2.60. Comparing the retention of *sec*-amylbenzene and *tert*-amylbenzene, the former is found to be more highly retained than the latter. This is because *sec*-amylbenzene has an *n*-butyl chain in its structure and it is known that *n*-butyl chains are a factor in increased solute retention, although the reason is not clear yet. The big difference in retention between *n*-butylbenzene and *tert*-butylbenzene indicates the important contribution of the *n*-butyl chains in this respect. Molecular length alone is not responsible for this effect: *p*-diethylbenzene, which has the same length as *n*-butylbenzene, is not retained as much as is shown in Figures 2.59 and 2.60. Moreover, *o*- and *m*-diethylbenzenes have retention similar to that of *n*-butylbenzene.

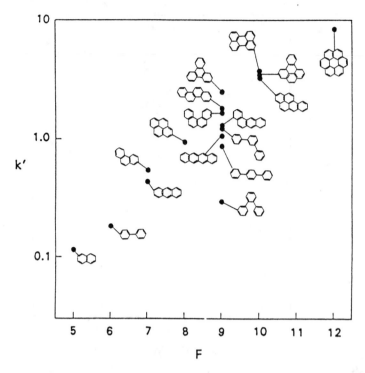

Figure 2.56 Relationship between log k' of standard PAHs and F number. Conditions are the same as in Figure 2.54, except the mobile phase was 100% methanol [66]. Reproduced from K. Jinno et al. in *J. Microcolumn Sep.* **5**, 517 (1993) by permission of John Wiley and Sons, New York, NY, USA.

2.6 C_{60} Bonded Silica Phase

We have just discussed the introduction of the fine powdery solid C_{60} as the stationary phase in microcolumn LC, and the results of a study of its separation performance and characteristics for various isomeric PAHs were presented. The results suggested that the fullerene C_{60} could be useful as an LC stationary phase, and it offers unique shape selectivity for PAHs compared to conventional ODS phases. Because of the good stability of the silica-based stationary phase compared to that of compound itself, a newly synthesized C_{60} bonded silica was deployed in the investigation of the retention behavior of PAHs, as described in Sections 2.6.1 and 2.6.2.

2.6.1 Bonded Phase Synthesis

The synthetic procedure of the silanization reagent of the C_{60} bonded silica phase, $C_{60}(H)[CH_2SiCl(CH_3)_2]$ (**1**) is described elsewhere [70]. Compound **1** was then

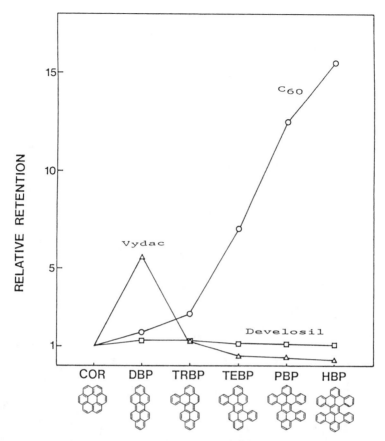

Figure 2.57 Relative retention as a function of structure for six large PAHs on three stationary phases. Conditions are the same as in Figure 2.54, except the mobile phase was 50:50 methanol/dichloromethane [66]. Reproduced from K. Jinno et al. in *J. Microcolumn Sep.* **5**, 517 (1993) by permission of John Wiley and Sons, New York, NY, USA.

treated with silica (Shiseido, Yokohama, Japan; particle diameter, $\approx$ 5μm; average pore size, 120 Å; surface area, $\approx$ 270 m^2/g) in the presence of pyridine at 100°C for 12 hours. After filtration, the brown solid materials were washed with toluene, chloroform, and methanol to give the C_{60} bonded silica phase **2**. Figure 2.61 shows the synthetic scheme of the phase. The synthesized bonded phase was then packed using a slurry method into 150 mm × 0.32 mm i.d. fused-silica capillaries [71].

2.6.2 Retention Behavior of PAHs

Table 2.22 shows the retention data of various PAH solutes with the C_{60} bonded silica stationary phase, as well as the data with C_{60} itself as the stationary phase (from Section 2.5.2). It was found that the powdery C_{60} stationary phase showed different selectivity for isomeric PAHs from general ODS phases, although ba-

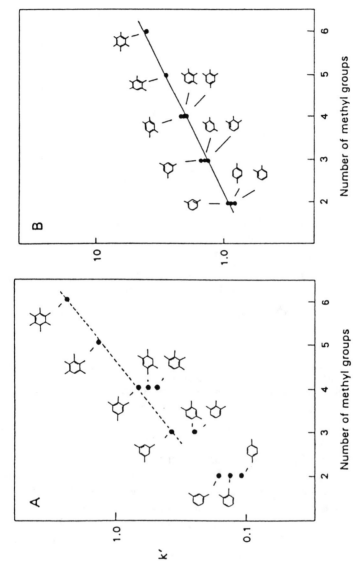

Figure 2.58 Relationship between log k' of polymethyl-substituted benzenes and the number of methyl groups in the compound structures. (A) 100 mm × 0.32 mm i.d. column on C_{60}; 90:10 methanol/water mobile phase, 2 μL/min flow rate; UV detection at 254 nm. (B) 150 mm × 0.53 mm i.d. column on monomeric ODS; 90:10 methanol/water mobile phase, 2 μL/min flow rate; UV detection at 254 nm [66]. Reproduced from K. Jinno et al. in *J. Microcolumn Sep.* **5**, 517 (1993) by permission of John Wiley and Sons, New York, NY, USA.

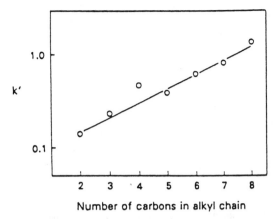

Figure 2.59 Relationship between log k' of normal alkylbenzenes and carbon number of the alkyl chains in their structures. Conditions are the same as in Figure 2.58 [66]. Reproduced from K. Jinno et al. in *J. Microcolumn Sep.* **5**, 517 (1993) by permission of John Wiley and Sons, New York, NY, USA.

sically linear relationships between log k' and F number with the C_{60} phase were observed, as in the case of typical monomeric ODS phases.

With chemically bonded C_{60} silica phase, almost the same trend has been obtained: that is, the retention increases with increasing solute F number. However,

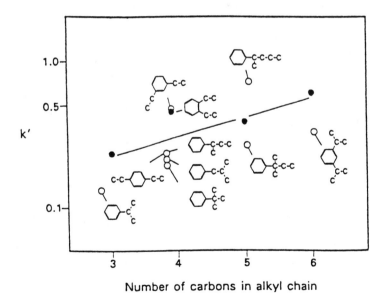

Figure 2.60 Relationship between log k' of isomeric alkylbenzenes and carbon number of the alkyl chains in their structures. Conditions are the same as in Figure 2.58; solid circles indicate the retention of *n*-alkylbenzenes [66]. Reproduced from K. Jinno et al. in *J. Microcolumn Sep.* **5**, 517 (1993) by permission of John Wiley and Sons, New York, NY, USA.

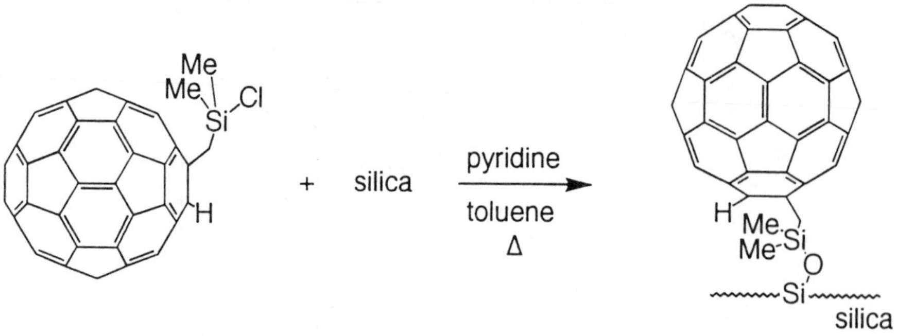

Figure 2.61 Synthetic scheme of the C_{60} bonded silica stationary phase **2**. Other characteristics of the C_{60} bonded phase are as follows: carbon content, 8.39%; surface coverage, 0.411 μmol/m^2 [71]. Reproduced from Y. Saito et al. in *J. High Resolu. Chromatogr. Chromatogr. Commun.* **18**, 569 (1995) by permission of Huethig Verlag, Heidelberg, Germany.

the elution order of isomeric PAHs differs somewhat from that of the ODS phases. For example, the elution order of planar four-ring PAH isomers on the C_{60} bonded phase and on the C_{60} phase is naphthacene, benz[a]anthracene, chrysene, and triphenylene, while the order with ODS phases is triphenylene, benz[a]anthracene, chrysene, and naphthacene.

Table 2.22   RETENTION DATA FOR PAHs WITH C_{60} BONDED SILICA AND THE C_{60} PHASES[a]

Solute	F	Retention Factor, k'	
		C_{60}-Bonded Phase	C_{60}
Naphthalene	5	0.179	0.116
Biphenyl	6	0.252	0.181
Phenanthrene	7	0.640	0.544
Anthracene	7	0.500	0.433
Pyrene	8	0.889	0.930
o-Terphenyl	9	0.392	0.298
m-Terphenyl	9	1.43	1.21
p-Terphenyl	9	1.02	0.868
Naphthacene	9	1.63	1.04
Benz[a]anthracene	0	1.68	1.29
Chrysene	9	1.81	1.78
Benzo[c]phenanthrene	9	2.13	1.68
Triphenylene	9	2.35	2.46
Benzo[a]pyrene	10	3.93	3.27
Benzo[e]pyrene	10	4.54	3.47
Perylene	10	4.76	3.78

[a]   Mobile phase, methanol.

Source: Ref. 71.

Similar trends have also been observed in the case of phenanthrene and anthracene, and five-ring PAHs. Upon careful examination of the foregoing results, it appears that the C_{60} bonded phase possesses specific selectivity for PAH isomers having a structure that is partially similar to C_{60} molecule as found with the C_{60} phase. The good correlation between the retention data obtained with the C_{60} bonded silica phase and data from the C_{60} phase is seen in Figure 2.62, where a high correlation coefficient ($r = 0.986$) was obtained.

The separation factors between triphenylene and o-terphenyl for the C_{60} bonded phase and the C_{60} phase are 5.99 and 8.26, respectively (larger than those with typical ODS and other aromatic stationary phases). The results can be explained by noting that the nonplanarity of o-terphenyl is very large, and this large curvature of the molecule is disturbed upon interaction with the C_{60} moiety, and thus the retention of o-terphenyl is reduced. Planar PAHs having a structure that is partially similar to C_{60} molecule, (e.g., triphenylene and perylene) interact effectively with the bonded moiety by π-π interaction, whereas slightly nonplanar PAHs such as benzo[c]phenanthrene are retained longer. From the foregoing observations, one can derive a scheme for the retention model of PAH solutes on the C_{60} bonded silica phase, as shown in Figure 2.63. In the truly interesting approach described in this section, even the molecule C_{60} can be modified on the silica surface and the singular structure of the bonded phase can give unusual and unique selectivity for the molecular shape and planarity of PAHs.

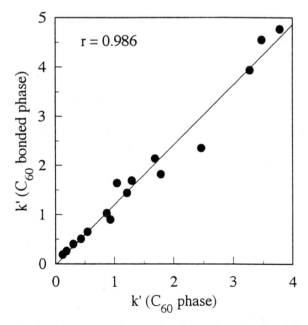

Figure 2.62 Relationship between the retention data with the C_{60} bonded silica phase and those with the C_{60} phase [71]. Reproduced from Y. Saito et al. in *J. High Resolu. Chromatogr. Chromatogr. Commun.* **18**, 569 (1995) by permission of Huethig Verlag, Heidelberg, Germany.

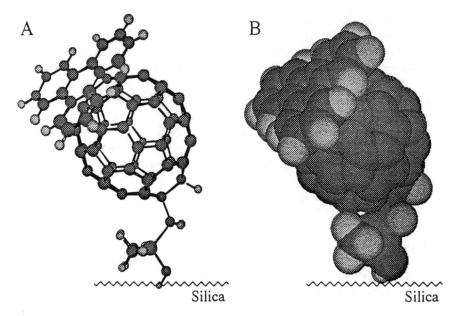

Figure 2.63 Molecular modeling schemes of the proposed π-π interaction between triphenylene and the C_{60} bonded stationary phase: (A) "ball and stick" model and (B) "space-filling" model. Diagrams created with Chem $3D^+$ software (Cambridge Scientific Computing).

2.7 Looking to the Future

It has been shown that the molecular shape recognition mechanism in LC is only the small part of overall separation mechanisms in this separation method. The concept described in this chapter, however, will open new views and possibilities for LC to be widely used to in the solution of separation problems so difficult that they are now regarded as impossible. This is the most important and promising prediction that can be made at the end of this chapter.

Acknowledgments

The author sincerely thanks the following people who contributed to this work and discussed the results with deep insight: Dr. J.C. Fetzer and Dr. W.R. Biggs, Chevron Research and Technology Company; Professor J.J. Pesek, Department of Chemistry, San Jose State University; and Dr. Y.-L. Chen, J&W Scientific, all of California; and Dr. N. Nagae, Nomura Chemicals, Seto, Japan. He also thanks Y. Saito, H. Ohta, T. Uemura, and K. Nakagawa, who were research students in his group at Toyohashi University of Technology. Financial support from the Japanese Ministry of Education and Culture as scientific research grants Nos. 05233108 and 06640781 is also acknowledged.

References

1. R.G. Harvey, Ed., *Polycyclic Hydrocarbons and Carcinogenesis,* ACS Symposium Series No. 283. American Chemical Society, Washington DC, 1985.

2. L.C. Sander and S.A. Wise, *Anal. Chem.* **56,** 504 (1984).

3. S.A. Wise and L.C. Sander, *High Resolution Chromatogr., Chromatogr. Commun.* **8,** 248 (1985).

4. L.C. Sander and S.A. Wise, *Adv. Chromatogr. (N.Y.),* **25,** 139 (1986).

5. K. Jinno, H. Shimura, N. Tanaka, K. Kimata, J.C. Fetzer, and W.R. Biggs, *Chromatographia,* **27,** 285 (1989).

6. R.K. Gilpin and M.E. Gangoda, *Anal. Chem.* **56,** 1470 (1984).

7. P. Shah, T.-B. Tsu, and L.B. Rogers, *J. Chromatogr.* **396,** 31 (1987).

8. M. Gangoda and R.K. Gilpin, *J. Magn. Resonance,* **74,** 134 (1987).

9. D.W. Sindorf and G.E. Maciel, *J. Am. Chem. Soc.* **105,** 3767 (1983).

10. D.E. Leydon, D.S. Kendall, L.W. Burggraf, F.J. Pern, and M. DeBello, *Anal. Chem.* **54,** 101 (1982).

11. R.K. Gilpin, *Anal. Chem.* **57,** 1465A (1985).

12. G.S. Garavajal, D.E. Leydon, G.R. Quinting, and G.E. Maciel, *Anal. Chem.* **60,** 1776 (1988).

13. A.K. Palmer and G.E. Maciel, *Anal. Chem.* **54,** 2194 (1982).

14. R.K. Gilpin and M.E. Gangoda, *J. Chromatogr. Sci.* **21,** 352 (1983).

15. D.W. Sindorf and G.E. Maciel, *J. Am. Chem. Soc.* **105,** 1848 (1983).

16. S.S. Yang and R.K. Gilpin, *J. Chromatogr.* **449,** 115 (1988).

17. J.K. Ohler, D.B. Chase, R.D. Farlee, A.J. Vega, and J.J. Kirkland, *J. Chromatogr.* **352,** 275 (1986).

18. C.R. Yonker, T.A. Zwier, and M.R. Burke, *J. Chromatogr.* **241,** 257 (1982).

19. P. Shah, L.B. Rogers, and J.C. Fetzer, *J. Chromatogr.* **388,** 411 (1987).

20. G.E. Maciel, R.C. Zeugler, and R.K. Taft, in D.E. Leydon, Ed., in *Silanes, Surface and Interfaces.* Gordon and Breach, New York, 1986, pp. 413–429.

21. K. Jinno, *J. Chromatogr. Sci.* **27,** 729 (1989).

22. M.E. McNally and L.B. Rogers, *J. Chromatogr.* **331,** 23 (1985).

23. K. Jinno, T. Ibuki, N. Tanaka, M. Okamoto, J.C. Fetzer, W.R. Biggs, P.R. Griffiths, and J.M. Olinger, *J. Chromatogr.* **461,** 209 (1989).

24. R.K. Gilpin and M.E. Gangoda, *J. Magn. Resonance,* **64,** 408 (1985).

25. E. Bayer, A. Paulus, B. Peters, G. Laupp, J. Reiners, and K. Albert, *J. Chromatogr.* **364,** 25 (1986).

26. M. Gangoda, R.K. Gilpin, and B.M. Fung, *J. Magn. Resonance,* **74,** 134 (1987).

27. E. Bayer, K. Albert, J. Reiners, M. Nieder, and D. Muller, *J. Chromatogr.* **264,** 197 (1983).

28. L.C. Sander, J.B. Callis, and L.R. Field, *Anal. Chem.* **55,** 1068 (1983).

29. D. Morel, K. Tabar, J. Serpinet, P. Claudy, and J.M. Letoffe, *J. Chromatogr.* **395,** 73 (1987).

30. J.C. Fetzer and W.R. Biggs, *J. Chromatogr.* **322,** 275 (1985).

31. H.L. Casal, D.G. Cameron, and H.H. Mantsch, *Can. J. Chem.* **61**, 1736 (1983).

32. M.L. Lee, R.C. Kong, C.L. Woolley, and J.S. Bradshaw, *J. Chromatogr. Sci.* **22**, 136 (1984).

33. B.A. Jones, J.S. Bradshaw, M. Nishioka, and M.L. Lee, *J. Org. Chem.* **49**, 4947 (1984).

34. J.S. Bradshaw, C. Schregenberger, K.H.C. Chang, K.E. Markides, and M.L. Lee, *J. Chromatogr.* **358**, 95 (1986).

35. K.A. Dill, *J. Phys. Chem.* **91**, 1980 (1987).

36. Z. Witkiewicz, Z. Suprynowicz, J. Wojcik, and R. Dabrowski, *J. Chromatogr.* **152**, 323 (1978).

37. F. Jassen, *Anal. Chem.* **51**, 2163 (1979).

38. R.J. Laub, W.L. Roberts, and C.A. Smith, *J. High Resolution Chromatogr., Chromatogr. Commun.* **3**, 355 (1980).

39. Z. Witkiewicz, *J. Chromatogr.* **251**, 311 (1982).

40. Z. Witkiewicz, *LC GC,* **8(3),** 224 (1990).

41. P.J. Taylor and P.L. Sherman, *J. Liquid Chromatogr.* **2**, 1271 (1979).

42. A.A. Aratskova, Z.P. Vetrova, and Y.I. Yashin, *J. Chromatogr.* **365**, 27 (1986).

43. J. Pesek and T. Cash, *Chromatographia,* **27**, 559 (1989).

44. H.-C.K. Chang, K.E. Markides, J.S. Bradshaw, and M.L. Lee, *J. Chromatogr. Sci.* **26**, 280 (1985).

45. K.E. Markides, H.-C.K. Chang, C.M. Schregenberger, B.J. Tarbet, J.S. Bradshaw, and M.L. Lee, *J. High Resolution Chromatogr., Chromatogr. Commun.* **8**, 516 (1985).

46. M.A. Apfel, H. Finemann, G.M. Janini, R.J. Laub, B.H. Luehmann, A. Price, W.L. Roberts, T.J. Shaw, and C.A. Smith, *Anal. Chem.* **57**, 651 (1985).

47. K.E. Markides, M. Nishioka, B.J. Tarbet, J.S. Bradshaw, and M.L. Lee, *Anal. Chem.* **57**, 1296 (1985).

48. S.A. Wise, L.C. Sander, H.-C.K. Chang, K.E. Markides, and M.L. Lee, *Chromatographia,* **25**, 473 (1988).

49. Y. Saito, K. Jinno, J.J. Pesek, Y.-L. Chen, G. Luehr, J. Archer, J.C. Fetzer, and W.R. Biggs, *Chromatographia,* **38**, 295 (1994).

50. J.F. Schabron, R.J. Hurtubise, and H.F. Silver, *Anal. Chem.* **49**, 2253 (1977).

51. K. Jinno and K. Kawasaki, *Chromatographia,* **17**, 445 (1983).

52. K. Jinno and K. Kawasaki, *J. Chromatogr.* **316**, 1 (1984).

53. K. Jinno and K. Kawasaki, in S. Ahuja (Ed.), Chromatography and Separation Chemistry, ACS Symposium Series No. 297, American Chemical Society, Washington DC, 1986.

54. N. Tanaka, Y. Tokuda, K. Iwaguchi, and J. Araki, *J. Chromatogr.* **239**, 761 (1982).

55. K. Jinno, N. Anazawa, M. Okamoto, N. Tanaka, J.C. Fetzer, and W.R. Biggs, *J. Chromatogr.* **402**, 173 (1987).

56. K. Jinno, T. Nagoshi, N. Tanaka, M. Okamoto, J.C. Fetzer, and W.R. Biggs, *J. Chromatogr.* **436**, 1 (1988).

57. S.A. Wise, W.J. Bonnett, F.R. Guenther, and W.E. May, *J. Chromatogr. Sci.* **19**, 457 (1981).

58. K. Jinno, K. Yamamoto, H. Nagashima, T. Ueda, and K. Itoh, *J. Chromatogr.* **517**, 193 (1990).

59. J.J. Kirkland, J.C. Glajch, and R.D. Farlee, *Anal. Chem.* **61**, 2 (1989).

60. H.J. Lempka, S. Obenland, and W. Schmidt, *Chem. Phys.* **96**, 349 (1985).

61. K. Jinno, H. Shimura, J.C. Fetzer, and W.R. Biggs, *J. High Resolution Chromatogr., Chromatogr. Commun.* **11**, 673 (1988).

62. K. Jinno, H. Shimura, J.C. Fetzer, and W.R. Biggs, *Polycyclic Aromat. Comp.* **1**, 151 (1990).

63. D.L. Stalling, C. Guo, K.C. Kuo, and S. Saim, *J. Microcolumn Sep.* **5**, 223 (1993).

64. D.L. Stalling, C.Y. Guo, and S. Saim, *J. Chromatogr. Sci.* **31**, 265 (1993).

65. K. Jinno, K. Yamamoto, J.C. Fetzer, and W.R. Biggs, *J. Microcolumn Sep.* **4**, 187 (1992).

66. K. Jinno, K. Fukuoka, J.C. Fetzer, and W.R. Biggs, *J. Microcolumn Sep.* **5**, 517 (1993).

67. W. Kratschmer, L.D. Lamb, K. Fostiropoulos, and D.R. Huffman, *Nature*, **347**, 3 (1990).

68. W. Kratschmer, L. Lamb, K. Fostiropoulos, and D. Huffman, *Chem. Phys. Lett.* **170**, 160 (1990).

69. D.S. Bethune, G. Meijer, W.C. Tang, H.J. Rosen, W.G. Golden, H. Seki, C.A. Brown, and M.S. deVries, *Chem. Phys. Lett.* **179**, 181 (1991).

70. H. Nagashima, T. Terasaki, Y. Saito, K. Jinno, and K. Itoh, *J. Org. Chem.* **60**, 4966 (1995).

71. Y. Saito, H. Ohta, H. Terasaki, H. Nagashima, K. Jinno, and K. Itoh, *J. High Resolution Chromatogr., Chromatogr. Commun.* **18**, 569 (1995).

CHAPTER
3

Molecular Recognition for Fullerenes in Liquid Chromatography

Kiyokatsu Jinno

3.1 Introduction

The history of fullerenes is highlighted by three important scientific findings:

- In 1970 the presence of a round, hollow, geodesic sphere–shaped molecule consisting of 60 carbon atoms was predicted by Osawa[1].
- In 1985 this stable carbon molecule was observed in the mass spectrum of laser-ablated graphite by Kroto et al. [2].
- However, before Kratschmer et al. [3] finally developed a method of preparing microscopic quantities by the resistive heating of graphite under an inert atmosphere, there was no evidence of the existence on our planet of this compound consisting of carbon atoms arranged to form 12 isolated pentagons and 20 hexagonal rings.

The substance was named "buckminsterfullerene" or "fullerene," and the distinctive molecules were called "Bucky balls," honoring Buckminster Fuller, the engineer and philosopher who is perhaps best known for designing geodesic dome structures. At almost the same time, Kratschmer et al. [3] found the second most abundant molecule formed by resistive heating of graphite, namely, C_{70}. Since this discovery, fullerenes, with their unique structures, have attracted the attention of many chemists, physicists, and materials scientists in a wide variety of fields. The characteristics of these carbon molecules have been studied by means of instrumental analytical techniques such as mass spectrometry (MS) [4–6], nuclear magnetic resonance

(NMR) spectrometry [7–10], infrared (IR) spectrometry [11–13], ultraviolet spectrometry [14], and other spectrometric measurements [15,16]. The key to obtaining sufficiently accurate spectrometric information was the isolation and purification of compounds from mixtures of carbonsoot. Thus various separation methods for the separation and purification of fullerenes and their derivatives were proposed. High performance liquid chromatography (HPLC) is clearly the most promising approach of those proposed methods, but it requires good stationary phases for improved separation performance.

The first several separations of fullerenes were performed on neutral alumina and silica gels [4,7,17,18], and subsequently several commercially available phases such as octadecylsilicas (ODS) were examined [19–28]. In addition, many other unique phases were developed solely for the purpose of fullerene separation. The demand for the preparative-scale separation of fullerenes has especially led to an additional challenge in developing new stationary phases that can permit such large-scale separations with toluene-based mobile phase systems. Hawkins and co-workers [6] used an aromatic stationary phase, since they thought the support's π-acidic dinitrobenzamide in 3,5-dinitrobenzoylphenylglysine (DNBPG) would interact with the π-basic groups of the aromatic soccer ball–like structures. After this success in using aromatic phases for the efficient separation of fullerenes, several new stationary phases based on the same concept were proposed; these include dinitroanilinopropyl (DNAP) [29] and multilegged phenyl phases [30]. Another highly successful approach for the separation of fullerenes with a toluene-rich mobile phase was the use of the Pirkle column, in which the tripodal ligand phase, called "Buckyclutcher I," provided the highest selectivity compared to other phases [31,32]. Other chiral phases such as cyclodextrins were also evaluated [33].

Thus the design of novel stationary phases that offer enhanced selectivity for HPLC separation of fullerenes is one of the most important tasks for separation scientists. Demands for such stationary phases to solve fullerene separation problems are huge, and to solve these problems, we must understand the nanoscale molecular–molecular interactions between stationary phases and fullerenes.

This chapter introduces three basic approaches to the development of a novel stationary phase capable of providing excellent resolution and selectivity for fullerene separations. The steps are very similar to the evaluation of various phases for the recognition at the molecular level of PAHs as described in Chapter 2. The difference lies in the solutes, two-dimensional molecules or bulky steric molecules. First we evaluated easily available ODS stationary phases for the separation of fullerenes, and the results imparted the basic understanding and knowledge needed for designing novel phases. Second, stationary phases different from ODS phases were evaluated. This experimentally obtained information can serve as the basis for designing and synthesizing novel stationary phases, to promote an understanding of the experimental results obtained by means of ODS and other stationary phases. After the confirmation of results in which selectivity enhancements for fullerene separations are seen, it became possible to design novel stationary phases that promise excellent selectivity for the separation of fullerenes.

3.2 ODS Stationary Phases for Fullerene Separation

In Chapter 2 we introduced the monomeric type of ODS phase, which can be synthesized using monochlorosilane as a starting material, and the polymeric type of ODS phase, which can be produced from trichlorosilane. These two types of phase differ in molecular shape and size recognition capability, especially for polycyclic aromatic hydrocarbons (PAHs) [34–37], as discussed in Chapter 2. These differences should be found in the separation of fullerenes as well. The section that follows describes the separation of fullerenes with monomeric and polymeric ODS phases using n-hexane or toluene/acetonitrile as the mobile phase.

3.2.1 Separation of C_{60} and C_{70}

The first step in the evaluation of monomeric ODS stationary phases, the separation of C_{60} and C_{70} with an n-hexane mobile phase, was carried out using various ODS phases that differ in surface coverage and in pore size. As shown in Table 3.1 Develosil ODS-5 contains about 20% carbon, with 16% for ODS-N5 and 11% for ODS-P5. The effect of the surface coverage (carbon content) of the stationary phases on the separation of C_{60} and C_{70} can be examined using the retention data in Table 3.1. The results suggest that the higher carbon content of the ODS phase, the better the separation factors obtained. Moreover, decreased surface coverage serves to shorten retention for both fullerenes.

The effect of silica pore size on separation power can be ascertained by comparing data for Develosil ODS-5 and Develosil 300 ODS-5. The separation factors with two columns are nearly the same as shown in Table 3.1, but retention with the 300 Å pore size was very poor. It is questionable, however, whether differences in pore size or surface coverage affect the separation power differential, even though the results clearly indicate that the stationary phase with high surface coverage gives

Table 3.1 RETENTION DATA FOR C_{60} AND C_{70} WITH VARIOUS ODS STATIONARY PHASES

Stationary Phase[a]	Pore Size (Å)	Carbon Content (%)	Retention Factor, k'		Separation Factor, α
			C_{60}	C_{70}	$k'_{C_{70}}/k'_{C_{60}}$
Develosil ODS-5	100	20	0.87	1.61	1.85
Develosil 300 ODS-5	300	11	0.48	0.89	1.85
Develosil ODS-N5	100	16	0.81	1.32	1.73
Develosil ODS-P5	100	11	0.44	0.64	1.45
Vydac 218-TPB5	300		0.22	0.45	2.05
Wakosil-II 5C18AR	120	18	0.25	0.53	2.11

[a] Develosil ODSs are produced by Nomura Chemicals in Japan as the monomeric type of ODS phase. Vydac 218, made by Separations Group in the United States, and Wakosil, made by Tokyo Kasei in Japan, are polymeric ODS phases.

better separation. It can be concluded that high-loaded monomeric ODS phases with high specific surface area are the best choice for the separation of C_{60} and C_{70} with an n-hexane mobile phase. The polymeric ODS phases are less effective at separating C_{60} and C_{70} than the monomeric ODS phases, as seen in Table 3.1.

3.2.2 Higher Fullerene Separation

To separate higher fullerenes by LC, the first task is to find the optimum mobile phase composition. Although n-hexane is commonly chosen, the low solubility of the higher fullerenes limits its use. A toluene-based mobile phase should be useful for solubility; therefore toluene/methanol and toluene/acetonitrile were evaluated. With Develosil ODS-5, the monomeric ODS, two mobile phases were examined, and the results are shown in Figure 3.1 for toluene/acetonitrile and toluene/methanol. Although the solvent strength does not explain the elution order if the retention is induced by a typical reversed-phase mechanism, both systems give relatively good separations for fullerenes higher than C_{70}. One can choose the countersolvent in toluene on the following basis: (1) if a better separation for higher fullerenes is desired, acetonitrile is the choice; (2) if a faster analysis with a high resolution column is required, methanol should be used.

Experiments to find the optimum composition of toluene in acetonitrile or methanol showed 50–60% acetonitrile and 45–55% methanol in the mobile phase give the best combination of resolution, analysis time, and solubility of fullerenes.

For separation of higher fullerenes, various ODS phases were evaluated, and retention factors of higher fullerenes were calculated. The data are summarized in Table 3.2A. Table 3.2B lists the retention data in terms of the retention of higher fullerenes relative to that of C_{60}. Because this work focuses on getting better separations of higher fullerenes, the mobile phase of 45:55 toluene/acetonitrile was selected. The peak assignments are tentatively made by referring to the published UV–visible spectra [27,38–40]. Several ODS phases cannot resolve C_{76}, C_{78}-$C_{2v'}$, and two isomers of C_{78}-C_{2v} and -D_3. Upon quick examination of the data, it is found that the polymeric ODS can resolve several isomers of C_{78} and the monomeric ODS cannot. Elution order also differs depending on the functionality of ODS phases.

The effect of the surface coverage (carbon content) of the stationary phases on the separation of the higher fullerenes can be examined by means of retention data for Develosil ODS-5, ODS-N5, and ODS-P5, which have carbon contents of 20, 16, and 11%, respectively. The chromatograms obtained are shown in Figure 3.2. The relative retention data of fullerenes higher than C_{60} can be found in Table 3.2B. Clearly, the higher carbon content of the ODS phases gave better separation, and the decrease surface coverage led to smaller retention. Though Develosil ODS-5 can resolve C_{76} and C_{78}-$C_{2v'}$, and C_{78}-C_{2v}, ODS-P5, with its lower carbon content, cannot separate the isomers of C_{78}. Surface coverage appears to be one of the important factors in obtaining high resolution of the higher fullerenes with ODS phases. The use of Develosil ODS-P5 and 300 ODS-5 permits the detailed evaluation of the effect of the pore size of the base silicas on the separation of higher fullerenes. The pore sizes are 100 and 300 Å, respectively, with other characteristics

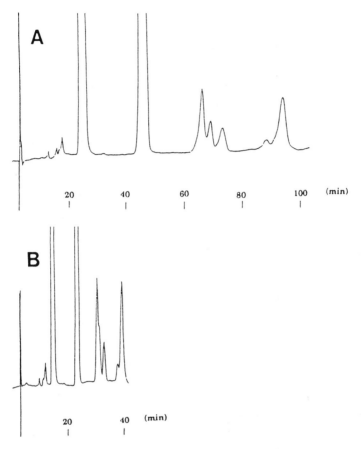

Figure 3.1   Room temperature chromatograms of fullerene mixtures on a Develosil ODS-5 column; flow rate, 1 mL/min; UV detection at 325 nm. Mobile phases: (A) toluene/acetonitrile (50:50) and (B) toluene/methanol (50:50) [25]. Reproduced from K. Jinno et al. in *Anal. Chem.* **65**, 2650 (1993) by permission of the American Chemical Society, Washington, D.C., USA.

the same. The chromatograms shown in Figure 3.3 indicate that resolution and retention with ODS-P5 were relatively poor. Silica pore size is a very good index of the effectiveness of separation of higher fullerenes. However, relative retentions of fullerenes higher than C_{60} with 300 ODS-5 are nearly the same as those with ODS-5, as shown in Table 3.2B. It can be concluded that high-loaded monomeric ODS phases with high specific surface area are the best choice for the separation of higher fullerenes with toluene-based mobile phase systems. Similar trends have been found for C_{60} and C_{70} separations.

Develosil ODS-5 gave the best results of all monomeric ODS systems evaluated; of the polymeric ODS phases, Wakosil-II 5C18 AR gave the best separation of higher fullerenes. To study the retention mechanism for fullerene separations, these two columns should be evaluated in detail with different functionalities. The chro-

Table 3.2A RETENTION FACTORS OF HIGHER FULLERENES ON VARIOUS STATIONARY PHASES[a]

Stationary Phase	$k'_{C_{60}}$	$k'_{C_{70}}$	$k'_{C_{76}}$	$k'_{C_{78}}-C_{2''}$	$k'_{C_{78}}-C_{2'}$	$k'_{C_{78}}-D_3$	$k'_{C_{82}}$	$k'_{C_{84}}$
1. Develosil ODS-5	12.20	24.98	37.03	39.18	41.33	41.33	51.09	54.96
2. Develosil ODS-N-5	12.78	24.40	34.60	37.17	38.46	38.46	47.22	51.67
3. Develosil ODS-P-5	5.42	9.13	12.02	12.90	13.14	13.14	15.62	16.85
4. Develosil 300 ODS-5	5.38	11.13	16.57	17.57	18.57	18.57	23.20	24.89
5. Develosil N.P. ODS-5								
6. Develosil 300 C30-5	9.15	18.75	27.68	30.25	31.25	31.25	39.68	43.76
7. Develosil N. P. C30-5								
8. Develosil 100								
9. Vydac 201 TPB 5	1.33	2.69	4.05	4.05	4.41	4.71	5.37	5.37
10. Vydac 218 TPB 5	1.69	3.34	4.89	4.89	5.38	5.58	6.40	6.40
11. Wakosil-II 5C18 AR	3.12	6.79	10.83	10.46	11.90	12.91	14.02	14.02
12. Capcell Pak C18 AG120	5.60	9.96	13.56	14.21	14.85	14.85	17.62	18.75
13. Capcell Pak C18 SG120	4.57	8.05	10.96	10.96	10.96	10.96	15.09	15.09
14. Capcell Pak C18 SG300	2.48	4.56	6.35	6.94	6.94	6.94	8.25	8.77
15. Capcell Pak C18	7.23	13.09	18.01	18.93	19.78	19.78	23.71	25.40
16. Shiseido ODS	6.81	12.35	17.02	17.96	18.76	18.76	22.43	24.03
17. Ashahipak ODP-50	9.35	18.45	26.93	30.04	30.04	30.04	37.40	40.78
18. YMC-Pack AM-302 S-5 120A ODS	9.27	18.00	25.56	27.24	28.47	28.47	34.83	37.73
19. Bondashere 5μ C18-100A	9.43	18.35	26.17	20.07	29.18	29.18	36.10	39.00
20. Spelcosil LC-DP	3.57	4.15						
21. Nova-pak HR C18	19.38	37.97	55.16	61.37	61.37	61.37	74.96	80.96

[a] Stationary phases 5, 7, and 8 did not give any separations of C_{60} and C_{70}. For this work the mobile phase was 45:55 toluene/acetonitrile; UV detection at 325 mn.

Source: Ref. 24.

Table 3.2B RELATIVE RETENTION OF HIGHER FULLERENES TO C_{60}[a]

Stationary Phase	C_{70}	C_{76}	C_{78}-$C_{2v'}$	C_{78}-C_{2v}	C_{78}-D_3	C_{82}	C_{84}
1. Develosil ODS-5	2.048	3.305	3.211	3.388	3.388	4.188	4.505
2. Develosil ODS-N5	1.909	2.707	2.908	3.009	3.009	3.695	4.043
3. Develosil ODS-P5	1.685	2.218	2.380	2.424	2.424	2.882	3.109
4. Develosil 300 ODS-5	2.069	3.080	3.266	3.452	3.452	4.312	4.626
5. Develosil N.P. ODS-5							
6. Develosil 300 C30-5	2.049	3.025	3.306	3.415	3.415	4.337	4.783
7. Develosil N. P. C30-5							
8. Develosil 100							
9. Vydac 201 TPB 5	2.023	3.045	3.045	3.316	3.541	4.038	4.038
10. Vydac 218 TPB 5	1.976	2.893	2.893	3.183	3.302	3.787	3.787
11. Wakosil-II 5C18 AR	2.176	3.471	3.353	3.814	4.138	4.494	4.494
12. Capcell Pak C18 AG120	1.779	2.421	2.538	2.652	2.652	3.146	3.348
13. Capcell Pak C18 SG120	1.176	2.398	2.398	2.398	2.398	3.302	3.302
14. Capcell Pak C18 SG300	1.839	2.560	2.798	2.798	2.798	3.327	3.536
15. Capcell Pak C18	1.811	2.491	2.618	2.736	2.736	3.279	3.513
16. Shiseido ODS	1.814	2.499	2.637	2.755	2.755	3.294	3.529
17. Ashahipak ODP-50	1.973	2.880	3.213	3.213	3.213	4.000	4.361
18. YMC-Pack AM-302 S-5 120A ODS	1.942	2.757	2.939	3.071	3.071	3.757	4.070
19. Bondashere 5μ C18-100A	1.946	2.775	2.977	3.094	3.094	3.828	4.136
20. Spelcosil LC-DP	1.162						
21. Nova-pak HR C18	1.959	2.846	3.167	3.167	3.167	3.868	4.178

[a] Other conditions as follows: mobile phase, 45:55 toluene/acetonitrile; UV detection at 325 mn.

Source: Ref. 24.

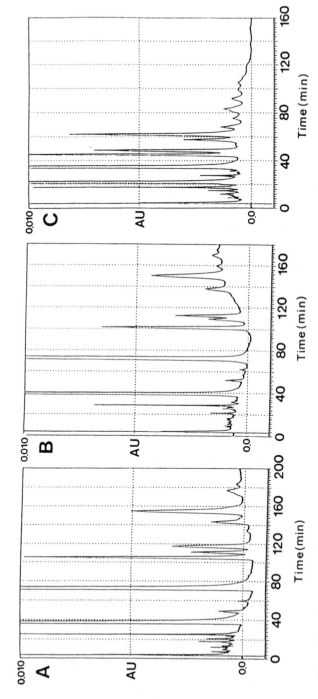

Figure 3.2 Chromatograms for the separation of higher fullerenes with three ODS phases: (A) Develosil ODS-5. (B) Develosil ODS-N5. and (C) Develosil ODS-P5. Mobile phase, toluene/acetonitrile (45:45): flow rate, 1 mL/min; UV detection at 325 nm; column temperature. 30°C.

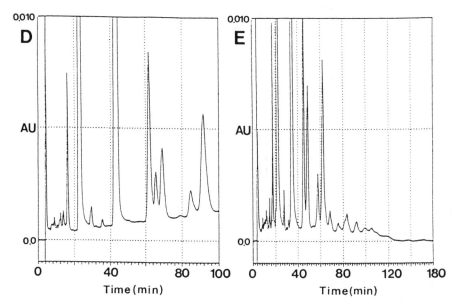

Figure 3.3   Chromatograms for higher fullerenes with two pore size ODS phases: (A) Develosil 300 ODS-5 and (B) Develosil ODS-P5. Other conditions as in Figure 3.2.

matogram of Figure 3.4, showing the separation of fullerenes with the monomeric ODS phase, contains several peaks later than the C_{70} retention time. The UV–visible spectra of these peaks are summarized in Figure 3.5. The results obtained by spectral identification are as follows: peak A, C_{60}; B, C_{70}; C, C_{76}; D, C_{78}-$C_{2v'}$; and E, C_{78}-$2C_{1}$. No reference data were found for F and G, but these peaks are probably C_{82} and C_{84}, respectively.

For the comparative study of the monomeric and the polymeric ODS phases, similar experiments were performed with the Wakosil ODS and the chromatogram obtained is shown in Figure 3.6. The UV–visible spectra of the peaks are summarized in Figure 3.7. It was confirmed that peaks J, K, L, and N are assigned to the same solutes of peaks D, C, E, and G, respectively, in Figure 3.4. Therefore the elution order of C_{76} and C_{78}-$C_{2v'}$ for the monomeric ODS is reversed with the polymeric ODS phase when the same mobile phase is used. The solute in peak M in Figure 3.6 cannot be seen in Figure 3.4, and its spectrum is very similar to that of C_{78}-D_{3}, as observed by Kikuchi et al. [40]. The right shoulder of peak N (Z in the Figure 3.6) gives a spectrum similar to peak F in Figure 3.4, although some overlapping of peaks distorts its spectrum. The results indicate that the polymeric ODS phase gives a different elution order compared to that with the monomeric ODS phase; such differences have also been found in the study of the elution characteristics of planar and nonplanar polycyclic aromatic hydrocarbons (PAHs) by Wise and Sander [34,35] and by Jinno et al. [36,37]. The explanation for these

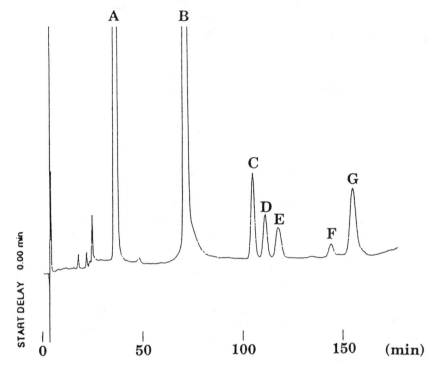

Figure 3.4 Room temperature separation of fullerene mixture with Develosil ODS-5. Other conditions as in Figure 3.2 [25]. Reproduced from K. Jinno et al. in *Anal. Chem.* **65**, 2650 (1993) by permission of the American Chemical Society, Washington, D.C., USA.

discrepancies lies in higher molecular planarity recognition capability of the polymeric phase, compared to the monomeric ODS phase.

Consideration of the shape of the isomers of higher fullerenes can explain the retention differences. Among the isomers of C_{78}, C_{78}-$C_{2v'}$ has a bulky shape, C_{78}-C_{2v} has a somewhat different symmetry, and D_3 has a narrower shape than the other two isomers. The polymeric ODS elutes bulkier $C_{2v'}$ first and the more cylindrical D_3 later by virtue of its planarity recognition capability. The reversed elution order of C_{84} and C_{82} with the polymeric ODS [as tentatively assigned, peak F in Figure 3.4 and the right shoulder of peak N (Z in Figure 3.6)] can be also explained in the same way. The C_{82} shape is considered to be longer and narrower than C_{84} (although theoretically at least three isomers of this fullerene isomer can exist [41]). This difference can induce such an elution order with the polymeric phase because of its planarity recognition capability. Although the sample used contained three C_{78} isomers, it remained unclear why the D_3 isomer of C_{78} could not be found with the monomeric ODS phase. Two isomers of C_{78}, C_{78}-C_{2v} and D_3, could not be resolved with the monomeric ODS phases under these separation conditions, and peak E may be assigned to the C_{78}-C_{2v} + D_3.

The high-loaded monomeric ODS phases cannot completely resolve three iso-

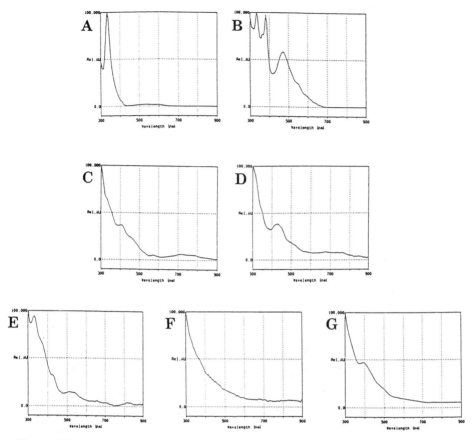

Figure 3.5 UV–visible spectra of the peaks in the chromatogram in Figure 3.4; letter labels of spectra correspond to peaks in the chromatogram [25]. Reproduced from K. Jinno et al. in *Anal. Chem.* **65**, 2650 (1993) by permission of the American Chemical Society, Washington, D.C., USA.

mers of C_{78}. We have also tried to separate three C_{78} isomers by consecutively combining a column with other columns: Develosil ODS-5 + ODS-N-5 + 300 C30-5 + YMC-Pack. These four ODS columns gave high separation factor values for C_{60} and C_{70} separation in the preliminary experiments. Toluene/methanol was used as the mobile phase in consideration of analysis time. In the chromatograms of Figure 3.8, the peak identification has been performed by measuring UV–visible spectra. However, the three isomers of C_{78} cannot be resolved completely even when four connecting columns are used.

3.2.3 Separation and Identification of Fullerenes Higher than C_{84}

There are a number of isomers of fullerenes far higher than C_{84}, although they are difficult to isolate and identify. Several papers focusing on the separation and

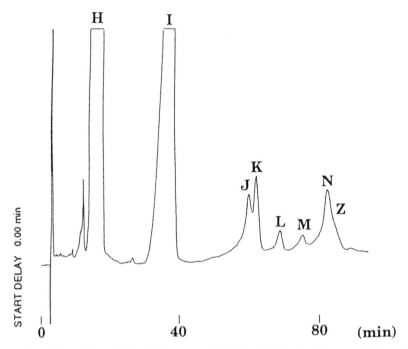

Figure 3.6   Separation of higher fullerenes with Wakosil ODS. Conditions as in Figure 3.4 [25]. Reproduced from K. Jinno et al. in *Anal. Chem.* **65**, 2650 (1993) by permission of the American Chemical Society, Washington, D.C., USA.

identification of large fullerenes by LC methods have been published, and the most recent results indicate that large fullerene fractions, at least to C_{90}, should exist and should be extractable from soot [25,42]. To confirm the existence of these large fractions, one needs to obtain multidimensional information from one chromatographic run. Hyphenated techniques are the most convenient and suitable analytical methods here; the LC-MS approach in particular would be the most powerful way to accomplish this purpose [42,44]. LC-MS is the combination of LC hardware with a mass spectrometer via unique interfacing devices that prevent any disturbances in mass spectral measurements due to the mobile phase solvents used for the LC separations. Typical interfacing devices feature electrospray ionization (ESI), atmospheric ionization (API), and fast atom bombardment (FAB) ionization; Sections 3.2.3.1 through 3.2.3.4 describe the LC-MS approach using the ESI interface for isolation and identification of higher fullerenes. The results show that this approach is powerful enough to identify large fullerenes in soot extracts, and the largest fullerene found in this work is C_{98}. Interesting elution behavior of C_{98} in LC separations using polymeric ODS as the stationary phase is also found and can be interpreted by means of the molecular planarity recognition mechanism of the polymeric ODS phase as described in Chapter 2.

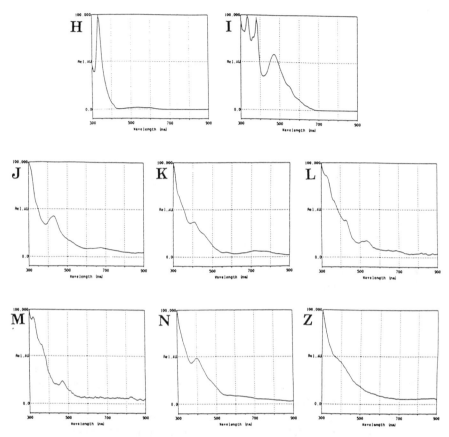

Figure 3.7 UV–visible spectra of the peaks in the chromatogram in Figure 3.6; letter labels of spectra correspond to peaks in the chromatogram [25]. Reproduced from K. Jinno et al. in *Anal. Chem.* **65**, 2650 (1993) by permission of the American Chemical Society, Washington, D.C., USA.

3.2.3.1 Analytical Procedure

The soot, obtained by arc discharge in an inert gas environment, was extracted first by toluene, then the residue was extracted by 1,2,4-trichlorobenzene, yielding a high concentration of fullerenes larger than C_{70}. The trichlorobenzene solution was vaporized by nitrogen gas, and the solvent was replaced by toluene/methanol (55:45), which served as the mobile phase for preparative-scale separations. The solution was then filtrated by Teflon membrane filter, and the solution obtained served as the sample for the preparative separations. The large fullerenes have been isolated and identified using two-step LC separations: first the preparative-scale separation to get large fullerene fractions, and then analytical-scale separations to identify large fullerenes in each fraction by LC-MS. A monomeric ODS was chosen as the stationary phase for preparative separations because the monomeric type

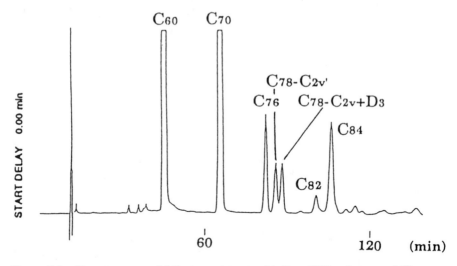

Figure 3.8 Chromatogram of fullerene mixtures with four ODS columns serially combined: column, Develosil ODS-5 + Develosil ODS-N5 + Develosil 300C30-5 + YMC-Pack; mobile phase; toluene/methanol (50:50). Other conditions as in Figure 3.4.

elutes fullerenes by their molecular weight. For analytical separations a polymeric ODS was used as the separation medium, since the polymeric type elutes fullerenes by their shape and structure, as discussed herein. Thus by using the monomeric ODS phase, one can collect fractions that contain fullerenes having different molecular weights. Each collected fraction then can be separated to identify fullerene isomers by applying the polymeric ODS phase to the task of recognizing shapes and structures.

Figure 3.9 presents a typical chromatogram for this preparative separation. The fraction collection was done manually, and five fractions as indicated in Figure 3.9 were accumulated 39 times. Table 3.3 gives information about each fraction and the assumed components, listing as well the amounts collected (by volume). The solvent was then evaporated to reduce the total sample amounts, and the reduced amounts of each fraction also are included in Table 3.3.

Analytical-scale separations for 10 μL samples of three fractions (nos. 3, 4, and 5) were done by LC-MS using an ESI interface. The MS conditions were as follows: mass range measured, 600–1600 amu, capillary voltage, 3.4 kV; ion source temperature, 70°C; split ratio, 1:20.

3.2.3.2 Analysis of Fraction No. 3

The third fraction should contain fullerenes C_{76} and C_{78}. Recent publications [19,22,25,38,40–42] indicate that C_{76} has a chiral isomer and C_{78} has at least three isomers. It is hard to separate each C_{76} chiral isomer, but three C_{78} isomers have

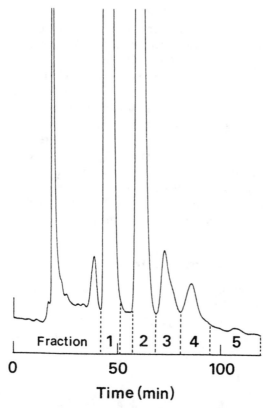

Figure 3.9 Chromatogram of the soot extract: preparative-scale separation. The five fractions were collected after 39 injections of 40 mL each on a Develosil Lop ODS column (50 mm i.d. × 500 mm long), 30 μm monomeric ODS. The mobile phase was toluene/methanol (55:45), and the flow rate was 40 mL/min [43]. Reproduced from K. Jinno et al. in *Chromatographia* **41**, 353 (1995) by permission of Vieweg Publishing, Wiesbaden, Germany.

Table 3.3   FRACTION COLLECTED BY PREPARATIVE-SCALE LC SEPARATIONS

Fraction	Time Range (min)	Desired Fullerenes for This Fraction	Collected Amount (mL)[a]	Final Amount (mL)	Sample No. for Analysis
1	43.0–51.75	C_{60}	13,650	45	
2	58.0–68.0	C_{70}	15,600	50	
3	68.0–79.25	$C_{76}-C_{78}$	17,550	30	3
4	79.25–95.5	$C_{82}-C_{86}$	25,350	35	4
5	95.5–120.5	$>C_{86}$	39,000	7	5

[a] After 39 accumulations.

been well characterized, namely: C_{78}-$C_{2v'}$, C_{78}-C_{2v}, and C_{78}-D_3. By using the polymeric ODS phase, the elution order of those fullerenes is found to be different from that with the monomeric ODS phase [11,12,16,17,22,23]. This is not surprising, however, since the monomeric ODS basically elutes fullerenes by their molecular weight but the polymeric phase uses molecular shape and structure as the basis for elution. The chromatograms measured by UV at 325 nm and MS by total ion monitoring included four clear peaks, one of which has a small shoulder at the early eluting part as shown in Figure 3.10. Therefore five positions in the retention time were referred to their mass spectra, and the results (Figure 3.11) allow the five peaks to be assigned as C_{60}, C_{78}, C_{76}, and two C_{78} isomers. The elution order found for fraction no. 3 with the polymeric ODS is determined as C_{78}-C_{2v}, C_{76}, C_{78}-C_{2v}, and C_{78}-D_3 by using results obtained earlier, identifying the isomers of C_{78} by photodiode array UV–visible detection. LC-MS information clearly indicated that our earlier findings on the elution order with the polymeric ODS were correct.

3.2.3.3 Analysis of Fraction No. 4

The fourth fraction should contain C_{82}, C_{84}, and some larger fullerenes. The UV and MS chromatograms (Figure 3.12) contain three major peaks, one of which has a broad shoulder. Thus we refer to peaks 1, 2, 3, and 4, observing that 2 and 3 constitute the same peak. The identification of each peak by mass spectra as shown

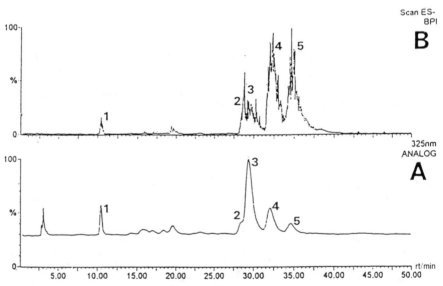

Figure 3.10 Chromatograms of fraction no. 3 by LC-MS measurements: (A) UV chromatogram monitored at 325 nm and (B) MS chromatogram by total ion-scanning mode (TIM). Wakosil II 5C18AR column (4.6 mm i.d. × 250 mm); 5 μm polymeric ODS; mobile phase, acetonitrile/toluene (55:45); flow rate, 1 mL/min. MS measurement conditions: VG Biotech platform with ESI interface; mass range, 600–1600, capillary voltage, 3.4 kV.

Sample No.3(C76,C78), Toluene/Acetonitrile(45/55), Wakosil-II 5C18AR, 30C

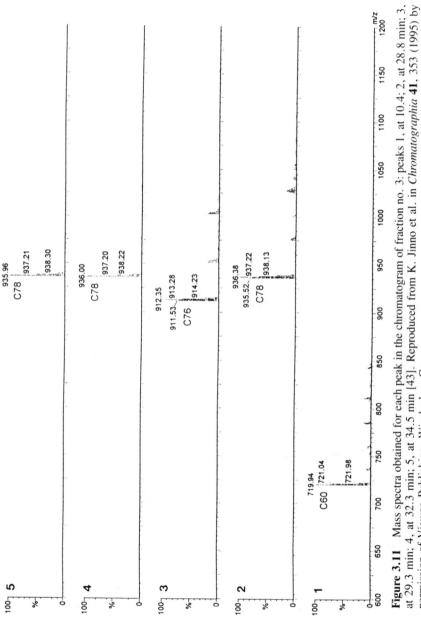

Figure 3.11 Mass spectra obtained for each peak in the chromatogram of fraction no. 3: peaks 1, at 10.4; 2, at 28.8 min; 3, at 29.3 min; 4, at 32.3 min; 5, at 34.5 min [43]. Reproduced from K. Jinno et al. in *Chromatographia* **41**, 353 (1995) by permission of Vieweg Publishing, Wiesbaden, Germany.

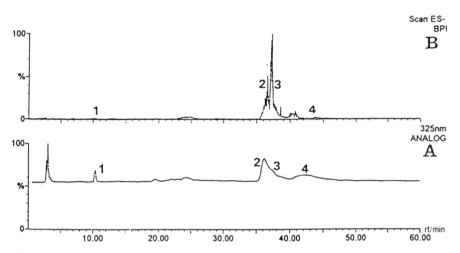

Figure 3.12   Chromatograms of fraction no. 4 by LC-MS measurements: (A) UV chromatogram monitored at 325 nm and (B) MS chromatogram by TIM. Other conditions as in Figure 3.10.

in Figure 3.13 is very easy. The elution order of C_{84} and C_{82} is interesting because the larger molecular weight fraction is eluted faster than C_{82} with the polymeric ODS phase. This in turn is because the polymeric ODS recognizes molecular planarity so well and the nonplanar PAH generally elutes faster than its planar counterpart. In shape and structure, C_{82} is less bulky than C_{84}, and the former elutes later than the latter. Isomers of C_{84} are not found in this fraction. Significantly, C_{86} *is* found in this fraction and elutes later than C_{84}. Thus the C_{86} structure is not unusual—if it were anomalous, it should elute faster or much later than C_{84}.

3.2.3.4 Analysis of Fraction No. 5

The most exciting results were obtained for the fifth fraction, which should contain fullerenes larger than C_{86}. The analysis was performed with 50:50 toluene/acetonitrile as the mobile phase instead of the 45:55 mixture used for the analysis of fractions nos. 3 and 4. Elution with 45:55 solvent was abandoned because of the long analysis time required and because the separations were not good enough to permit recognition of each peak, given the lower solubility of the sample into the mobile phase. The chromatograms obtained for UV and for MS monitoring are shown in Figure 3.14. Fifteen peaks are seen in both chromatograms, and the main mass numbers for the peaks are determined by referring to the mass spectrum at each peak position. The mass spectra for all 15 peaks are summarized in Figure 3.15. The results can help to identify the components, as summarized in Table 3.4.

Table 3.4 highlights some interesting facts. First, fraction no. 5 still contains

Sample No.4(C82,C84), Toluene/Acetonitrile(45/55), Wakosil-II 5C18AR, 30C

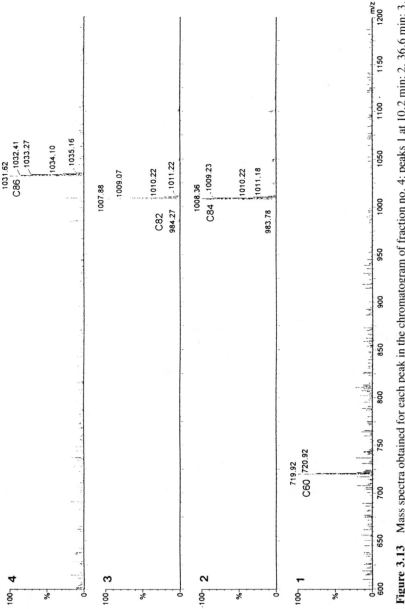

Figure 3.13 Mass spectra obtained for each peak in the chromatogram of fraction no. 4: peaks 1 at 10.2 min; 2, 36.6 min; 3, 37.3 min; 4, 42.5 min [43]. Reproduced from K. Jinno et al. in *Chromatographia* **41**, 353 (1995) by permission of Vieweg Publishing, Wiesbaden, Germany.

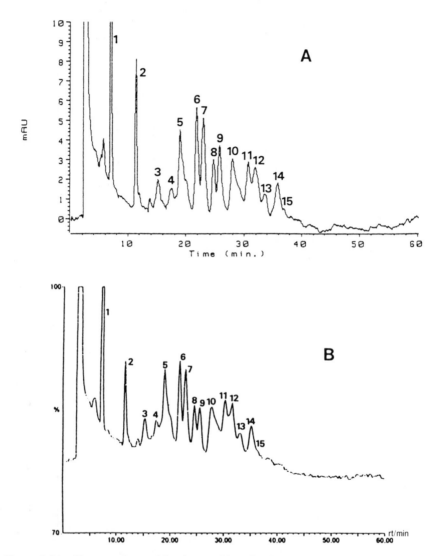

Figure 3.14 Chromatograms of fraction no. 5 by LC-MS measurements: (A) UV chromatogram monitored at 325 nm and (B) MS chromatogram by TIM. Other conditions as in Figure 3.10 [43]. Reproduced from K. Jinno et al. in *Chromatographia* **41**, 353 (1995) by permission of Vieweg Publishing, Wiesbaden, Germany.

C_{60}, C_{70}, and C_{84}, which should have been excluded by the preparative separations, since the concentration difference between those fullerenes and higher fullerenes is so huge. Second, it appears that C_{84} has at least two isomers, C_{86} has two isomers, C_{88} has also two isomers, C_{90} has three isomers, and C_{98} has at least two isomers; but C_{92} and C_{94} have only one isomer according to these results. Kikuchi et al. [45]

Sample No.5(>C86), Toluene/Acetonitrile(50/50), Wakosil-II 5C18AR, 30C

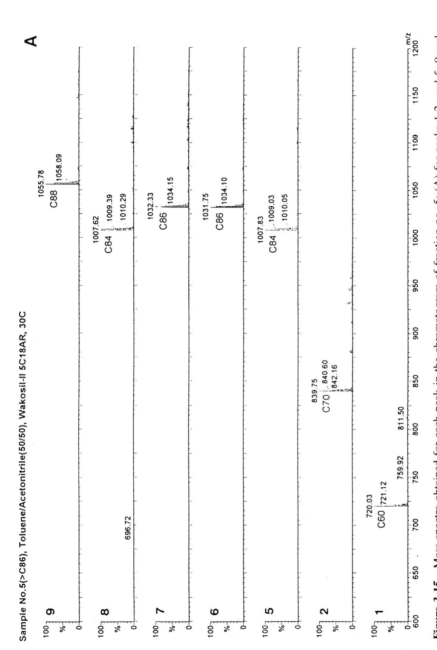

Figure 3.15 Mass spectra obtained for each peak in the chromatogram of fraction no. 5: (A) for peaks 1.2. and 5–9 and (B) for peaks 3, 4, and 10–15 [43]. Reproduced from K. Jinno et al. in *Chromatographia* **41**, 353 (1995) by permission of Vieweg Publishing, Wiesbaden, Germany.

Sample No.5(>C86), Toluene/Acetonitrile(50/50), Wakosil-II 5C18AR, 30C

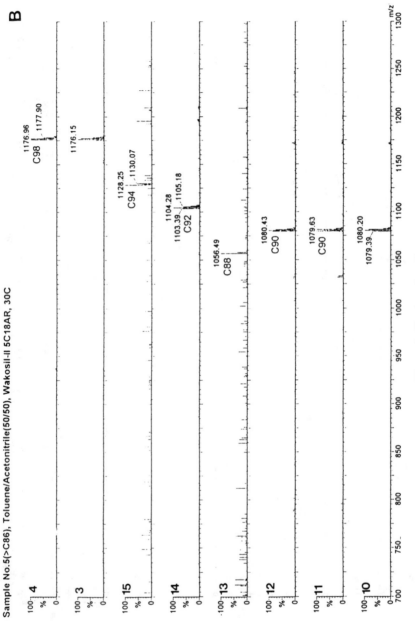

Figure 3.15 *(continued)*

Table 3.4 PEAK ASSIGNMENTS FOR THE ANALYSIS OF
FRACTION NO. 5 BY LC-MS

Peak No.	Retention Time (min)	Major Mass Number	Fullerenes Identified
1	7.2	720	C_{60}
2	11.6	840	C_{70}
3	15.3	1176	$C_{98\text{-}1}$
4	17.5	1176	$C_{98\text{-}2}$
5	19.1	1008	$C_{84\text{-}1}$
6	21.8	1032	$C_{86\text{-}1}$
7	23.0	1032	$C_{86\text{-}2}$
8	24.0	1008	$C_{84\text{-}2}$
9	25.7	1056	$C_{88\text{-}1}$
10	27.7	1080	$C_{90\text{-}1}$
11	30.6	1080	$C_{90\text{-}2}$
12	31.5	1080	$C_{90\text{-}3}$
13	32.5	1056	$C_{88\text{-}2}$
14	35.9	1104	C_{92}
15	36.6	1128	C_{94}

separated and identified two isomers of C_{84}, C_{90}, C_{92}, C_{94}, and C_{96} and one isomer for C_{88} using an expensive commercial preparative column (Buckyprep, Nacalai Tesque, Kyoto, Japan) by off-line MS measurements for the collected fractions. Our results are not totally consistent with Kikuchi's results, but on-line LC-MS information should be more accurate than off-line measurements. It should be noted that differences in soot sources and production methods contribute to the discrepancies in results. Further work is required to confirm the number of isomers for higher fullerenes with LC-MS on-line detection.

Another important observation stemming from these results entails the elution time for two isomers of C_{98}, which are eluted with unusual rapidity—faster than C_{84}. Since the monomeric phase elutes fullerenes by molecular weight and the polymeric phase elutes them on the basis of shape and structure, the two C_{98} isomers should be very bulky and unusual in structure. This type of information will help to characterize the higher fullerenes, and further developments of new stationary phases in LC can give such information on shape and structure based on differential molecular recognition capability.

To obtain the UV–visible spectra of higher fullerenes identified by LC-MS measurements photodiode array detection was again used to obtain LC separation. The UV/visible spectra for each peak obtained by the measurements are summarized in Figure 3.16. Spectra like those of Figure 3.16 will be very valuable in further attempts to isolate and identify higher fullerenes because LC–photodiode array detection is much easier to perform than LC-MS measurements in the average laboratory.

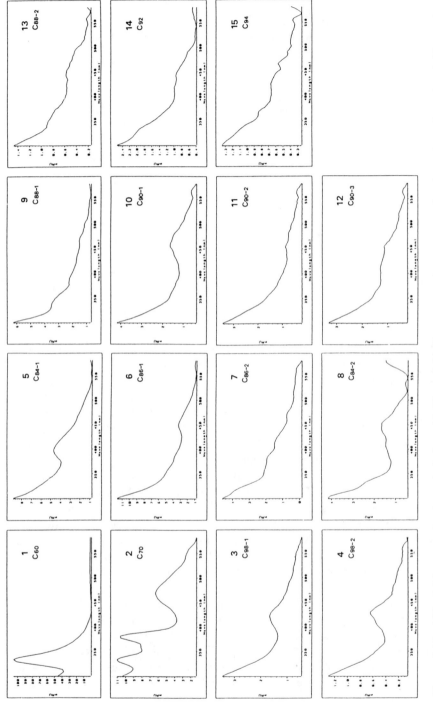

Figure 3.16 UV–visible spectra for the peaks indicated in Figure 3.14. Spectrum numbers correspond to the peak numbers given in Figures 3.14 and 3.15 [43]. Reproduced from K. Jinno et al. in *Chromatographia* **41**, 353 (1995) by permission of Vieweg Publishing, Wiesbaden, Germany.

3.2.4 Effects of Temperature in the Separation of Higher Fullerenes

Figure 3.17 shows chromatograms in which three monomeric ODS phases, with *n*-hexane as the mobile phase, induce separations at column temperatures between 20 and −70°C. Not only does the higher surface coverage give better separation performance, but decreasing the temperature leads to an improvement in resolution. In Figure 3.17, it is apparent that the best separation of fullerene isomers is achieved at −20°C with the monomeric ODS bonded phase, Develosil ODS-5 (20C%). A peak of the isomers of C_{78} (peaks 2, 3, and 4 for $C_{2v'}$, C_{2v}, and D_3, respectively) above 20°C is separated into three peaks at lower temperature and at −20°C the best resolution is obtained with an *n*-hexane mobile phase. At the temperature lower than

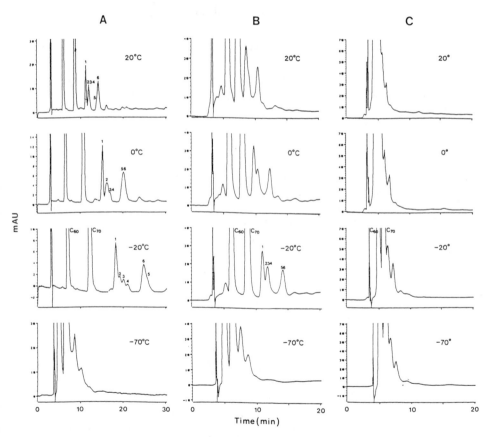

Figure 3.17 Separations of higher fullerenes with monomeric ODS phases at various temperatures: (A) Develosil ODS-5, 20 C%; (B) Develosil ODS-N5, 16 C%; and (C) Develosil ODS-P5, 11 C%. Mobile phase, *n*-hexane; flow rate, 1 mL/min^{-1}; detection at 312 nm. Peaks: 1, C_{76}; 2, C_{78}-$C_{2v'}$; 3, C_{78}-C_{2v}; 4, C_{78}-D_3; 5, C_{82}; 6, C_{84} [26]. Reproduced from K. Jinno et al. in *Chromatographia* **39**, 453 (1994) by permission of Vieweg Publishing, Wiesbaden, Germany.

$-20°C$—that is, $-70°C$, the peak of C_{78}-C_{2v} almost merges into the peak of C_{76} (peak 1 in the figure), since C_{78}-C_{2v} elutes closely with C_{76}. Thus the retention time of C_{78}-C_{2v} tends to be shorter than that of C_{76} as a function of decreasing in temperature.

A change in elution order of C_{82} (peak 5) and C_{84} (peak 6) is also observed when the temperature is decreased. In the preceding section we showed that the elution order of the higher fullerenes is controlled by the order of their molecular size and shape with monomeric and polymeric ODS bonded phases using a toluene/acetonitrile mobile phase. The polymeric ODS phase resolves the three C_{78} isomers because one of them is eluted faster than C_{76}, while the monomeric ODS cannot resolve two of the three C_{78} isomers. This behavior is attributed to the higher shape recognition ability of the polymeric ODS phase.

It has also been asserted [46] that the difference between the surface structural characteristics of the ODS bonded phases revealed by both NMR and FTIR explains why the functionality of these phases can result in different molecular planarity recognition capabilities. Sander and Wise [35] reported that at low temperatures phase selectivity is more "polymeric-like" for both ODS phase types. Therefore the best separation of the three isomers of C_{78} at $-20°C$ is also due to the increase of shape recognition ability of the monomeric ODS bonded phase. Contrary to expectation below $-20°C$ a decrease in the temperature induces a decrease in the resolution in retention. It is assumed that lower column temperature changes the structure of the stationary phase to rigid and ordered; therefore, with a more "slotlike" structure on the surface, a better separation of higher fullerenes is achieved at $-20°C$. At this stage, the monomeric phase behaves like a polymeric phase. Below that temperature the bonded phases tend to become more rigid and ordered, and finally the phases can hardly retain the bulky and globular fullerenes, with the result that retentions for whole fullerenes were decreased.

This situation of the bonded phase can be thought as resembling the case of slotlike surface structures, which are much harder than the structure of polymeric phases at normal temperature. The same tendency was observed with Develosil ODS N5 (16C%) and Develosil ODS P5 (11C%) in Figure 3.17. In addition, increasing the carbon contents of the monomeric phases serves to increase selectivity. In this investigation, however, it was also observed that as the surface coverage increases, the temperature for the maximum retention of fullerenes will be higher. Figure 3.18, which plots $\log k'$ for C_{60} and C_{70} with three monomeric ODS phases versus temperature, illustrates this point. The approximate maximum retention temperatures for C_{60} and C_{70} are -10 to $-20°C$ for Develosil ODS-5, $-30°C$ for Develosil ODS N5, and $-50°C$ for Develosil ODS P5, respectively. It is reasonable to explain the variation of maximum retention temperature difference with the carbon content of the bonded phases as follows: with high surface coverage, the average distance between each C_{18} bonded phase on the silica surface is small and a rigid structure can be easily obtained at relatively high temperature ($\approx -20°C$). With lower surface coverage the distance is larger, prohibiting a rigid structure of the bonded phases at a similar temperature; thus a rigid situation can be made at lower temperature such as -30 and $-50°C$ for phases N5 and P5, respectively.

Figures 3.19 and 3.20 present chromatograms with the polymeric ODS phases

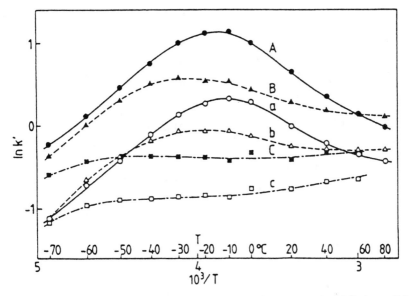

Figure 3.18   Ln k' versus temperature for C_{60} and C_{70} for monomeric ODS phase of three different carbon contents: a-○, C_{60}, Develosil ODS-5 (20 C%); A-●, C_{70}, Develosil ODS-5; b-△, C_{60}, Develosil ODS-N5 (16 C%); B-▲, C_{70}, Develosil ODS-N5; c-□, C_{60}, Develosil ODS-P5 (11 C%); C-■, C_{70}, Develosil ODS-P5 [26]. Reproduced from K. Jinno et al. in *Chromatographia* **39**, 453 (1994) by permission of Vieweg Publishing, Wiesbaden, Germany.

and the plots of log k' versus temperature, respectively. The retention with the polymeric ODS phases is smaller than with the monomeric ODS phases. With Wakosil ODS, the elution order can be determined at 0 or −10°C, and the order is quite different from the monomeric phases: C_{78}-$C_{2v'}$, then C_{76}, C_{78}-C_{2v}, C_{84}, C_{78}-D_3, and C_{82}. This order, which can be explained by the strong molecular shape recognition power of the polymeric phase, is unusual compared to that with the monomeric phases at the same temperature around 0°C. The maximum retention temperature of fullerenes with the polymeric ODS phases is higher than that of the three monomeric ODS phases. The average distance of C_{18} bonded phases in the aggregated situation of the polymeric phase is smaller than that in monomeric phase. For comparison between monomeric and polymeric ODS phases, relationships between the separation factors for C_{60} and C_{70} with four different stationary phases and column temperature are illustrated in Figure 3.21. Upon increasing the surface coverage of the monomeric phases, the separation factor for C_{60} and C_{70} is found to be improved. With Develosil ODS-5, the separation for C_{60} and C_{70} above 50°C is similar to that of other monomeric ODS phases; however below −20°C the differences in separation factor disappear. The result suggests that the high surface coverage monomeric ODS bonded phase is "polymeric-like" as a result of the increase in surface coverage and the decrease in column temperature.

   In an attempt to explain the specific phenomenon of the separation of fullerenes, a mixture of PAHs and C_{60} was separated with the ODS phases. Figure 3.22A

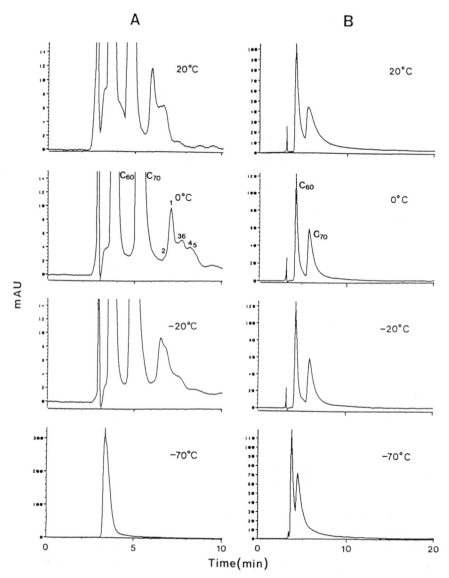

Figure 3.19 Separations of higher fullerenes on polymeric ODS phases at various temperatures: (A) Wakosil, 20 C%, and (B) Vydac; mobile phase, *n*-hexane; flow rate, 1 mL/min; detection at 312 nm. Peaks: 1, C_{76}; 2, C_{78}-$C_{2v'}$; 3, C_{78}-C_{2v}; 4, C_{78}-D_3; 5, C_{82}; 6, C_{84} [26]. Reproduced from K. Jinno et al. in *Chromatographia* **39**, 453 (1994) by permission of Vieweg Publishing, Wiesbaden, Germany.

shows the chromatograms illustrating the effect of temperature when using Develosil ODS-5, a monomeric ODS bonded phase. The difference between the retention behavior of PAHs and C_{60} is dramatic. The behavior is clearly seen in the plots of log k' for C_{60} and the main three PAHs versus temperature (Figure 3.23). The retention of PAHs increased linearly with decreasing temperature; the plot of the re-

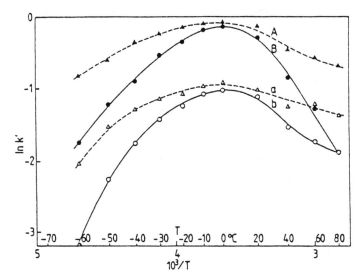

Figure 3.20 Ln k' versus temperature for C_{60} and C_{70} for two polymeric ODS phases: a-△, C_{60}, Vydac (8–9 C%); A-▲, C_{70}, Vydac; b-○, C_{60}, Wakosil (18 C%); B-●, C_{70}, Wakosil [26]. Reproduced from K. Jinno et al. in *Chromatographia* **39**, 453 (1994) by permission of Vieweg Publishing, Wiesbaden, Germany.

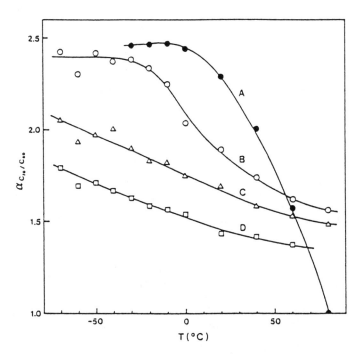

Figure 3.21 Separation factors versus column temperatire for C_{60} and C_{70} for four stationary phases: A-●, Wakosil (polymeric type); B-○, Develosil ODS-5 (monomeric type, 20 C%); C-△, Develosil ODS-N5 (monomeric type, 16 C%); D-□, Develosil ODS-P5 (monomeric type, 11 C%) [26]. Reproduced from K. Jinno et al. in *Chromatographia* **39**, 453 (1994) by permission of Vieweg Publishing, Wiesbaden, Germany.

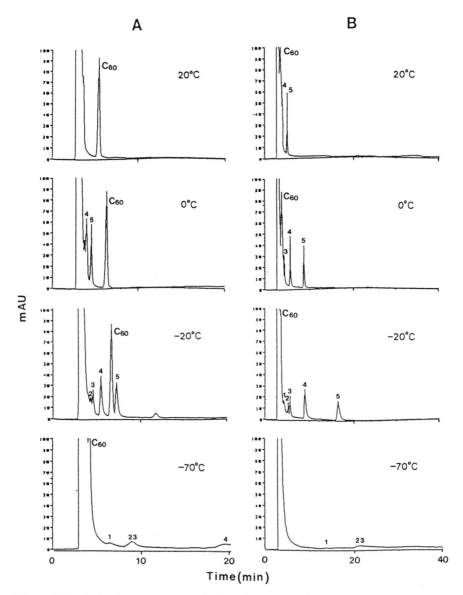

Figure 3.22 Separations of C_{60} and PAHs with monomeric ODS and polymeric ODS phases at various temperatures: (A) Develosil ODS-5 and (B) Wakosil. Mobile phase, *n*-hexane; flow rate, 1 mL/min; detection, 254 nm. Peaks: 1, phenanthrene; 2, anthracene; 3, pyrene; 4, triphenylene; 5, chrysene [26]. Reproduced from K. Jinno et al. in *Chromatographia* **39**, 453 (1994) by permission of Vieweg Publishing, Wiesbaden, Germany.

tention for C_{60}, however, revealed that lower temperature did not give longer retention for the fullerene. The reason for this difference may lie in the different molecular shapes of the PAHs and C_{60}; PAHs are relatively two-dimensional molecules, whereas C_{60} is bulky and spherical. With the polymeric ODS bonded phase (Wako-

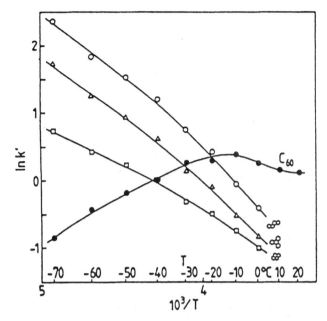

Figure 3.23 Ln k' versus temperature for C_{60} and PAHs with monomeric ODS phases (Develosil ODS-5): ●, C_{60}, ○, chrysene; △, triphenylene; □, pyrene [26]. Reproduced from K. Jinno et al. in *Chromatographia* **39**, 453 (1994) by permission of Vieweg Publishing, Wiesbaden, Germany.

sil ODS) as seen in Figure 3.22B, a similar evaluation was made (Figure 3.24). The strange retention behavior of C_{60} with the monomeric ODS phases was not observed with the polymeric phase. However the high shape recognition ability of PAHs was clearly seen with the polymeric ODS phase. It is difficult with the polymeric ODS phase to retain the bulky and spherical fullerene C_{60} at low temperature, since the rigid and ordered bonded phase excludes the molecule. In any event, the investigation just described clearly indicates that decreasing the temperature induces the increase of rigid and ordered "slotlike" structures on the ODS surface.

The results can be easily explained by reference to Figure 3.25, using the concept of different interligand distances among the bonded C_{18} functional groups. At ambient and/or higher temperature, the C_{18} functional groups, being easily relaxed and movable, can interact with the solute C_{60} without restriction. With a high surface coverage phase, the distance between each C_{18} group is about 7.6 Å, which is very similar to the diameter of C_{60} molecule; thus very efficient interactions are induced. However, decreasing the surface coverage of the ODS phase results in an increase in the distance between C_{18} functional groups, and in less interaction with the C_{60} molecule. At high temperatures, all the C_{18} ligands of the three bonded phases have freedom to move, and there is not much difference in C_{60} recognition among these phases due to such different intervals between the alkyl chains; however, the possibility of interaction is dominant. The C_{18} ligands become more rigid and ordered at lower temperatures and the ligand distance becomes very crucial, enabling them to interact with the C_{60} molecule. Any mismatch of ligand distance

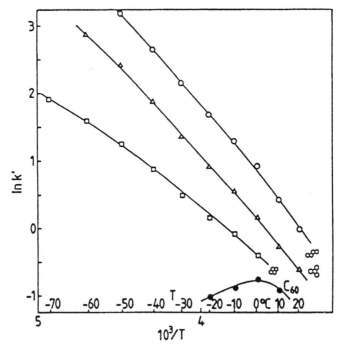

Figure 3.24   Ln k' versus temperature for C_{60} and PAHs with polymeric ODS phases (Wakosil): ●, C_{60}; ○, chrysene; △, triphenylene; □, pyrene [26]. Reproduced from K. Jinno et al. in *Chromatographia* **39**, 453 (1994) by permission of Vieweg Publishing, Wiesbaden, Germany.

and diameter of the C_{60} molecule results in a weak interaction and similar retentions. Thus the ODS–ligand distance between adjacent pairs is the most important aspect in obtaining improved selectivity for fullerenes in LC separations.

The foregoing conclusion can be confirmed by measuring NMR spectra of the phases under conditions of changing temperature and noting the consistency between NMR and chromatographic information with temperature change.

Figure 3.26 shows NMR spectral changes as a function of temperature for the ODS-5 phase. Drastic spectral changes are induced by temperature variations, especially the signals around 30 ppm, which are typically assigned to methylene (CH_2) groups in the middle part of the alkyl chains [46–50]. These initial results suggested that a more careful examination of NMR spectra would yield insights into the configuration changes of the alkyl chains with temperature, which would provide an explanation of the chromatographic behavior observed earlier.

Figure 3.27 shows a typical CP-MAS spectrum obtained from ODS-P5 over the temperature range studied. In curve A, the experimental spectrum of the P5 phase at 30°C (303 K), peaks at 30.0, 32.0, and 33.6 ppm can be easily identified. Curve B is the composite simulated spectrum, and curve C shows the individual components of the simulated spectrum. The match between the experimental and simulated

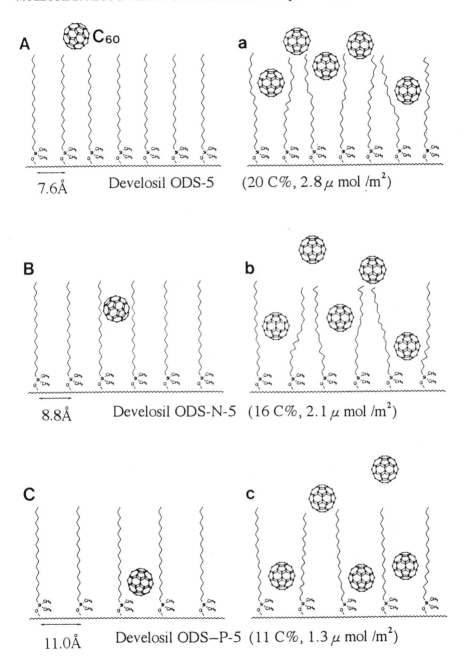

Figure 3.25 Schematic diagram of the interaction between C_{60} and ODS bonded phases: (A)–(C) under low temperatures and (a)–(c) under relatively higher temperatures.

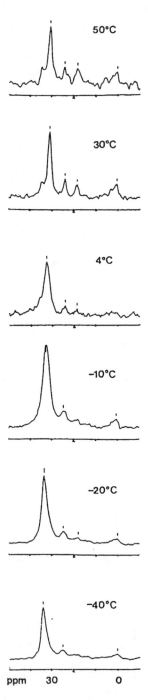

Figure 3.26 ^{13}C NMR CP-MAS spectra of Develosil ODS-5 at different temperatures [50]. Reproduced from H. Ohta et al. in *Chromatographia* **40**, 507 (1995) by permission of Vieweg Publishing, Wiesbaden, Germany.

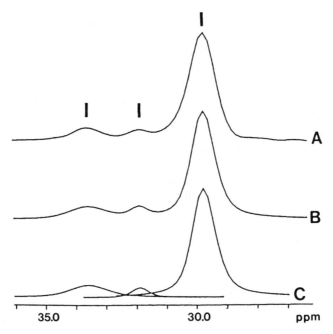

Figure 3.27 ^{13}C NMR CP-MAS data for Develosil ODS-P5 around the 30 ppm region: (A) measured spectrum, (B) composite simulated spectrum, and (C) individual components of the simulated spectrum [50]. Reproduced from H. Ohta et al. in *Chromatographia* **40**, 507 (1995) by permission of Vieweg Publishing, Wiesbaden, Germany.

spectra is excellent. In all cases (i.e., at each temperature for all three phases), the line width of the 30 ppm peak was used for the calculation of the relaxation time. The line width is related to the spin–spin relaxation time by

$$\Delta W_{\text{(line width)}} = \frac{1}{\pi T_2} \tag{3.1}$$

where T_2 is the relaxation time. A general expression for T_2 can be written as follows:

$$\frac{1}{T_2^*} = \frac{1}{T_2} + \frac{1}{T_{2(\text{inhomo})}} + \frac{1}{T_{2(\text{diff})}} \tag{3.2}$$

Where $T_{2(\text{inhomo})}$ and $T_{2(\text{diff})}$ are, respectively, contributions to relaxation from an inhomogeneous magnetic field and from diffusion effects. The value obtained is an apparent spin–spin relaxation time ($T_2^* = \pi T_2$), which can include effects other than molecular motion such as field inhomogeneities. Since the lines are already quite broad, it is unlikely that inhomogeneity contributions will be significant. The molecules are not freely diffusing, which minimizes the contributions of $T_{2(\text{diff})}$. Since the samples are treated in the same manner, any additional contributions to relaxation not specifically identified should be uniform. Therefore, the change we observe for T_2^* is taken to be generated by temperature-dependent differences in molecular motion.

Figure 3.28 plots T_2^* versus temperature for the three phases using all three methods of line width determination. The three curves are similar in general shape. A large decrease in T_2^* occurs as the temperature is lowered from 30°C to -50°C; then T_2^* remains approximately constant as the temperature is lowered further to -50°C. The major decrease in the relaxation time T_2^* occurring between 30 and -50°C is most likely due to restrictions in molecular motion as the temperature is lowered, and the minimum attained at certain temperatures may reflect a stable configuration (e.g., an association between adjacent C_{18} chains) assumed by the bonded moieties. In all cases the measurements are reversible; that is, whether the temperature has been raised or lowered, the same value for line width is obtained when we return to the original temperature. The curves all show a decrease in relaxation time from higher to lower temperature, with a minimum reach at some point of the curve. In the higher temperature region, the T_2^* values obtained for the packings are in the order: ODS-P5 > ODS-N5 > ODS-5. The temperatures at which a minimum in T_2^* is reached are ordered ODS-5 > ODS-N5 > ODS-P5. Both these observations parallel the loading density of C_{18} bonded phase (ODS-5 > ODS-N5 > ODS-P5). Since the ODS-5 phase has the highest loading, it would be expected to have greater association in the higher temperature range than the other phases. Therefore T_2^* should be the least for this phase, followed by N5, and P5 should be the highest. It would also be expected that a stable configuration, or maximum association, would be retained at a higher temperature for ODS-5 relative to N5. Similarly, ODS-N5 should retain maximum association at a higher temperature than ODS-P5. These are exactly the results observed.

Comparison of the NMR data and the LC data shows a remarkable correlation. The retention k' of the fullerene compounds (particularly C_{70}) goes through a maximum at about the same temperature that the relaxation time T_2^* is at minimum. This correlation can be explained by assuming that increased contact (interaction) occurs between the solute and the bonded moiety as the temperature is lowered, until a point is reached at which association of the stationary phase with itself begins to exclude solute. Considering the relative average distance between the bonded C_{18} chains, large molecules such as C_{60} and C_{70} would be particularly susceptible to such an exclusion mechanism. A related observation is that all three phases possess similar structure (i.e., nearly identical accessible surface for solute interaction) at this low temperature. Thus we would predict that retentions of C_{60} and (particularly because of its larger size) C_{70} would be quite similar below the T_2^* minimum temperature for each phase. That prediction is supported by experimental results.

Finally, the exclusion effect of a bonded phase for bulky solutes such as fullerenes should be more clearly seen when using C_8 as the stationary phase, since a shorter alkyl chain length more easily induces an association of the bonded phase moieties at temperatures higher than for C_{18}. Figure 3.29 shows two-dimensional elution profiles of chromatograms generated at various temperatures with a C_8 phase and C_{60} solute. The C_{60} retention monotonically decreases with column temperature, without showing the maximum retention behavior observed with C_{18} phases. Below -20°C the retention time observed for C_{60} is shorter than the solvent peak

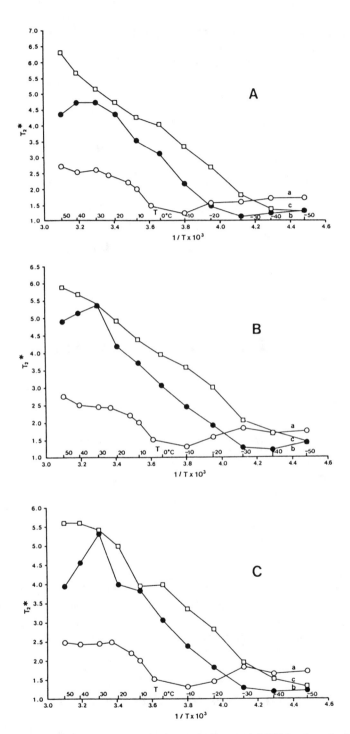

Figure 3.28 Plots of T_2^* versus: (A) manual calculation, (B) initial Linesim calculation, and (C) final Linesim calculation. Curves a, ODS-5; b, ODS-N5; c, ODS-P5 [50]. Reproduced from H. Ohta et al. in *Chromatographia* **40**, 507 (1995) by permission of Vieweg Publishing, Wiesbaden, Germany.

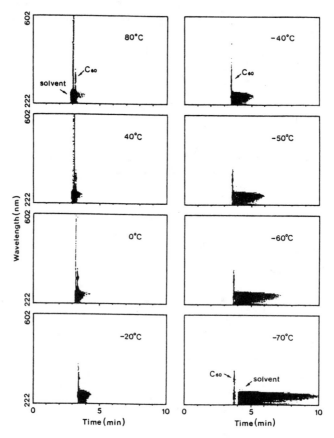

Figure 3.29 Two-dimensional elution profiles of C_{60} and solvent (toluene) at various temperatures; C_8 stationary phase. The mobile phase was n-hexane and the flow rate was 1 mL/min [50]. Reproduced from H. Ohta et al. in *Chromatographia* **40**, 507 (1995) by permission of Vieweg Publishing, Wiesbaden, Germany.

(toluene), a condition most clearly evident in the profile obtained at $-70°C$. In this case, a total exclusion effect dominates the retention of C_{60}. The results obtained for the C_8 phase clearly support the model described earlier for the ODS bonded phase, where bonded phase association and the configurational change with temperature are the most important factors in controlling the retention of solutes in LC. Large bulky solutes such as fullerenes apparently can magnify this effect.

3.3 Alkyldiphenyl Phases for Fullerene Separation

Another important factor at the surface of the bonded phases is the length of the alkyl chain in conjunction with other functional groups. When phenyl groups are present, they are expected to form strong π-π interactions with fullerenes. Therefore several novel phases have been synthesized (Figure 3.30:[51]). These phases consist of a monoalkyl chain such as ODS, octyl (C_8), and butyl (C_4) phases with two phenyl or two methyl groups.

Figure 3.31 shows chromatograms of the separation of C_{60} and C_{70} fullerenes with three alkyldiphenyl bonded silica phases under the same conditions. As seen in Table 3.5, the retention of the C_{18}Diph bonded phase is larger than those of the C_8Diph and C_4Diph phases. This pattern suggests the contribution of the long alkyl chains in the bonded phase to fullerene retention, since surface coverage values of these three phases are comparable, as shown in Table 3.5. To clarify this behavior, C_{60} and C_{70} were separated by means of various bonded phases shown in Figure 3.30. These retention data are summarized in Table 3.5. The C_{18}Diph phase gives the best separation for fullerenes.

A comparison of the retention data for C_{18}Diph and Develosil ODS-P5 suggests that two phenyl rings in the bonded phase structure contribute significantly to the retention values for fullerenes. Similar trends are also observed by comparing for the C_4Diph and C_4 phases. In spite of the lower surface coverage of the C_{18}Diph phase versus the C_{18} phase, the former phase retained the fullerene molecules longer than the latter. Those data also indicate the contribution of the phenyl rings.

Comparison of the separation factors for C_{18} (120 Å pore size, 9.08 C%) and Develosil ODS-N5 (100 Å pore size, 16 C%), where both phases have similar surface coverage, indicates that the differences in the retention factors for fullerenes with these two bonded phases are due mainly to the different silica gel properties.

On the other hand, the effect of alkyl chain length can be seen in the C_{18}Diph, C_8Diph, and C_4Diph, and corresponding alkyldimethyl phases. With the C_{18}Diph phase, larger retention factors and better separation factors were found for fullerenes in comparison to those observed with C_8Diph and C_4Diph phases. Thus the length of the alkyl chains in the bonded phases appears to make a large contribution

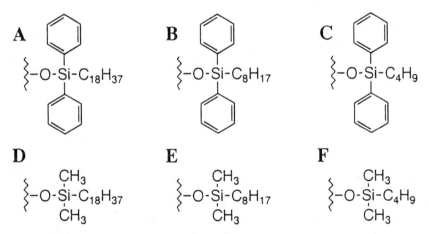

Figure 3.30 Structures of various bonded stationary phases evaluated in this work: (A) C_{18} Diph, (B) C_8 Diph, (C) C_4 Diph, (D) C_{18}, (E) C_6, and (F) C_4 [51]. Reproduced from Y. Saito et al. in *J. Liq. Chromatogr.*, **18**, 1897 (1995) by permission of Marcel Dekker Inc., New York, NY, USA.

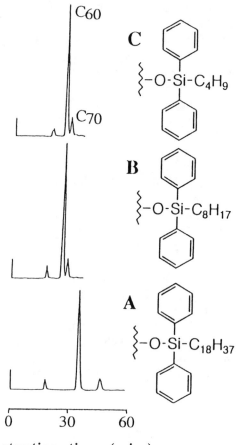

Figure 3.31   Chromatograms for the separation of C_{60} and C_{70} with three alkyldiphenyl bonded phases: (A) C_{18}Diph, (B) C_8Diph, and (C) C_4Diph. Microcolumn, 0.53 mm i.d. × 150 mm; mobile phase, *n*-hexane; flow rate, 2 µL/min; detection at 325 mm; temperature, 30°C [51]. Reproduced from Y. Saito et al. in *J. Liq. Chromatogr.*, **18**, 1897 (1995) by permission of Marcel Dekker Inc., New York, NY, USA.

to fullerene retention, and C_8 and C_4 chains are too short to have effective interactions with fullerene molecules.

Table 3.6 shows the retention data for the separation of C_{60} and C_{70} with the C_{18}Diph phase at various column temperatures. The corresponding van't Hoff plots are shown in Figure 3.32. For both C_{60} and C_{70}, linear van't Hoff plots are obtained over the temperature range investigated (from 0 to 80°C); when the column temperatures are elevated, however, the separation factor of these fullerenes decreases more slowly than is the case with Develosil ODS-P5. Since the Develosil ODS-P5 phase possesses a surface coverage almost identical to that of the C_{18}Diph phase, this phenomenon could be explained by the uniformity of the average ligand interval

Table 3.5   RETENTION DATA FOR THE SEPARATION[a] OF C_{60} AND C_{70} WITH VARIOUS BONDED PHASES

Bonded Phase	Carbon Content (%)	Surface Coverage (μmol/m^2)	Retention Factor		Separation Factor, α
			$k'_{C_{60}}$	$k'_{C_{70}}$	$(k'_{C_{70}}/k'_{C_{60}})$
C_{18}Diph	8.48	1.54	1.02	1.67	1.64
C_8Diph	5.89	1.56	0.42	0.57	1.36
C_4Diph	4.15	1.35	0.41	0.56	1.37
C_{18}	9.08	2.52	0.52	0.84	1.63
C_8	5.08	2.68	0.086	0.124	1.44
C_4	1.73	1.50	0.061	0.087	1.43
C_{18}(Develosil ODS-5)[b]	20	2.8	0.87	1.60	1.84
C_{18}(Develosil ODS-N5)[b]	16	2.1	0.81	1.32	1.64
C_{18}(Develosil ODS-P5)[b]	11	1.6	0.44	0.64	1.45

[a]   Mobile phase, *n*-hexane; column temperature, 30°C.
[b]   Commercial ODS phases from Nomura Chemicals, Seto, Japan.

Source: Ref. 50. Reproduced from H. Ohta et al. in *Chromatographia* **40**, 507 (1995) by permission of Vieweg Publishing, Wiesbaden, Germany.

($\approx$11.2 Å). The C_{18}Diph phase interacts more effectively with fullerenes than the ODS-P5 phase because the former can show better uniformity of the ligand interval. More detailed investigations are required if we are to make firmer conclusions on these phases.

Figure 3.33 shows two-dimensional chromatograms for C_{60}, C_{70}, and the solvent (toluene) at various temperatures with C_{18} and C_{18}Diph stationary phases. It was observed that the C_{18} phase gave better separation of C_{60} and C_{70} with decreasing

Table 3.6   RETENTION DATA FOR THE SEPARATION OF C_{60} AND C_{70} WITH C_{18}Diph AT VARIOUS COLUMN TEMPERATURES WITH AN *n*-HEXANE MOBILE PHASE

Column Temperature (°C)	Retention Factor, k'		Separation Factor, α $(k'_{C_{70}}/k'_{C_{60}})$	
	C_{60}	C_{70}		
0	1.24	2.13	1.72	[1.54][a]
20	1.09	1.81	1.66	[1.45][a]
30	1.02	1.67	1.64	
40	0.971	1.58	1.63	[1.42][a]
50	0.916	1.48	1.62	
60	0.852	1.38	1.62	[1.38][a]
70	0.827	1.32	1.60	
80	0.767	1.23	1.60	

[a]   Data with Develosil ODS-P5.

Source: Ref. 51. Reproduced from Y. Saito et al. in *J. Liq. Chromatogr.*, **18**, 1897 (1995) by permission of Marcel Dekker Inc., New York, NY, USA.

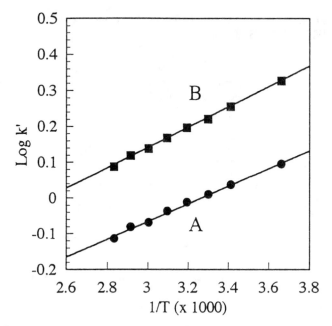

Figure 3.32   van't Hoff plots for C_{60} and C_{70} fullerenes with the C_{18} Diph phase: (A) C_{60} and (B) C_{70} [51]. Other conditions as in Figure 3.18. Reproduced from Y. Saito et al. in *J. Liq. Chromatogr.*, **18**, 1897 (1995) by permission of Marcel Dekker Inc., New York, NY, USA.

column temperature until about $-30°C$, when a maximum in retention was achieved. The retention of these solutes then decreased upon further lowering of the column temperature. In contrast, the retention of the solvent (toluene) increases monotonically with decreasing column temperature. The elution order change among the three components is clearly seen at about $-50°C$ from one order (solvent, C_{60}, C_{70}) to another (C_{60}, C_{70}, solvent). However, this kind of elution order change was not observed with the C_{18}Diph phase. Rather, C_{60} and C_{70} gave the expected elution behavior with decreasing column temperature (80 to $-70°C$), that is, a monotonic increase in retention for the two solutes.

It was postulated in the preceding discussion that a configurational change of the ligand as a function of temperature is the most important factor in controlling fullerene retention in LC and that this change is caused by an exclusion effect of the rigid alkyl chains, which do not have enough space for fullerenes to fit in between adjacent bonded moieties. Large bulky solutes such as fullerenes can apparently magnify this temperature exclusion effect. This means that bonded phases having sufficient space and possessing a shape to match that of the fullerenes should retain such molecules to a greater extent even at lower temperatures.

Figures 3.34 and 3.35 show two-dimensional chromatograms for C_8/C_8Diph and C_4/C_4Diph stationary phases, respectively. It appears that none of these phases are able to separate fullerenes well, and the elution order change is seen at a higher temperature than in the case of the C_{18} phase. The shorter alkyl chains induce an

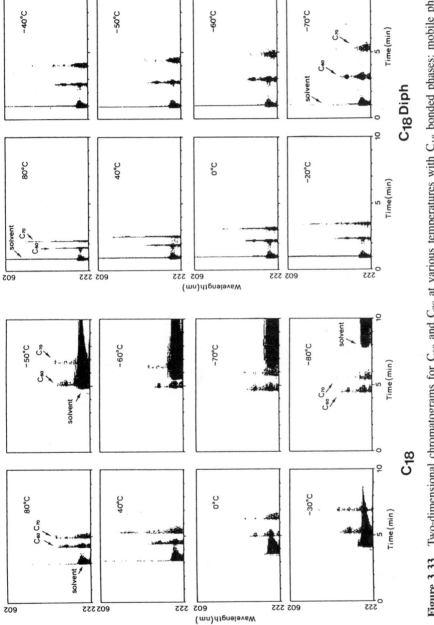

Figure 3.33 Two-dimensional chromatograms for C_{60} and C_{70} at various temperatures with C_{18} bonded phases: mobile phase, n-hexane; solvent, toluene; flow rate, 1 mL/min [54]. Reproduced from H. Ohta et al. in *Chromatographia* **42**, 56 (1996) by permission of Vieweg Publishing, Wiesbaden, Germany.

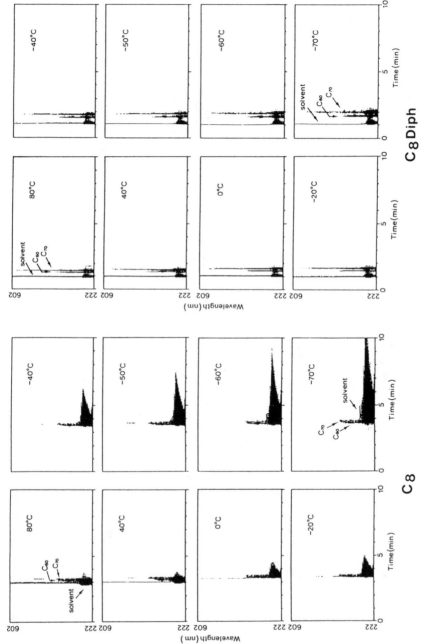

Figure 3.34 Two-dimensional chromatograms for C_{60} and C_{70} at various temperatures with C_8 bonded phases: mobile phase, *n*-hexane; solvent, toluene; flow rate, 1 mL/min [54]. Reproduced from H. Ohta et al. in *Chromatographia* **42**, 56 (1996) by permission of Vieweg Publishing, Wiesbaden, Germany.

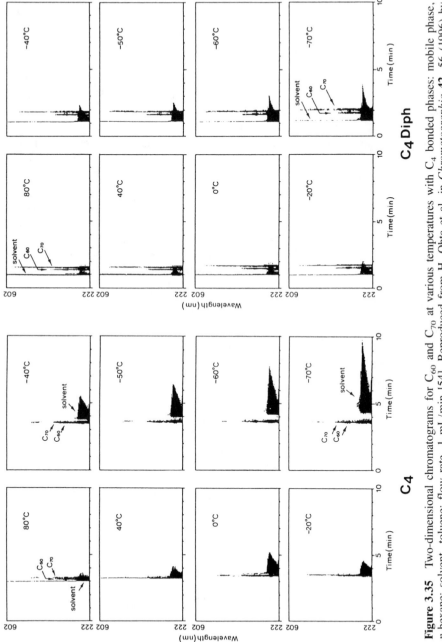

Figure 3.35 Two-dimensional chromatograms for C_{60} and C_{70} at various temperatures with C_4 bonded phases: mobile phase, *n*-hexane; solvent, toluene; flow rate, 1 mL/min [54]. Reproduced from H. Ohta et al. in *Chromatographia* **42**, 56 (1996) by permission of Vieweg Publishing, Wiesbaden, Germany.

exclusion effect for fullerenes at a relatively high temperature because they lack the depth needed to capture and hold with the ligand space molecules as bulky as fullerenes. If the stationary phases have two phenyl groups at the bottom part of the bonded moieties, the contribution of the aromatic groups to fullerene retention becomes significant. It appears that fullerene separation can be improved by using shorter alkyl chain stationary phases if these phases have phenyl groups near the bonding position.

The phenomena just described can be readily seen in Figure 3.36, where the retention ratio between C_{60} and the solvent is plotted against temperature for each phase. C_{60} followed the expected behavior with decreasing column temperature (80 to $-70°C$): that is, monotonic increase of retention is seen for the C_{18}Diph, C_8Diph, and C_4Diph stationary phases, whereas an elution order change between C_{60} and the solvent was observed with the C_{18}, C_8, and C_4 stationary phases at -50, -20, and $60°C$, respectively.

Solid state NMR spectroscopy should be a very useful tool for acquiring a better understanding of the specific phenomena that affect the separation of fullerenes because the spin–spin relaxation time (T_2^*) has been shown to correlate with the chromatographic behavior of fullerenes in LC, especially in the low temperature region. The general trend of the T_2^* values when plotted as a function of $1/T$— namely, for this relaxation time to decrease with a decrease in temperature—is

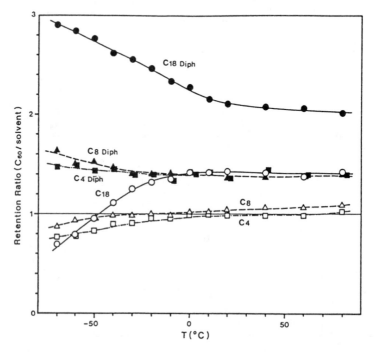

Figure 3.36 Plots of the retention ratio of C_{60} to the solvent (toluene) versus temperature [54]. Reproduced from H. Ohta et al. in *Chromatographia* **42**, 56 (1996) by permission of Vieweg Publishing, Wiesbaden, Germany.

readily rationalized by our awareness that molecular motion decreases as the temperature is lowered, generally leading to shorter relaxation times for samples in the limited-motion regime on the NMR time scale. Zeigler and Maciel [52] have shown that the real spin–spin relaxation time differs from T_2^* determined according to the following equation by a factor that can vary from 2.6 to 48 for C_{18} silica depending on the loading of the phase and the position of the carbon in the alkyl chain:

$$\Delta W = \frac{1}{T_2^*} \tag{3.3}$$

The difference between the natural T_2^* and the one calculated from the line width has been attributed to variations in the surface topography of the silica, which in turn lead the alkyl chains to assume a variety of conformations and interactions that result in an inhomogeneous broadening of the line (usually referred to as chemical shift dispersion). Therefore it should be possible to let this inhomogeneous broadening act as a probe of the surface, one that might be sensitive to structural changes in the bonded moiety as a function of an experimental variable such as temperature. That is, the increased interaction of neighboring alkyl chains should facilitate the detection of a more ordered structure on the surface. Conclusions consistent with the preceding explanation were made after a series of C_{18} bonded phases had been studied by variable temperature CP-MAS NMR as described earlier. The T_2^* values decreased with decreasing temperature up to a certain point. At a temperature that is different for each phase and dependent on the carbon loading, the T_2^* value reached a minimum and became relatively constant. This was interpreted as the formation of a more ordered configuration. The more ordered structure would lead to a decrease in the surface inhomogeneity, hence should bring the observed line width closer to that of the natural line width. The latter condition could be detected experimentally by inspecting the line width, which should either remain constant or perhaps even decrease. In terms of the T_2^* values, line widths would remain constant or perhaps become slightly longer.

Figure 3.37 shows the relationships between the spin–spin relaxation time T_2^* and the temperature for the C_{18} and C_{18}Diph phases. The C_{18} phase shows a decrease in T_2^* until -30 to $-40°C$, when the value becomes constant. The C_{18}Diph phase shows no such plateau, indicating that the bonded alkyl chain had not achieved any ordered structure at the lowest temperature measured. It is very interesting to note that the relaxation time change for the C_{18} phase correlates well with the plots of log k' for C_{60} and the three PAHs versus temperature, as seen in Figure 3.38, where the retention behavior of the PAHs is shown to differ greatly from that of C_{60}. For the C_{18} phase, the retention of the PAHs increases linearly with decreasing temperature, but there is a curvature at about 30°C.

In addition, the retention of C_{60} displays a drastic decrease below about $-30°C$. It was suggested that the major decrease in T_2^* occurring between 30 and $-30°C$ is most likely due to the restriction of molecular motion as the temperature is lowered, and the minimum attained at around $-30°C$ may be due to the assumption by the bonded moieties of a stable configuration (e.g., an association between adjacent C_{18}

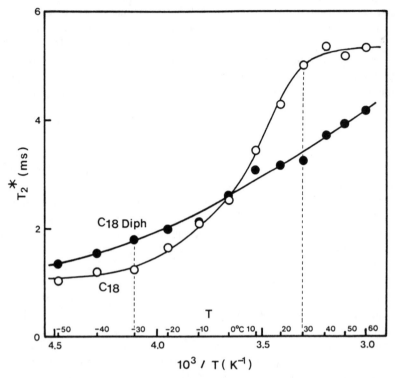

Figure 3.37   Plots of spin–spin relaxation time T_2^* versus temperature for the C_{18} and C_{18} Diph phases [54]. Reproduced from H. Ohta et al. in *Chromatographia* **42**, 56 (1996) by permission of Vieweg Publishing, Wiesbaden, Germany.

chains). Then, it was assumed that the rigidity and orderliness of the ligand at low temperature greatly reduced the retention of the bulky fullerenes and promoted the retention of the PAHs, which are relatively planar. On the other hand, the relaxation time on the C_{18}Diph phase decreased linearly with decreasing temperature. C_{60} and the PAHs gave a monotonic increase in retention with decreasing temperature, leading to the supposition that the lower rigidity and orderliness of the ligand of the C_{18}Diph material at low temperature greatly facilitated the retention of the bulky fullerenes because the phase has enough depth to trap and hold bulky fullerenes in the space between bonded moieties at low temperature, whereupon the contribution of the phenyl groups to retention becomes significant.

To substantiate the foregoing explanation, investigators examined data obtained by LC and CP-MAS NMR for a C_{30} alkyl bonded phase, which should be more rigid than C_{18} phases at normal temperatures. Typical CP-MAS NMR spectra of the C_{30} phase at various temperatures are summarized in Figure 3.39. The two major peaks in the spectra at 30 and 32 ppm were also reported by Albert et al. [53]. These researchers concluded that the 30 ppm peak is the more mobile configuration, while the 32 ppm peak represents a more rigid structure. The T_2^* values for the two peaks

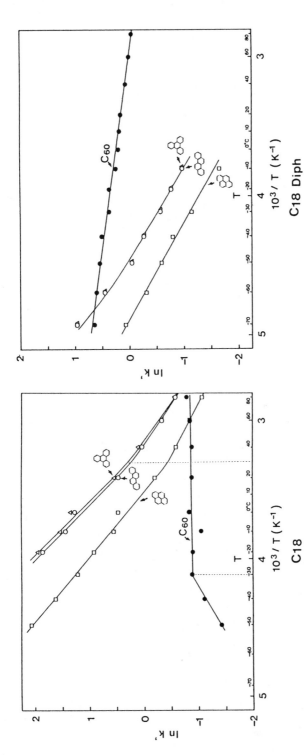

Figure 3.38 Plots of ln k' for C_{60} and three PAHs versus temperature for the C_{18} and C_{18} Diph phases [54]. Reproduced from H. Ohta et al. in *Chromatographia* **42**, 56 (1996) by permission of Vieweg Publishing, Wiesbaden, Germany.

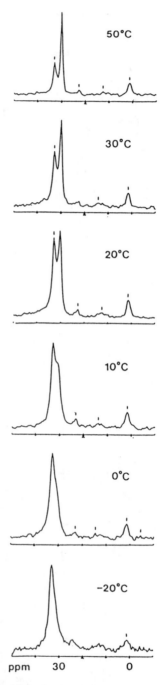

Figure 3.39 CP-MAS spectra of the C_{30} bonded phase at various temperatures [54]. Reproduced from H. Ohta et al. in *Chromatographia* **42**, 56 (1996) by permission of Vieweg Publishing, Wiesbaden, Germany.

were calculated, and the results are plotted versus the temperature in Figure 3.40. The relaxation time for the 32 ppm peak seems to decrease monotonically with decreasing temperature, but the relaxation time for the 30 ppm peak has a different behavior as a function of temperature. Below about 10°C, the relaxation time is practically constant down to −30°C, which indicates that the rigidity and orderliness of the alkyl chains is not much changed by lowering the temperature below 10°C. This NMR spectral information is quite consistent with the LC retention data of Figure 3.41. The retention of C_{60} is almost constant above 20°C and decreases with decreasing temperature from 10°C. This temperature is the same one found in the NMR data, where the relaxation time changes from a monotonic decrease to a constant value. Therefore the chromatographic behavior of fullerenes on the C_{30} phase can be explained by NMR spectroscopic data in terms of changes in the configuration of the alkyl chain.

From these results, it appears that the spin–spin relaxation time is a good indicator of the LC retention behavior of fullerenes [54]. The strong correlation between the changes in spin–spin relaxation time and retention indicate that the exclusion effect observed for bulky fullerenes with alkyl bonded silica stationary phases is

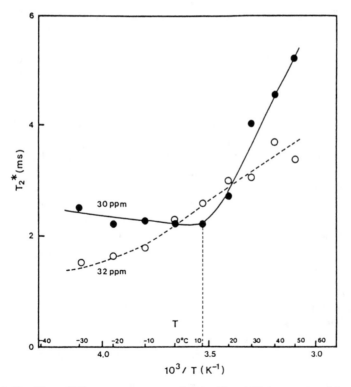

Figure 3.40 Plots of T_2^* versus temperature for the 30 and 32 ppm peaks of the C_{30} phase [54]. Reproduced from H. Ohta et al. in *Chromatographia* **42**, 56 (1996) by permission of Vieweg Publishing, Wiesbaden, Germany.

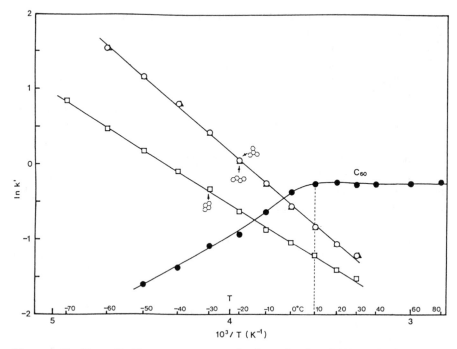

Figure 3.41   Plots of ln k' versus temperature for C_{60} and various PAHs on the C_{30} bonded phase [54]. Reproduced from H. Ohta et al. in *Chromatographia* **42**, 56 (1996) by permission of Vieweg Publishing, Wiesbaden, Germany.

dependent on the rigidity and orderliness of the ligand as well as the interval between adjacent moieties. A low surface coverage phase (i.e., large interval between adjacent ligands) does not exhibit such an exclusion effect. Bulky fullerenes that possess the right size and shape are able to provide this important insight into the retention mechanism in LC for alkyl bonded silica stationary phases. Without the phenyl group at the base of the bonded moiety, there is not enough space between adjacent ligands to allow fullerene penetration at low temperatures. Although further consideration about the retention mechanism of fullerenes with chemically bonded stationary phases is needed, we can make three conclusions from the systematic evaluations just described.

1. Two phenyl rings at the bottom part of the bonded phase contribute to effective interaction between the bonded phase ligands and fullerene molecules.
2. The phenyl rings bonded to the silicon atom induce uniformity in the interval between alkyl chains; and therefore, good retention can be obtained, because the phenyl rings work as the spacer of the bonded phase ligands on the silica support.
3. There may be a critical chain length for effective interaction with fullerenes, and longer may be better.

3.4 Multimethoxyphenylpropyl Bonded Phases for Fullerene Separation

It has been found that phenyl groups in the bonded moiety contribute greatly to the retention of fullerene, and to the enhancement of their selectivity. This section, therefore, evaluates 4-methoxy-, 3,4-dimethoxy-, and 3,4,5-trimethoxyphenyl-propyl bonded silica (MPP, DMPP, and TMPP, respectively) phases for the separation of fullerenes as an extension of the preceding discussion. A phenylpropyl (PP) bonded phase, synthesized for purposes of comparison, is described as well.

Since microcolumns are ideal for studies in which only limited amounts of laboratory-made, experimental stationary phases are available, this technique has been applied in almost all the evaluations of new stationary phases. Figure 3.42 shows chromatograms of the separation of C_{60} and C_{70} on new methoxyphenyl-propyl bonded phases synthesized in the laboratory [55,56]. In these separations, n-hexane was used as the mobile phase, and the column temperature was maintained at 30°C. The retention times of both C_{60} and C_{70} were increased, and their separation was improved significantly on the bonded phases having methoxy functional group(s), compared to those without. These results are summarized in Table 3.7. The retention data obtained on two commercial monomeric ODS bonded phases were also tabulated for comparison. In contrast to the PP phase, methoxy bonded phenylpropyl phases possessed greater retention for C_{60} and C_{70} than ODS phases, especially in the case of DMPP and TMPP, although the separation factors obtained on these methoxy bonded phases are comparable or superior to the commercial ODS bonded phases listed in Table 3.7.

Our basic studies revealed that a solute having a chemical structure complementary to the preferred conformation of the bonded phase interacts strongly with the stationary phase. This was also indicated when C_{60} was used as a stationary phase in Chapter 2. In such solute–stationary phase interactions, the stationary phase provides so-called docking sites that possess the appropriate dimensions and functional groups to accommodate the solute(s) of interest, and to produce strong interactions. The retention behavior of the fullerenes on methoxyphenylpropyl bonded phases can be explained by arguments such as those shown in Figure 3.43: these conformations were determined by computerized, energy-minimized molecular modeling calculations by using MM2PP software.

In the case of the DMPP phase, two methoxy groups at the top of the aromatic ring are forced to some extent into a planar conformation to the aromatic ring of the phase. This realignment is due to the contribution of the steric hindrance and, possibly, hydrogen bonding between the methoxy groups. Therefore, the fullerenes could interact effectively with the phenyl rings of the DMPP phase, and the interactive sites of the phase could have a curved conformation to complement the curvature of fullerene molecules, especially for C_{70}.

The PP bonded phase with no methoxy group interacts weakly with fullerenes because its interaction area is smaller than those of the phases having methoxy

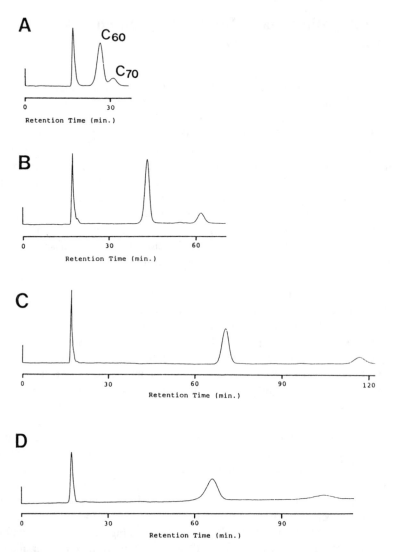

Figure 3.42   Chromatograms for the separation of C_{60} and C_{70} with various multimethoxy-phenylpropyl bonded phases: (A) PP, (B) MPP, (C) DMPP, and (D) TMPP. Column size, 0.53 mm i.d. × 200 mm; mobile phase, n-hexane; flow rate, 2 μL/min; UV detection at 280 nm; temperature, 30°C [55]. Reproduced from K. Jinno et al. in *J. Microcolumn Sep.* **5**, 135 (1993) by permission of John Wiley and Sons, New York, NY, USA.

groups. The MPP and TMPP phases should produce smaller retention values, and have less recognition capability for C_{60} and C_{70} than the DMPP bonded phase, because the conformational fitting of these bonded phase groups with the C_{60} and C_{70} solutes is worse than in the case of the DMPP phase.

Table 3.7   RETENTION DATA FOR THE SEPARATION OF C_{60}
AND C_{70} WITH VARIOUS BONDED PHASES

Bonded Phase	Retention Factor, k'		Separation Factor, α $(k'_{C_{70}}/k'_{C_{60}})$
	C_{60}	C_{70}	
PP	0.58	0.83	1.43
MPP	1.49	2.56	1.72
DMPP	3.09	5.77	1.87
TMPP	2.75	4.90	1.78
Develosil ODS-5	0.87	1.60	1.84
Capcell Pak C18 SG120[a]	0.36	0.56	1.56

[a]   A conventional monomeric ODS column with polymer-coated silica (4.6 mm i.d. ×
250 mm; Shiseido, Tokyo, Japan); flow rate, 1 mL/min; mobile phase, *n*-hexane; column
temperature, 30°C.

Source: Ref. 55.

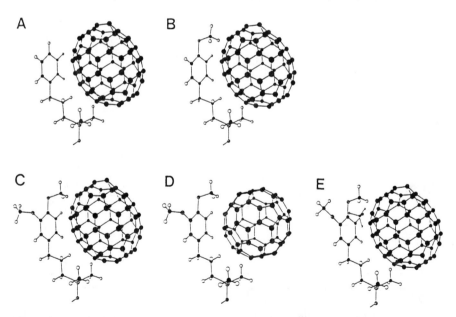

Figure 3.43   Molecular modeling diagrams of interactions between fullerenes and bonded
phases: (A) PP, (B) MPP, (C) DMPP, (E) TMPP with C_{70}, and (D) DMPP with C_{60} [55].
Reproduced from K. Jinno et al. in *J. Microcolumn Sep.* **5**, 135 (1993) by permission of John
Wiley and Sons, New York, NY, USA.

3.5 Multiphenyl Bonded Phases for Fullerene Separation

In further studies of novel stationary phases for the separation of fullerenes, we evaluated stationary phases of three types: triphenyl (Triph), diphenyl (Diph), and monophenyl (Monoph) bonded silica phases [57].

Figure 3.44 shows chromatograms for the separation of C_{60} and C_{70} on Triph, Diph, and Monoph bonded phases prepared with silica gels having different pore size. These retention data are also summarized in Table 3.8. Of the phases having a 70 Å pore size, the Triph phase exhibited the best retention power and separation performance, although good separation was obtained with all phases, as shown in Figure 3.44A. When the pore size was 150 Å (Figure 3.44B), retentions of C_{60} and C_{70} were shorter than those with the 70 Å phases. The decrease in retention is due to the lower carbon loadings on the wide pore silica surfaces.

The results presented in Figure 3.44 and Table 3.8 indicate that the three phenyl rings of Triph bonded phases possess a strong retention based on the π-π interaction between the bonded phase and fullerene molecules. Furthermore, the chromatographic characteristics of these three phases (i.e., Triph, Diph, Monoph) are most evident for the 70 Å silica gel phases.

The Triph 70 Å phase has a much higher retention power than commercial ODS phases, although the retention of the triphenyl phase is somewhat less than can be obtained with the DMPP bonded phase.

The chromatograms in Figure 3.45 show separation of C_{60} and C_{70} with the Triph

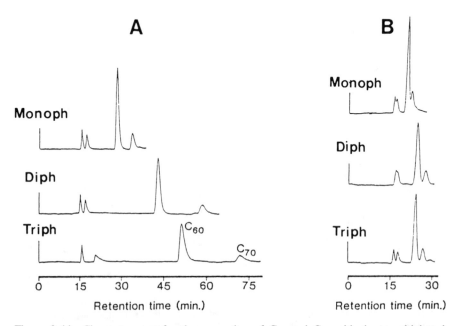

Figure 3.44 Chromatograms for the separation of C_{60} and C_{70} with three multiphenyl bonded phases: (A) 70 Å silica pore size and (B) 150 Å silica pore size [57]. Other conditions as in Figure 3.42. Reproduced from Y. Saito et al. in *J. Liq. Chromatogr.*, **17**, 2359 (1995) by permission of Marcel Dekker Inc., New York, NY, USA.

Table 3.8 RETENTION DATA FOR THE
SEPARATION[a] OF C_{60} AND C_{70}
WITH VARIOUS BONDED PHASES

Bonded Phase	Retention Factor, k'		Separation Factor, α $(k'_{C_{70}}/k'_{C_{60}})$
	C_{60}	C_{70}	
70 Å			
Triph	2.28	3.59	1.58
Diph	1.85	2.89	1.56
Monoph	0.86	1.24	1.44
150 Å			
Triph	0.45	0.63	1.39
Diph	0.45	0.64	1.41
Monoph	0.27	0.37	1.36

[a] Mobile phase, n-hexane; column temperature, 30°C.

Source: Ref .57. Reproduced from Y. Saito et al. in *J. Liq. Chromatogr.*, **17**, 2359 (1995) by permission of Marcel Dekker Inc., New York, NY, USA.

70 Å bonded phase at different column temperatures. These chromatograms indicate that with the triphenyl phase, temperature has little influence on the retention. For comparison, retention data for five PAHs were also measured (see Table 3.9). While the retention factors of PAHs decrease with increasing column temperature, only a slight variation in retention for C_{60} and C_{70} is found, and this is understood to be a phenomenon specific to fullerenes.

Figure 3.46 displays a summary of separation factors of C_{60} and C_{70} with various phases at different temperatures, as well as some additional data. The separation factors with ODS phases, especially in the case of the polymeric phase, decrease significantly when the column temperature is elevated. The conformational rigidity of the DMPP phase, however, causes this phase to exhibit a relatively smaller dependence on temperature than the ODS phases. Therefore, it can be assumed from the foregoing results that the Triph 70 Å phase also possesses a conformational rigidity that is not witnessed when the ODS phases are used.

3.6 Liquid Crystal Bonded Phases for Fullerene Separation

Although certain ODS phases studied show good separation of fullerene mixtures, some novel stationary phases having phenyl ring(s) indicate stronger retention and better separation power than ODS stationary phases. Those investigations revealed the following trend: with phenyl-derivatized bonded phases, the effective π-π interaction between the phenyl ligand(s) of the bonded phases and fullerene molecules is dominant in the retention mechanism; the conformational fitting between the ligand of the bonded phases and fullerene molecules also has an influence on retention,

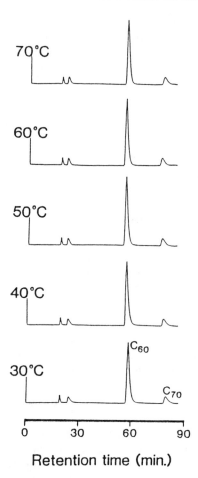

Figure 3.45   Chromatograms for the separation of C_{60} and C_{70} with the Triph 70 Å phase at different column temperatures [57]. Other conditions as in Figure 3.42. Reproduced from Y. Saito et al. in *J. Liq. Chromatogr.*, **17**, 2359 (1995) by permission of Marcel Dekker Inc., New York, NY, USA.

however. Thus our studies suggest that the retention behavior of fullerenes and PAHs can be explained by the same concept. Since fullerene compounds can be regarded as very large PAHs with specific features and dimensions, it is natural for the retention mechanisms for fullerenes and PAHs to be very similar.

Pesek et al. and Jinno et al. introduced liquid crystal bonded phases to the separation of PAHs and investigated the retention behavior in HPLC [37,58,59] and supercritical fluid chromatography (SFC) [60]. Their results clearly confirm the excellent separation performance of the liquid crystal phases for isomeric PAHs and also suggest the possibility of applying these phases for the separation of fullerenes because of their good molecular shape recognition capability.

In this section, we introduce liquid crystal bonded silica phases as stationary phases for the LC separation of C_{60} and C_{70} fullerenes [61].

Table 3.9   RETENTION FACTORS FOR C_{60}, C_{70}, AND FIVE PAHs WITH Triph 70 Å PHASE AT DIFFERENT COLUMN TEMPERATURES

	Retention Factor, k'				
	30°C	40°C	50°C	60°C	70°C
Naphthalene	0.76	0.71	0.63	0.58	0.54
Phenanthrene	1.52	1.38	1.20	1.07	0.971
o-Terphenyl	2.32	2.07	1.77	1.57	1.40
Pyrene	1.66	1.54	1.41	1.30	1.21
Triphenylene	3.02	2.77	2.44	2.22	2.02
C_{60}	1.98	1.97	1.94	1.93	1.93
C_{70}	3.10	3.09	3.03	3.02	3.02

Source: Ref. 57. Reproduced from Y. Saito et al. in *J. Liq. Chromatogr.*, **17**, 2359 (1995) by permission of Marcel Dekker Inc., New York, NY, USA.

3.6.1   Separation of C_{60} and C_{70} with Two Liquid Crystal Bonded Silica Phases

To evaluate the basic separation performance of the liquid crystal phases, C_{60} and C_{70} fullerenes were chromatographically separated with *n*-hexane as the mobile

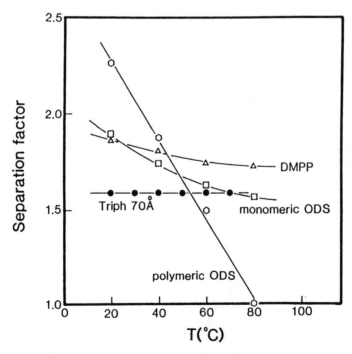

Figure 3.46   Relationships between separation factors for C_{60} and C_{70} with four different bonded phases versus column temperature [57]. Conditions as in Figure 3.45. Reproduced from Y. Saito et al. in *J. Liq. Chromatogr.*, **17**, 2359 (1995) by permission of Marcel Dekker Inc., New York, NY, USA.

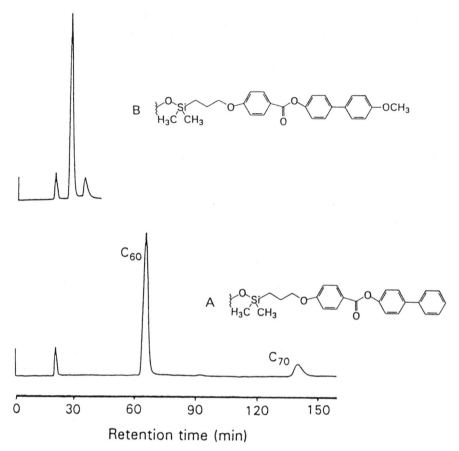

Figure 3.47 Chromatograms for the separation of C_{60} and C_{70} with (A) liquid crystal phase 1, and (B) liquid crystal phase 2. Column size, 0.53 mm i.d. × 200 mm; mobile phase, *n*-hexane; flow rate, 2 μL/min; detection, UV at 320 nm; temperature, 30°C [61]. Reproduced from Y. Saito et al. in *J. Microcolumn Sep.* **7**, 41 (1995) by permission of John Wiley and Sons, New York, NY, USA.

phase. Chromatograms and retention data are shown in Figure 3.47 and Table 3.10, respectively. Data obtained with a commercial ODS phase, Develosil ODS-5, are given in Table 3.10 for comparison.

The liquid crystal phase 1, whose structure appears in Figure 3.47, has excellent separation performance for C_{60} and C_{70} compared with various commercial ODS phases. Develosil ODS-5 was found to be the best stationary phase among a wide variety of ODS phases. With liquid crystal phase 2, whose structure is also shown in Figure 3.47, the separation factor between C_{60} and C_{70} is less than that of phase 1, and the value is almost comparable to those of commercial ODS bonded phases.

Section 3.2 describes a certain trend, namely, that an ODS phase having its ligand interval near to the molecular size of fullerenes possesses good retention for fullerene molecules. The respective ligand intervals of liquid crystal phase 1 and 2 are calculated to be about 6.9 and 5.9 Å based on the carbon content values of these

Table 3.10 BASIC CHARACTERISTICS OF BONDED PHASES AND RETENTION DATA
FOR THE SEPARATION[a] OF C_{60} AND C_{70} WITH THESE PHASES

Bonded Phase	Pore Size (Å)	Carbon Content (%)	Surface Coverage (μmol/m²)	Separation Factor, α		
				$k'_{C_{60}}$	$k'_{C_{70}}$	$(k'_{C_{70}}/k'_{C_{60}})$
Liquid crystal phase 1	300	9.7	3.6	2.82	8.46	3.00
Liquid crystal phase 2	300	13	4.7	0.42	0.79	1.88
Develosil ODS-5	120	20	2.8	0.87	1.60	1.84

[a] Mobile phase, n-hexane; column temperature, 30°C.

Source: Ref. 61. Reproduced from Y. Saito et al. in *J. Microcolumn Sep.* **7**, 41 (1995) by permission of John Wiley and Sons, New York, NY, USA.

phases and the specific surface area of the silica gel. The respective molecular dimensions of C_{60} and C_{70} are 7 Å × 7 Å and 7 Å × 9 Å. Although the effect of the methoxy group at the end of the bonded phase ligand on the retention of fullerenes should be further investigated, it is likely that the ligand interval of these liquid crystal phases also affects interactions between the solutes and the bonded phases.

From the results of the evaluation on the basic separation performances of the two bonded silica phases, liquid crystal phase 1 is used in the following discussions.

3.6.2 Effect of Mobile Phase Composition on Fullerene Separation

Figure 3.48a shows the chromatograms for the separation of C_{60} and C_{70} with liquid crystal phase 1 using an n-hexane/dichloromethane mobile phase system. These retention data are summarized in Table 3.11A. Smaller retention factors for these fullerenes were observed when the dichloromethane content in the mobile phase solvent was increased.

The corresponding chromatograms and the retention data with n-hexane/toluene as the mobile phase (Figure 3.48b and Table 3.11B, respectively) also indicate a decrease of retention values for C_{60} and C_{70} as observed in Figure 3.48a and Table 3.11A. A logarithmic decrease in the solutes' retentions with modifier concentration in the mobile phase was found. Although the effect of adding toluene to the mobile phase is similar effect to that of dichloromethane on the retention of fullerenes, the chromatographic peak shape observed with n-hexane/toluene mixtures is better than that seen when dichloromethane is in the mobile phase. Ruoff et al., who reported the solubility of C_{60} in various organic solvents, demonstrated that C_{60} is 10 times more soluble in toluene than in dichloromethane [62]. Therefore, our present observation in the chromatograms for the separation of C_{60} and C_{70} presumably reflects the increased solubility of fullerenes in the mobile phase containing toluene as the modifier.

The plots in Figure 3.49 indicate that the liquid crystal phase and a mobile phase containing toluene up to about 42% (v/v) can be used to obtain the same separation factor (i.e., $k'_{C_{70}}/k'_{C_{60}} = 1.84$) as Develosil ODS-5 using n-hexane as the mobile

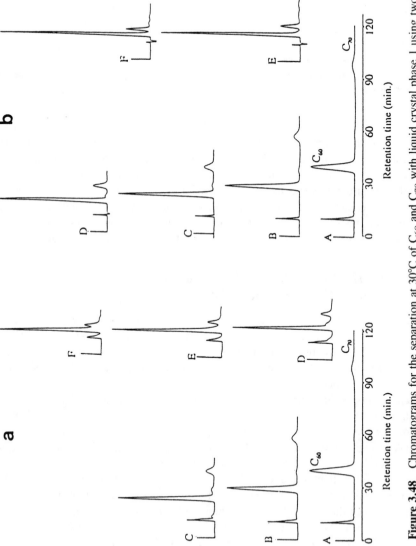

Figure 3.48 Chromatograms for the separation at 30°C of C_{60} and C_{70} with liquid crystal phase 1 using two mobile phase compositions: (a) n-hexane/dichloromethane and (b) n-hexane/toluene. Composition: A, 100:0; B, 90:10; C, 80:20; D, 70:30; E, 60:40; F, 50:50.

Table 3.11A   RETENTION DATA FOR THE SEPARATION OF C_{60} AND C_{70} WITH LIQUID CRYSTAL PHASE 1 USING DIFFERENT MOBILE PHASE COMPOSITIONS

Mobile Phase n-Hexane/Dichloromethane[a]	Retention Factor, k'		Separation Factor, α $(k'_{C_{70}}/k'_{C_{60}})$
	C_{60}	C_{70}	
100:0	2.82	8.46	3.00
90:10	1.86	4.72	2.54
80:20	1.22	2.80	2.30
70:30	0.79	1.64	2.08
60:40	0.54	1.03	1.90
50:50	0.39	0.69	1.75

[a]   Column temperature, 30°C.

Source: Ref. 61. Reproduced from Y. Saito et al. in *J. Microcolumn Sep.* **7**, 41 (1995) by permission of John Wiley and Sons, New York, NY, USA.

phase solvent. One of the most advantageous features of this phase is that better solubility results in greater sample loadability for preparative scale LC. In contrast to ODS phases, other good solvents in which fullerenes are very soluble can be employed as the mobile phase with the liquid crystal bonded silica phase.

3.6.3 Effect of Column Temperature on Fullerene Separation

Figure 3.50 shows typical chromatograms for the separation of C_{60} and C_{70} with liquid crystal phase 1 at different column temperatures. The retention data at various

Table 3.11B   RETENTION DATA FOR THE SEPARATION OF C_{60} AND C_{70} WITH LIQUID CRYSTAL PHASE 1 USING DIFFERENT MOBILE PHASE COMPOSITIONS

Mobile Phase n-Hexane/Toluene[a]	Retention Factor, k'		Separation Factor, α $(k'_{C_{70}}/k'_{C_{60}})$
	C_{60}	C_{70}	
100:0	2.82	8.46	3.00
90:10	1.82	4.68	2.57
80:20	1.22	2.80	2.30
70:30	0.83	1.72	2.07
60:40	0.60	1.12	1.86
50:50	0.44	0.76	1.74
40:60	0.34	0.54	1.59
30:70	0.24	0.36	1.49

[a]   Column temperature, 30°C.

Source: Ref. 61.

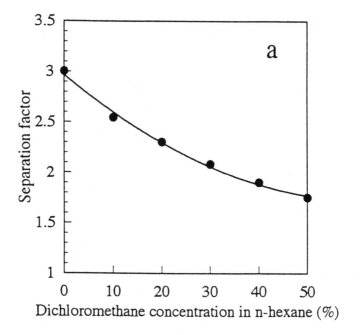

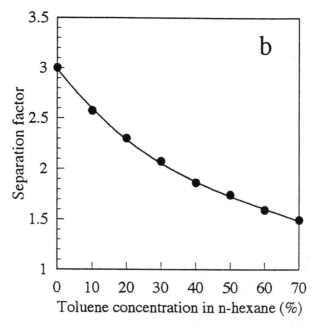

Figure 3.49 Plots of separation factors for C_{60} and C_{70} versus mobile phase composition for (a) *n*-hexane/dichloromethane and (b) *n*-hexane/toluene [61]. Reproduced from Y. Saito et al. in *J. Microcolumn Sep.* **7**, 41 (1995) by permission of John Wiley and Sons, New York, NY, USA.

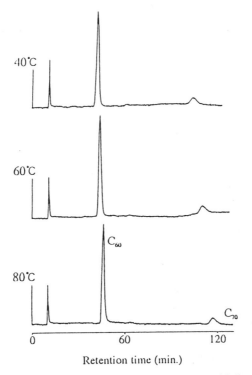

Figure 3.50 Chromatograms for the separation of C_{60} and C_{70} with liquid crystal phase 1 at different column temperatures using the *n*-hexane mobile phase.

temperatures are tabulated along with the data of some PAHs in Table 3.12. These results highlight an unusual increase in the retention of fullerenes with increasing column temperature, while all the PAHs studied behaved normally under the same conditions, as shown by the van't Hoff plots in Figure 3.51. The retention mechanism of fullerenes appears to differ from that of PAHs, because the enthalpy values for fullerenes and for PAHs have different signs. Pirkle and Welch also observed similar behavior for the retentions of C_{60} and C_{70} with a π-acidic chiral stationary phase [31].

The molecular shape recognition capabilities of the liquid crystal bonded phase in reversed-phase LC have been evaluated at various temperatures using PAHs as the sample probes for the interpretation of the retention mechanism for fullerenes. Those results clearly indicate that the liquid crystal phase possesses excellent planarity and shape recognition power compared to commercial ODS phases, and that the selectivity of the liquid crystal phase decreases significantly with increasing temperature.

The separation factors between planar and nonplanar PAHs at different temperatures are tabulated in Table 3.13A. A decrease in the planarity recognition capability of liquid crystal phase 1 is indicated with increasing temperature. Similar decreases

Table 3.12   RETENTION DATA FOR FOUR PAHs AND FULLERENES
WITH LIQUID CRYSTAL PHASE 1 AT VARIOUS COLUMN
TEMPERATURES AND AN *n*-HEXANE MOBILE PHASE

Solute	Retention Factor, k'			
	0°C	40°C	60°C	80°C
o-Terphenyl	0.20	0.13	0.11	0.10
Phenanthrene	0.42	0.29	0.25	0.22
Anthracene	0.62	0.37	0.30	0.25
Triphenylene	1.13	0.67	0.54	0.45
C_{60}	2.52	2.88	3.08	3.27
C_{70}	7.58	8.63	9.18	9.76
Separation factor, α^a	(3.01)	(3.00)	(2.98)	(2.98)

a   Between C_{70} and C_{60}.

Source: Ref. 61. Reproduced from Y. Saito et al. in *J. Microcolumn Sep.* **7**, 41 (1995) by permission of
John Wiley and Sons, New York, NY, USA.

in the molecular shape recognition power of the bonded phase are also observed for
other PAHs, where the temperature dependence of the shape selectivity between
PAHs having "rodlike" and "squarelike" shapes was investigated (Table 3.13B).
The more elongated PAHs show a larger decrease in retention with increasing
temperature. The results suggest that at elevated column temperatures, the phase
ordering of this liquid crystal bonded silica phase is reduced by the thermal mobility

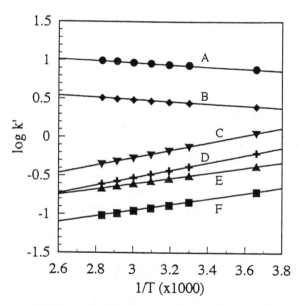

Figure 3.51   van't Hoff plots for fullerenes and four PAHs with liquid crystal phase 1: A,
C_{70}; B, C_{60}; C, triphenylene; D, anthracene; E, phenanthrene; F, *o*-terphenyl [61]. Reproduced from Y. Saito et al. in *J. Microcolumn Sep.* **7**, 41 (1995) by permission of John Wiley
and Sons, New York, NY, USA.

Table 3.13A   SEPARATION FACTORS BETWEEN PLANAR AND NONPLANAR PAHs WITH LIQUID CRYSTAL PHASE 1 AT DIFFERENT COLUMN TEMPERATURES AND AN n-HEXANE MOBILE PHASE

Solute Pair	Separation Factor, α $(k'_{planar}/k'_{nonplanar})$			
	0°C	40°C	60°C	80°C
Triphenylene/o-terphenyl	5.71	5.06	4.78	4.57
Benzo[a]pyrene/TBN	2.03	1.60	1.45	1.33
Dibenzo[ghi]perylene/dibenzo[c,g]phenanthrene	3.48	2.98	2.79	2.64
Coronene/PhPh	3.80	3.24	3.03	2.87

Source: Ref. 61. Reproduced from Y. Saito et al. in *J. Microcolumn Sep.* **7**, 41 (1995) by permission of John Wiley and Sons, New York, NY, USA.

of the bonded ligands. A similar observation was reported by Sander and Wise [35], who demonstrated with polymeric-type ODS phases a shape selectivity change similar to the present trend. In their theory, the bonded phase structure of ODS phases synthesized by a trifunctional silanization reagent in the presence of water are regarded as an ordered structure, a "slotlike" structure at ambient temperature. The ordering of these phases decreases with increasing column temperature, with the result that the selectivity of the polymeric phase becomes similar to that of monomeric ODS phases lacking the "slotlike" surface structure.

Finally, Figure 3.52 proposes two interaction models between solutes and liquid crystal phase 1. At ambient or lower temperatures, planar rodlike PAHs interact effectively with the bonded phases having an ordered structure (Figure 3.52A), while nonplanar molecules interact less well (Figure 3.52B). Fullerenes such as C_{60} can be retained longer than the PAHs at elevated column temperature, because the ligand interval of phase 1 (≈ 6.9 Å) is assumed to be smaller than the molecular size of fullerenes at ambient temperature. Since, however, the interval increases with increasing temperature, interaction between C_{60} and the bonded phase is easier, as shown schematically drawn in Figure 3.52C. Although further investigation will be

Table 3.13B   SEPARATION FACTORS BETWEEN RODLIKE AND SQUARELIKE PAHs WITH LIQUID CRYSTAL PHASE 1 AT DIFFERENT COLUMN TEMPERATURES AND AN n-HEXANE MOBILE PHASE

Solute Pair	Separation Factor, α $(k'_{rodlike}/k'_{squarelike})$			
	0°C	40°C	60°C	80°C
Anthracene/phenanthrene	1.47	1.27	1.20	1.13
Benz[a]anthracene/triphenylene	1.55	1.35	1.27	1.21
Chrysene/triphenylene	2.17	1.69	1.53	1.40
Naphthacene/triphenylene	16.6	8.32	6.30	4.93

Source: Ref. 61.

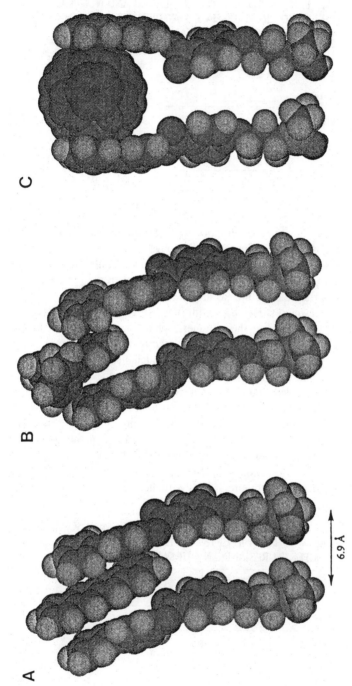

6.9 Å

Figure 3.52 Molecular modeling of the interactions between solutes and the bonded phase: (A) naphthacene with liquid crystal phase 1 having a slotlike ordered structure at lower temperature, (B) *o*-terphenyl with liquid crystal phase 1 having a slotlike ordered structure at lower temperature, and (C) C₆₀ with liquid crystal phase 1 at higher temperature [61]. Diagrams rendered by Chem 3D+ software. Reproduced from Y. Saito et al. in *J. Microcolumn Sep.* **7**, 41 (1995) by permission of John Wiley and Sons, New York, NY, USA.

required to identify the origin of these abnormal temperature effects, the current results indicate not only the possibility of effective preparative separation of fullerenes with this bonded phase using a toluene-rich mobile phase but also important information about the retention mechanism of the probes on this phase.

3.7 Multilegged Stationary Phases for Fullerene Separation

3.7.1 Basic Concept of Multilegged Phases

Another conclusion that can be drawn is that the phenyl group, in addition to having appropriate interligand distance and longer alkyl chain length, is a key ingredient in fullerene separations. All the stationary phases evaluated thus far have been vertical. Figure 3.53A shows a typical example of the vertical type of chemically bonded phase. We can design the phase from a different approach, as well, such that the bonded phase is situated horizontally relative to the silica surface (Figure 3.53 [63]). As described in the preceding section, the phenyl groups at the bottom part of the ODS bonded phase contribute to the retention of fullerenes. It seems to be necessary to insert phenyl groups in the bonded phase to enhance the retention and the selectivity for fullerene separations. The details of the "Multilegged" bonded phases chosen for the purposes of evaluation are discussed in Chapter 2. Each phenyl moiety is in the cavity-like structure lying parallel to the silica surface. The walls of the cavity can be made by several methyl groups attached to two or three silicon atoms, which are bonded to the silica surface and to phenyl groups such as phenyl, biphenyl, benzyl, and anthracene groups. BP, BMB, and BBB were made to investigate the contribution to the retention of the biphenyl group, methylene groups, and varying cavity size. An anthracene moiety was also added to investigate a different geometrical contribution to the retention.

TP was tried first for the separation of C_{60} and C_{70} because this stationary phase has been shown to be very effective for the separation of planar and nonplanar small PAHs (see Chapter 2). Moreover, its cavity size is the most suitable for the retention of nonplanar small PAHs. Compared to ODS phases, TP has been found to have a reversed elution order in o-terphenyl and triphenylene. This retention behavior is reasonably explained by the molecular modeling interpretation.

In accordance with the same concept described in Chapter 2, one can expect to encounter the slot-like or cavity-fit mechanism for a C_{60} and C_{70} separation with multilegged bonded phases. The molecular size of C_{60} is approximately 7 Å × 7 Å, with the carbon atoms arranged like a soccer ball, and C_{70} is approximately 9 Å × 7 Å with the carbons arranged like a rugby ball. Therefore, one would expect that the TP phase ($\approx$5 Å × 5 Å), where methyl groups and one phenyl ring form the cavity, would be too small to accept either of these fullerenes, and thus the retention behavior should be similar for both solutes. Experimental results thus far clearly agree with this hypothesis; no separation has been found for the fullerenes. With the BP phase, in contrast, the size of the cavity is about 9–10 Å, and the size of C_{70}

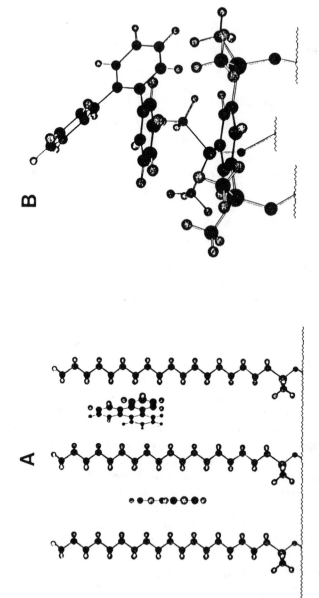

Figure 3.53 Schematic diagram of two different chemically bonded stationary phases: (A) vertical situation, typical bonded phase structure and (B) horizontal situation, multilegged bonded phase structure.

should make it a better fit to the cavity of the BP phase. The retention of C_{70} should be longer than that of C_{60}, and the separation should be improved. The result shown in Figure 3.54 supports this prediction. In addition, since the interaction between the BP phase and the solutes in the sense of molecular recognition results in the retention of both solutes, the BP phase has a large relative retention value for the solutes and large retention factors for both solutes. By the same token, this concept of molecular–molecular interaction is useful in the design of novel stationary phases that offer specific and/or selective separation of other fullerenes.

3.7.2 Retention Behaviors of Fullerenes with Multilegged Phases

Based on the foregoing concept, other multilegged phases (Figure 3.55) were also evaluated, and the results indicate that this mechanism works very well for fullerene separations. Of the phases studied, BMA has the smallest coverage because of its difficult bonding chemistry, which indices the lowest retention for C_{60} and C_{70} despite the high selectivity of this phase.

The retention data are summarized in Table 3.14. The BBB and BMB phases retain C_{60} and C_{70} more than either BP or ODS. The BMB phase shows the longest retention and good selectivity for C_{60} and C_{70}. The difference between BBB and BMB is in the location of the methylene groups in their structures. BMB contains a diphenyl group that can interact with fullerenes easily because movement of diphenyl is allowed. Two methylene groups between the diphenyl group and the silicon atom can work as a buffer for such movement, which is similar to the

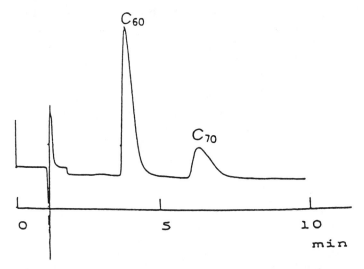

Figure 3.54   Chromatogram for the separation of C_{60} and C_{70} with the BP phase: column size, 4.6 mm i.d. × 150 mm; mobile phase, *n*-hexane; flow rate, 1 mL/min; detection, UV at 320 nm; temperature, 30°C.

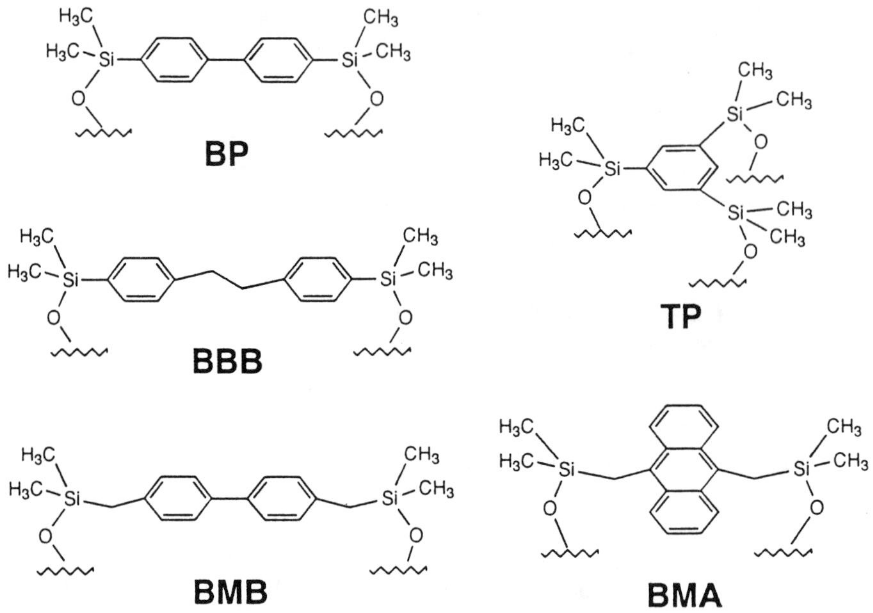

Figure 3.55   Structure of the multilegged bonded phases evaluated [63]. Reproduced from K. Jinno et al. in *J. Chromatogr. A.*, **691**, (1995) by permission of Elsevier Science-NL, Amsterdam, The Netherlands.

entrapment routine of a sea anemone. As a result, the phase can interact with fullerenes through π-π interaction. The benzyl group of the BBB phase also lacks the ability to move as freely because one edge is bonded to a silicon atom. The BP phase also does not have the freedom of movement expected in the case of the BMB phase. Therefore BMB is the most suitable stationary phase for fullerene separations. The mechanism can be seen in Figure 3.56. The BMA phase also has freedom

Table 3.14   RETENTION DATA FOR THE SEPARATION[a]
OF C_{60} AND C_{70} WITH VARIOUS
STATIONARY PHASES

| Stationary Phase | Retention Factor, k' | | Separation Factor, α $(k'_{C_{70}}/k'_{C_{60}})$ |
	C_{60}	C_{70}	
BP	0.75	1.22	1.63
BBB	2.36	4.26	1.80
BMB	5.36	12.0	2.23
BMA	0.24	0.96	2.21
Develosil ODS-5	0.87	1.60	1.84

[a] Mobile phase. *n*-hexane; column temperature. 30°C.

Source: Ref. 50. Reproduced from H. Ohta et al. in *Chromatographia* **40**. 507 (1995) by permission of Vieweg Publishing. Wiesbaden. Germany.

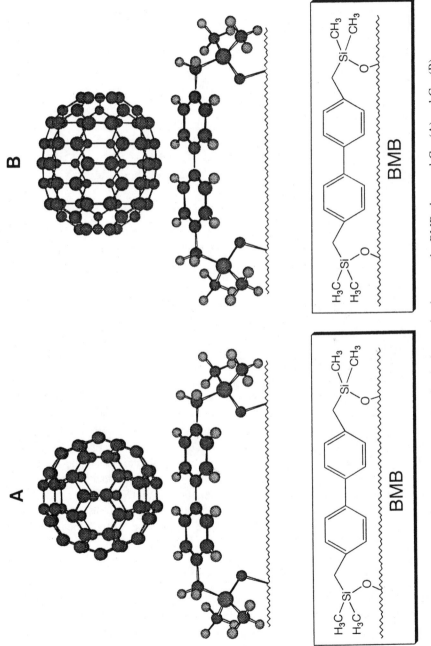

Figure 3.56 Molecular modeling for the interaction between the BMB phase and C_{60} (A) and C_{70} (B).

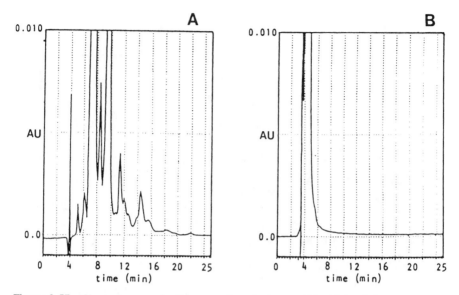

Figure 3.57 Chromatograms for the separation of mixture of the higher fullerenes wit the BMB and monomeric ODS phases: (A) BMB phase and (B) Develosil ODS-5 (both in 4.6 mm i.d. × 250 mm column); mobile phase, n-hexane/toluene (70:30) [50]. Reproduced from H. Ohta et al. in *Chromatographia* **40**, 507 (1995) by permission of Vieweg Publishing, Wiesbaden, Germany.

of the movement because the anthracene moiety is not connected to the silicon atom, and therefore this phase shows high selectivity. Its low surface coverage does not permit long retention, however.

Another practical possibility has been evaluated using BMB. Because fullerenes are only slightly soluble in most organic solvents, the mobile phase for preparative-scale separations can be selected from only a very few solvents (e.g., toluene). However, since pure toluene does not give a good separation, one needs to use a mobile phase with toluene as a modifier. To determine the practicality of using a BMB phase, the mobile phase of a toluene/n-hexane mixture was applied to BMB, and the two chromatograms shown in Figure 3.57 were obtained. With 70:30 n-hexane/toluene, BMB gave good separation and selectivity for higher fullerenes; therefore this phase can be used for preparative-scale separations. On the other hand, monomeric ODS (Develosil ODS-5) gave only one peak for all the fullerenes.

3.8 Copper–Phthalocyanine Phases

A new stationary phase with a tetraphenylporphyrin bonded stationary phase [64] that can induce π-π interaction between fullerenes and the bonded phase was shown to be very powerful when toluene was used as the major mobile phase constituent [63]. Stationary phases very similar in type to the tetraphenylporphyrin phase were evaluated for the separation of polycyclic aromatic compounds, and those phases should be examined to ascertain their performance for fullerene separation [65].

This section describes two kinds of copper–phthalocyanine (Cu-PCS) bonded phase for fullerene separation in LC.

Two kinds of Cu-PCS were synthesized in the laboratory [56, 57]. Figure 3.58 illustrates the synthetic procedure for Cu-PCS phases. The materials are based on a commercial aminopropyl-bonded phase Develosil-NH$_2$ (Nomura Chemicals, Seto, Japan). As shown in Figure 3.58 Cu-phthalocyanine is reacted with chlorosulfonic acid at 135–150°C for 2–3 hours. The Cu-phthalocyanine tetrasulfonylchloride obtained is then refluxed in dioxane with the aminopropyl silica, and the stationary phase Cu-PCS is produced. Next the Cu-PCS is subjected to the end capping process for the residual aminopropyl groups on the silica surface using *n*-butanoyl(butyryl) chloride and *n*-decanoylchloride, as shown in Figure 3.59. In this section, the former product is called Butanoyl Cu-PCS and the latter is Decanoyl Cu-PCS. Develosil ODS-5 (monomeric ODS phase) was also evaluated as the reference compound. All

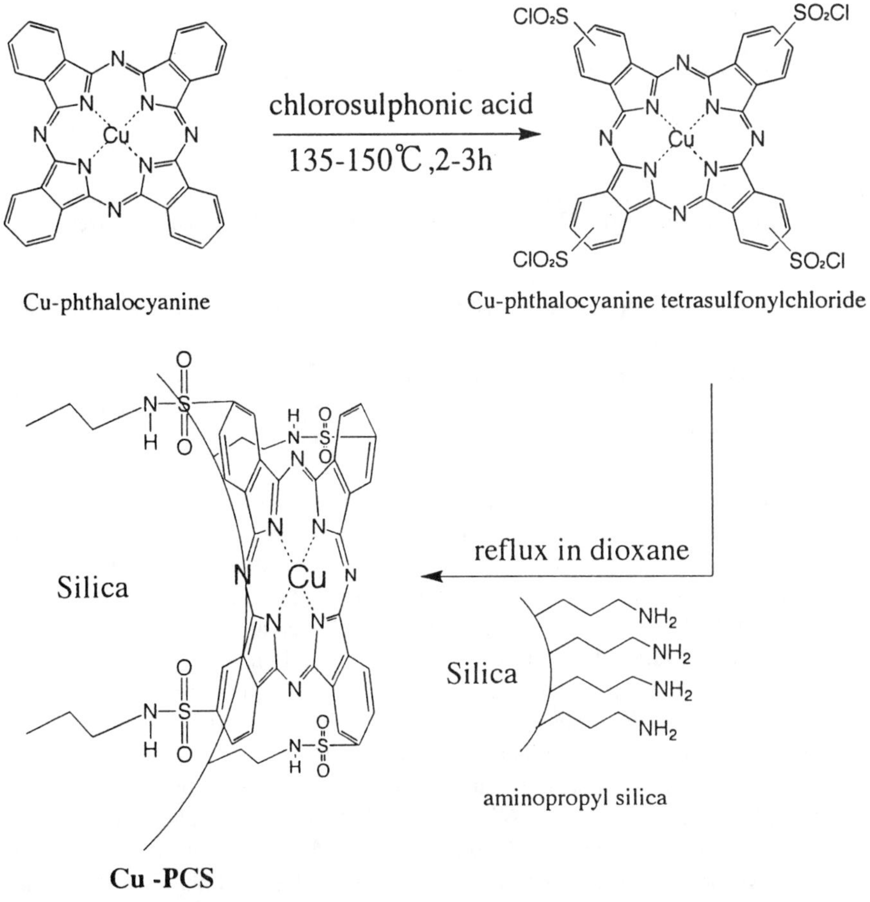

Figure 3.58 Synthetic process of Cu-PCS phases immobilized on an aminopropyl silica bonded phase (Develosil-NH$_2$) [67]. Reproduced from K. Jinno et al. in *J. Microcolumn Sep.* **8**, 13 (1996) by permission of John Wiley and Sons, New York, NY, USA.

R= CH$_3$CH$_2$CH$_2$COCl : Butanoyl chloride

CH$_3$(CH$_2$)$_8$COCl : Decanoyl chloride

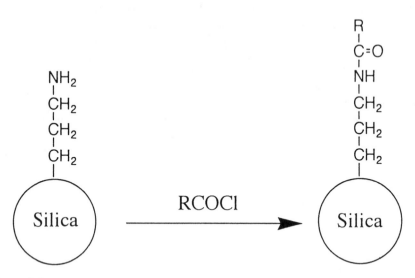

Figure 3.59 End capping process for the Cu-PCS phase [67]. Reproduced from K. Jinno et al. in *J. Microcolumn Sep.* **8**, 13 (1996) by permission of John Wiley and Sons, New York, NY, USA.

three stationary phases (Butanoyl Cu-PCS, Decanoyl Cu-PCS, and Develosil-ODS) were packed into a series of fused-silica capillaries, each 0.53 mm i.d. × 150 mm long, by means of the slurry technique for C$_{60}$ and C$_{70}$ separations.

The microcolumn LC system was used, and the mobile phases were *n*-hexane, *n*-hexane/toluene, and *n*-hexane/*cis*- or *trans*-decalin mixtures under isocratic conditions.

The separation of two basic fullerenes, C$_{60}$ and C$_{70}$, was examined with three phases, Decanoyl Cu-PCS, Butanoyl Cu-PCS, and ODS. Typical chromatograms with those stationary phases, and *n*-hexane as the mobile phase, are shown in Figure 3.60. The retention data, summarized in Table 3.15, indicate that Decanoyl Cu-PCS gives larger retention than the Butanoyl and ODS phases, and the separation factors with Butanoyl are larger than those with Decanoyl and ODS. However, it is also found that increasing toluene concentration in the mobile phase reduces the difference between two Cu-PCS phases. The results can be interpreted by examining the molecular structures of the stationary phases.

Figure 3.61 represents schematically the molecular interactions between C$_{60}$, C$_{70}$, and three stationary phases. With Develosil ODS the main interaction controlling the retention should be made between the alkyl chains and the solute based on so-called hydrophobic interaction. On the other hand, Cu-PCS phases offer mainly two interactions for fullerene retention: one is the same as the ODS phase "hydrophobic interaction" between the alkyl chains inserted into the separation system by

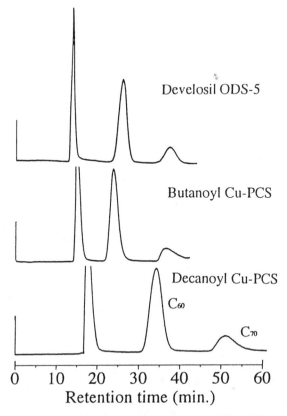

Figure 3.60 Chromatograms for the room temperature separation of C_{60} and C_{70} with three stationary phases: mobile phase, *n*-hexane; flow rate, 2 μL/min; detection, 320 nm.

the end capping process and the solute, and another is the most important "π-π interaction" between the phthalocyanine moiety and the fullerene. Since with Butanoyl Cu-PCS the end-capped alkyl chain length is shorter than that with Decanoyl Cu-PCS, the dominant interaction with the former is "π-π." With the latter, the stereo situation of the alkyl chains can enhance the interaction between the phthalocyanine moiety and the fullerene, because the alkyl chains around the Cu-PCS moiety can also interact with the solute more easily than the chains of the Butanoyl Cu-PCS phase. Therefore the retention values for fullerenes with Decanoyl are larger than those for Butanoyl, but Butanoyl Cu-PCS provides better separation for fullerenes because it can offer the pure "π-π interaction" for the retention; moreover, only C_{70} is able to interact with short Butanoyl alkyl chains around the Cu-PCS moiety. It has been convincingly demonstrated that the bonded phase structure and the associated characteristics can control the retention of fullerenes well and that this kind of concept can be applied to the design of much better stationary phases for the separation of fullerenes.

Table 3.15   EFFECT OF TOLUENE CONCENTRATION IN THE MOBILE PHASE ON THE RETENTION FACTORS OF C_{60} AND C_{70} AND THEIR SEPARATION FACTORS, α $(k'_{C_{70}}/k'_{C_{60}})^a$

Stationary Phase		n-Hexane/Toluene				
		100:0	97:3	95:5	93:7	90:10
Develosil	$k'_{C_{60}}$	0.811	0.612	0.575	0.483	0.379
ODS-5	$k'_{C_{70}}$	1.540	1.131	1.046	0.864	0.667
	α	1.90	1.85	1.82	1.79	1.76
Decanoyl	$k'_{C_{60}}$	0.940	0.802	0.683	0.598	0.512
Cu-PCS	$k'_{C_{70}}$	1.766	1.391	1.156	0.984	0.815
	α	1.88	1.73	1.69	1.64	1.59
Butanoyl	$k'_{C_{60}}$	0.757	0.640	0.555	0.483	0.408
CU-PCS	$k'_{C_{70}}$	1.742	1.262	1.018	0.850	0.698
	α	2.30	1.97	1.83	1.76	1.71

[a] Data obtained at room temperature except for data with Butanoyl Cu-PCS, which were obtained at 20°C.

To learn more about the performance of Cu-PCS phases for the separation of fullerenes, the effect of mobile phase compositions was evaluated. Because of the limited solubility of fullerenes in n-hexane, other mobile phases, including better solvents for fullerenes, should be used if preparative-scale LC is required. Ruoff et al. [62], who reported the solubility of fullerenes in various organic solvents, found that toluene and decalin are better solvents than n-hexane on the basis of solubility: C_{60} solubility (mg/mL) was 0.043 for n-hexane, 2.8 for toluene, 4.6 for decalin (cis/trans = 3/7), 2.2 for cis-decalin, and 1.3 for trans-decalin. Therefore adding toluene or decalin cis-, trans-, and mixture) to n-hexane has been tried. The results with toluene was summarized in Table 3.15, and Tables 3.16A–3.16C list those with decalins. From the experimental standing point, decalin is a better mobile phase solvent than toluene because there is less UV absorption of decalin than of toluene and because the background noise can be reduced. In this evaluation cis-decalin, trans-decalin, and two commercial mixtures of cis- and trans-decalin were used, and a 1:1 mixture was made in the laboratory. Typical chromatograms with Decanoyl Cu-PCS using those mobile phases are shown in Figure 3.62. The results in Tables 3.15 and 3.16 and Figure 3.62 indicate that decalin is a better solvent than toluene in separation performance, and trans-decalin is better than the cis isomer. The maximum amount of toluene or decalin that can be added to n-hexane is only 10% with the Cu-PCS phases, and even though the results are promising, preparative use is not yet feasible in terms of either cost or performance.

Another factor to be considered in attempts to improve the separation performance for fullerenes is temperature. It has been found that column temperature

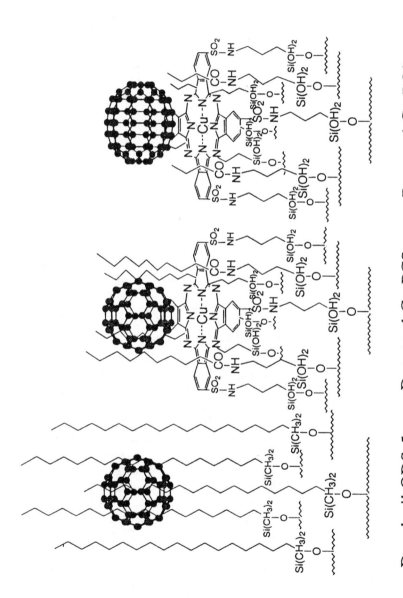

Develosil ODS-5   **Decanoyl Cu-PCS**   **Butanoyl Cu-PCS**

Figure 3.61 Schematic molecular interactions between three stationary phases, and C_{60} and C_{70} [67]. Reproduced from K. Jinno et al. in *J. Microcolumn Sep.* **8**, 13 (1996) by permission of John Wiley and Sons, New York, NY, USA.

Table 3.16A   EFFECT OF DECALIN CONCENTRATION IN THE MOBILE PHASE ON THE RETENTION FACTORS OF C_{60} AND C_{70} AND THEIR SEPARATION FACTORS, α $(k'_{C_{70}}/k'_{C_{60}})$, WITH DEVELOSIL ODS-5 AS THE STATIONARY PHASE

		n-Hexane/Decalin			
		100:0	95:5	90:10	85:15
Decalin[a]	$k'_{C_{60}}$	0.811	0.640	0.499	0.390
	$k'_{C_{70}}$	1.54	1.152	0.873	0.663
	α	1.90	1.80	1.75	1.70
cis-Decalin	$k'_{C_{60}}$		0.595	0.488	0.352
	$k'_{C_{70}}$		1.076	0.839	0.585
	α		1.81	1.72	1.66
trans-Decalin	$k'_{C_{60}}$		0.684	0.528	0.429
	$k'_{C_{70}}$		1.250	0.922	0.735
	α		1.83	1.75	1.71
cis/trans-Decalin (50:50)	$k'_{C_{60}}$		0.651	0.512	0.404
	$k'_{C_{70}}$		1.191	0.898	0.680
	α		1.83	1.75	1.68

[a] Commercial mixture of cis- and trans-decalins.

influences the separation of fullerenes strongly. Higher temperature offers the possibility of increasing the solubility of fullerenes into the mobile phase solvents. Therefore the effect of column temperature on the retention of C_{60} and C_{70} was examined, and the results are summarized in Table 3.17. The plots of the temperature and log k' for three phases with an n-hexane mobile phase appear in Figure 3.63. With ODS phase increasing the temperature the retentions of C_{60} and C_{70} are decreasing, while with Decanoyl and Butanoyl Cu-PCS phases the retentions are rather increasing with increasing the temperature. The chromatograms in Figure 3.64 also show better performance at higher column temperature with the Butanoyl Cu-PCS phase. Figure 3.65 presents the relationships between the separation factors for C_{60} and C_{70} with three phases and the column temperature. Cu-PCS phases gave higher selectivity with increasing temperature, whole ODS gave the opposite effect. This unusual behavior (increasing the temperature while increasing the retention) also was found by Pirkle and Welch with their column [31] for C_{60} and C_{70} separations, and in this case the mechanism can be interpreted as follows: Cu-PCS phases offer two retention mechanisms, one is "lipophilic" between the alkyl chains and fullerenes and the other is "π-π" between the phthalocyanine moiety and fullerenes,

MOLECULAR RECOGNITION FOR FULLERENES IN LIQUID CHROMATOGRAPHY   **227**

Table 3.16B   EFFECT OF DECALIN CONCENTRATION IN THE MOBILE PHASE ON THE RETENTION FACTORS OF C_{60} AND C_{70} AND THEIR SEPARATION FACTORS, α $(k'_{C_{70}}/k'_{C_{60}})$, WITH DECANOYL Cu-PCS AS THE STATIONARY PHASE

		n-Hexane/Decalin			
		100:0	95:5	90:10	85:15
Decalin 1[a]	$k'_{C_{60}}$	0.940	0.787	0.640	0.538
	$k'_{C_{70}}$	1.760	1.408	1.091	0.893
	α	1.83	1.79	1.71	1.66
Decalin 2[b]	$k'_{C_{60}}$		0.784	0.627	0.530
	$k'_{C_{70}}$		1.370	1.056	0.864
	α		1.74	1.68	1.63
cis-Decalin	$k'_{C_{60}}$		0.716	0.592	0.514
	$k'_{C_{70}}$		1.246	0.995	0.838
	α		1.74	1.68	1.63
trans-Decalin	$k'_{C_{60}}$		0.812	0.687	0.553
	$k'_{C_{70}}$		1.451	0.182	0.927
	α		1.79	1.72	1.66
cis/trans-Decalin (50:50)	$k'_{C_{60}}$		0.786	0.658	0.525
	$k'_{C_{70}}$		1.385	1.099	0.867
	α		1.76	1.69	1.65

[a]   Commercial mixture of cis- and trans-decalins.
[b]   Another commercial mixture of cis- and trans-decalins.

and increasing the temperature can induce an increase in the mobility of the alkyl chains around the phthalocyanine moiety, followed by a reduction in the interference with the interaction of fullerenes and the phthalocyanine. Butanoyl Cu-PCS can elicit such an effect better than the Decanoyl phase because the former has shorter end-capped alkyl chains. As a result, one can conclude that increasing the column temperature with Cu-PCS phases offers two big advantages for the preparative-scale LC separation of fullerenes: increasing the solubility into the mobile phase (by means of higher temperature) and enhancing separation performance (by means of higher separation factors).

To evaluate our conclusion, the effect of mobile phase compositions on separation performance was examined again, at 30 and 80°C with Butanoyl Cu-PCS. The results appear in Figure 3.66, which shows three chromatograms with mixtures of

Table 3.16C   EFFECT OF DECALIN CONCENTRATION IN THE MOBILE PHASE ON THE RETENTION FACTORS OF C_{60} AND C_{70} AND THEIR SEPARATION FACTORS, α $(k'_{C_{70}}/k'_{C_{60}})$, WITH BUTANOYL Cu-PCS AS THE STATIONARY PHASE

		n-Hexane/Decalin			
		100:0	95:5	90:10	85:15
Decalin[a]	$k'_{C_{60}}$	0.757	0.655	0.523	0.439
	$k'_{C_{70}}$	1.742	1.278	0.936	0.755
	α	2.30	1.95	1.81	1.74
cis-Decalin	$k'_{C_{60}}$		0.634	0.501	0.420
	$k'_{C_{70}}$		1.230	0.897	0.727
	α		1.94	1.79	1.73
trans-Decalin	$k'_{C_{60}}$		0.662	0.546	0.452
	$k'_{C_{70}}$		1.311	1.005	0.796
	α		1.98	1.84	1.76
cis/trans-Decalin (50:50)	$k'_{C_{60}}$		0.645	0.525	0.435
	$k'_{C_{70}}$		1.271	0.961	0.757
	α		1.97	1.83	1.74

[a] Commercial mixture of cis- and trans-decalins.

n-hexane, toluene, and decalin as the mobile phases. It appears that the separation with n-hexane/toluene (90/10) at 30°C is very similar to that with n-hexane/decalin (85/15) at 80°C. This means that the solubility in the mobile phase can be increased with similar separation performance at higher temperatures with the Butanoyl Cu-PCS phase.

3.9 C_{60} Bonded Phase

As discussed in Chapter 2, the C_{60} bonded phase has a unique ability to assist in the molecular shape and size recognition of PAHs [68]. Since the bonded phase moiety has a bulky, roundlike shape, it should give unusual retention behavior for fullerene separation. As an extension of earlier descriptions, the performance of this unique phase must be evaluated with respect to the separation of fullerenes. The separation has been tried using microcolumn technology.

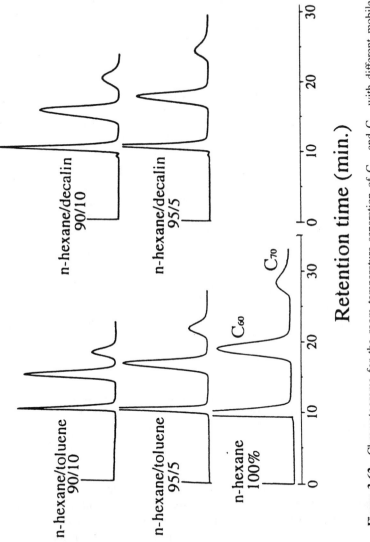

Figure 3.62 Chromatograms for the room temperature separation of C_{60} and C_{70} with different mobile phase systems: the decalin used is the one commercial *cis/trans* mixture cited in note *a*, Table 3.16A; stationary phase, Decanoyl Cu-PCS; flow rate, 4 µL/min; detection, 320 nm.

Table 3.17   EFFECT OF COLUMN TEMPERATURES
ON THE RETENTION FACTORS C_{60} AND
C_{70} AND THE SEPARATION FACTORS
FOR THREE STATIONARY PHASES[a]

Temp. (°C)	Retention Factor, k'		Separation Factor, α $(k'_{C_{70}}/k'_{C_{60}})$
	$k'_{C_{60}}$	$k'_{C_{70}}$	
Develosil ODS-5			
20	0.863	1.631	1.89
30	0.736	1.347	1.83
40	0.666	1.180	1.77
50	0.603	1.044	1.73
60	0.568	0.953	1.68
70	0.526	0.861	1.64
80	0.498	0.804	1.61
Decanoyl Cu-PCS			
20	0.922	1.696	1.84
30	1.017	1.917	1.89
40	1.037	2.034	1.96
50	1.135	2.257	1.99
60	1.192	2.409	2.02
70	1.270	2.608	2.05
80	1.299	2.720	2.09
Butanoyl Cu-PCS			
20	0.757	1.742	2.30
30	0.817	1.926	2.36
40	0.855	2.078	2.43
50	0.874	2.153	2.46
60	0.952	2.363	2.48
70	0.997	2.503	2.51
80	1.056	2.649	2.51

[a] Other conditions as follows: mobile phase, 100% *n*-hexane; flow rate, 2 μL/min; detection at 300 nm.

Figure 3.67 shows a typical chromatogram for the separation of C_{60} and C_{70} fullerene molecules with the C_{60} bonded phase; the retention data for *n*-hexane/ toluene mobile phase systems are summarized in Table 3.18. In spite of strong mobile phase conditions such as 50% toluene in *n*-hexane, the phase possesses excellent retentivity for both fullerenes, while with the same mobile phase compositions a typical ODS phase could not keep such large retention values for these solutes. Although further investigation should be carried out for the retention mechanism with the C_{60} bonded silica phase, it can be said that the C_{60} phase has exotic selectivity for PAHs and is potentially usable for other difficult separation problems entailing geometrical isomers.

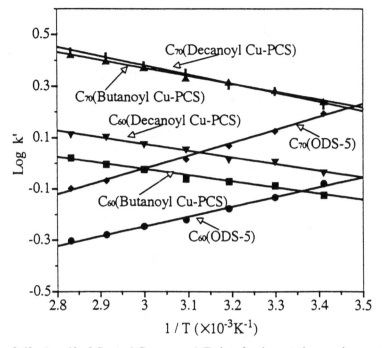

Figure 3.63 Log k' of C_{60} and C_{70} versus $1/T$ plots for three stationary phases at room temperature: mobile phase, n-hexane; flow rate, 2 μL/min; detection, 320 nm.

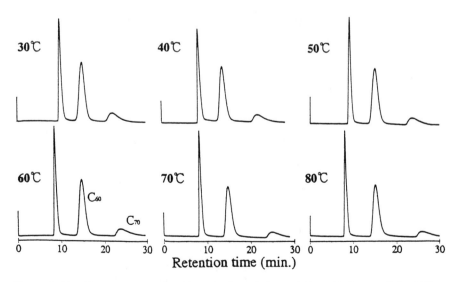

Figure 3.64 Chromatograms for the separation of C_{60} and C_{70} with Butanoyl Cu-PCS at various column temperatures: mobile phase, n-hexane; flow rates, 4 μL/min; detection, 320 nm [67]. Reproduced from K. Jinno et al. in *J. Microcolumn Sep.* **8**, 13 (1996) by permission of John Wiley and Sons, New York, NY, USA.

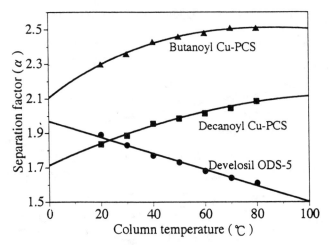

Figure 3.65 Relationships between the separation factors for C_{60} and C_{70} with three stationary phases and column temperature: mobile phase, *n*-hexane; flow rate, 2 μL/min; detection, 320 nm [67]. Reproduced from K. Jinno et al. in *J. Microcolumn Sep.* **8**, 13 (1996) by permission of John Wiley and Sons, New York, NY, USA.

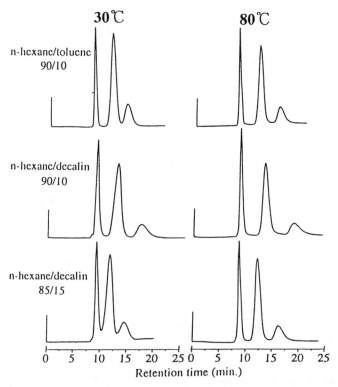

Figure 3.66 Effect of column temperature on the separation of C_{60} and C_{70} with Butanoyl Cu-PCS stationary phase using three mobile phase compositions at a flow rate of 4 μL/min; detection, 320 nm [67]. Reproduced from K. Jinno et al. in *J. Microcolumn Sep.* **8**, 13 (1996) by permission of John Wiley and Sons, New York, NY, USA.

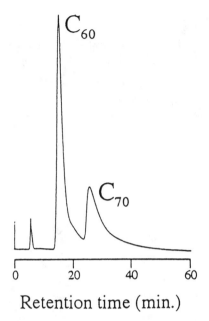

Retention time (min.)

Figure 3.67   Typical chromatogram for the separation of C_{60} and C_{70} with the C_{60} bonded silica stationary phase: mobile phase, n-hexane/toluene (60:40) [68]. Reproduced from Y. Saito et al. in *J. High Resolu. Chromatogr. Chromatogr. Commun.* **18**, 569 (1995) by permission of Huethig Verlag, Heidelberg, Germany.

3.10 Conclusion

Thus the design of a novel stationary phase for enhancing fullerene selectivity calls for a consideration of many points, as discussed here. These include:

1. If the bonded phase is a nonaromatic moiety, longer alkyl chains are preferable

Table 3.18   RETENTION DATA FOR THE SEPARATION OF C_{60} AND C_{70} WITH THE C_{60} BONDED SILICA PHASE USING DIFFERENT MOBILE PHASE COMPOSITIONS

Mobile Phase n-Hexane/Toluene	Retention Factor, k'		Separation Factor, α $(k'_{C_{70}}/k'_{C_{60}})$
	C_{60}	C_{70}	
90:10	6.04	17.6	2.91
80:20	4.00	10.3	2.58
70:30	2.34	5.42	2.32
60:40	1.63	3.56	2.18
50:50	1.18	2.36	2.00

Source: Ref. 68. Reproduced from Y. Saito et al. in *J. High Resolu. Chromatogr. Chromatogr. Commun.* **18**, 569 (1995) by permission of Huethig Verlag, Heidelberg, Germany.

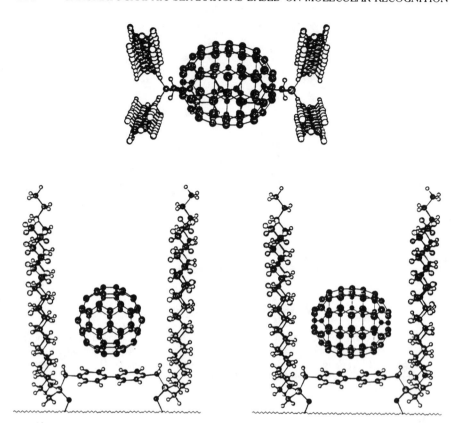

Figure 3.68   This novel stationary phase, proposed by the investigations described here, should be made from the silane compound 4,4′-bis(dioctadecylchlorosilylmethyl)biphenyl [63]. Reproduced from K. Jinno et al. in *J. Chromatogr. A.*, **691**, (1995) by permission of Elsevier Science-NL, Amsterdam, The Netherlands.

and the distance between each bonded moiety is the most important to enhancing selectivity.

2. If a separation mechanism based on molecular recognition (e.g., fit of size or shape) is expected, multilegged phases are the most highly preferred phases because they can produce a cavity-like structure on the silica surface. The diphenyl group should be included in the structure.

3. Combining points 1 and 2 will produce the most desirable novel phase, which has long alkyl chains and diphenyl groups at the bottom part of the cavity-like structure.

By incorporating these points into the structure of the bonded phase as shown in Figure 3.68, the most promising phase may be obtained. The long alkyl chain in this case is C_{18}, but a chain shorter than C_{18} probably could do a better job of catching fullerenes and form the cavity walls as well. Biphenyl at the bottom part of a multilegged structure can catch fullerenes by π-π interaction when the solutes enter the cavity. The biphenyl group and methylene group also work to control the distance between alkyl chains. After this novel phase has been synthesized in the laboratory, its performance should be confirmed by actual chromatographic experiments. In this way, the systematic approach for designing novel phases to solve difficult separation problems will be established, and the foregoing discussion will be recognized as the correct approach.

Acknowledgments

The author sincerely thanks the following people who contributed to work on the projects and discussed the results with deep insights: Dr. J.C. Fetzer and Dr. W.R. Biggs, Chevron Research and Technology Company; Professor J.J. Pesek, Department of Chemistry, San Jose State University; and Dr. Y.-L. Chen, J&W Scientific, all of California, and, in Japan, Dr. N. Nagae, Nomura Chemicals, Seto; Professor H. Nagashima, Toyohashi University of Technology, Toyohashi; Professor K. Itoh, Nagoya University, Nagoya; and Y. Saito, H. Ohta, T. Uemura, and K. Nakagawa, research students in the author's laboratory. Financial support from the Japanese Ministry of Education and Culture grant nos. 05233108 and 06640781 is also acknowledged.

References

1. E. Osawa, *Kagaku,* **25,** 854 (1970).

2. H.W. Kroto, J.R. Heath, S.C. O'Brien, R.F. Curl, and R.E. Smalley, *Nature,* **318,** 162 (1985).

3. W. Kratschmer, K. Fostiropoulos, and D.R. Huffman, *Chem. Phys. Lett.* **170,** 167 (1990).

4. R. Taylor, J.R. Hare, A.K. Abdul-Sada, and H.W. Kroto, *J. Chem. Soc., Chem. Commun.* 1423 (1990).

5. J.R. Heath, S.C. O'Brien, Q. Zhang, Y. Lin, R.F. Curl, H.W. Kroto, and R.E. Smalley, *J. Am. Chem. Soc.* **107,** 7779 (1985).

6. J.M. Hawkins, T.A. Lewis, S.D. Loren, A. Meyer, J.R. Heath, Y. Shibato, and R.J. Saykally, *J. Org. Chem.* **55,** 6520 (1990).

7. R.D. Johnson, G. Meijer, and D.S. Bethune, *J. Am. Chem. Soc.* **112,** 8983 (1990).

8. H. Aije, M.M. Alvarez, S.J. Ang, R.D. Beck, F. Diederich, K. Fostirofoulos, R.D. Huffman, W. Kratschmer, Y. Rubin, K.E. Schriver, K. Sensharma, and R.L. Whetten, *J. Phys. Chem.* **94,** 8630 (1990).

9. R.D. Johnson, G. Meijer, J.R. Salem, and D.S. Bethune, *J. Am. Chem. Soc.* **113,** 3619 (1991).

10. P.W. Fowler, P. Lazzetti, M. Malagoli, and R. Zanasi, *Chem. Phys. Lett.* **179,** 174 (1991).

11. C.I. Frum, R. Engleman, H.G. Hedderich, P.F. Bernath, L.D. Lamb, and D.R. Huffman, *Chem. Phys. Lett.* **176,** 504 (1991).

12. J.R. Hare, T.J. Dennis, H.W. Kroto, R. Taylor, A.W. Allaf, S.P. Balm, and D.R.M. Walton, *J. Chem. Soc., Chem. Commun.* **179,** 412 (1991).

13. D.S. Bethune, G. Meijer, W.C. Tang, H.J. Rosen, W.G. Golden, H. Seki, C.A. Brown, and M.S. de Vries, *Chem. Phys. Lett.* **179,** 181 (1990).

14. J.P. Hare, H.W. Kroto, and R. Taylor, *Chem. Phys. Lett.* **170,** 167 (1990).

15. C. Reber, L. Yee, J.I. McKieman, J.I. Zink, R. Williams, N.W. Tong, D.A.A. Dahlberg, R.L. Whetten, and F.N. Diederich, *J. Phys. Chem.* **95**, 2127 (1991).

16. D.S. Bethune, G. Meijer, W.C. Tang, and J.H. Rosen, *Chem. Phys. Lett.* **174**, 181 (1990).

17. P.M. Allenmand, A. Koch, and F. Wudl, *J. Am. Chem. Soc.* **113**, 1050 (1991).

18. J.C. Fetzer, private communication.

19. J.C. Fetzer and E.J. Gallegos, *Polycyclic Aromat. Comp.* **2**, 245 (1992).

20. M. Diach, R.L. Hettich, R.N. Compton, and G. Guiiochon, *Anal. Chem.* **64**, 2143 (1992).

21. R.C. Klute, H.C. Dorn, and H.M. McNair, *J. Chromatogr. Sci.* **30**, 438 (1992).

22. K. Jinno, T. Uemura, H. Nagashima, and K. Itoh, *Chromatographia,* **35**, 38 (1993).

23. K. Jinno, T. Uemura, H. Nagashima, and K. Itoh, *High Resolution Chromatogr.* **15**, 627 (1992).

24. K. Jinno, T. Uemura, H. Ohta, H. Nagashima, and K. Itoh, *Chim. Oggi,* **12(9)**, 19 (1994).

25. K. Jinno, T. Uemura, H. Ohta, H. Nagashima, and K. Itoh, *Anal. Chem.* **65**, 2650 (1993).

26. H. Ohta, Y. Saito, K. Jinno, H. Nagashima, and K. Itoh, *Chromatographia,* **39**, 453 (1994).

27. F. Diederich, R.L. Whetten, C. Thilgen, R. Ettyl, I. Chao, and M.M. Alvarez, *Science,* **254**, 1768 (1991).

28. J.F. Anacleto and M.A. Quilliam, *Anal. Chem.* **65**, 2236 (1993).

29. D.M. Cox, S. Behal, M. Disko, S.M. Gorun, M.G. Greaney, C.S. Hsu, E.B. Kollin, J. Miller, R.D. Robbins, R.D. Sherwood, and P. Tindall, *J. Am. Chem. Soc.* **113**, 2940 (1991).

30. K. Jinno, K. Yamamoto, T. Ueda, H. Nagashima, and K. Itoh, *J. Chromatogr.* **594**, 105 (1992).

31. W.H. Pirkle and C.J. Welch, *J. Org. Chem.* **56**, 6973 (1991).

32. C.J. Welch and W.H. Pirkle, *J. Chromatogr.* **609**, 89 (1992).

33. K. Cabrera, G. Wieland, and M. Schafer, *J. Chromatogr.* **644**, 396 (1993).

34. L.C. Sander and S.A. Wise, *Anal. Chem.* **59**, 2309 (1987).

35. L.C. Sander and S.A. Wise, *Anal. Chem.* **61**, 1749 (1989).

36. K. Jinno, T. Ibuki, N. Tanaka, M. Okamoto, J.C. Fetzer, W.R. Biggs, P.R. Griffiths, and J.M. Olinger, *J. Chromatogr.* **461**, 209 (1989).

37. K. Jinno, Y. Saito, R. Malhan née Chopra, J.J. Pesek, J.C. Fetzer, and W.R. Biggs, *J. Chromatogr.* **557**, 459 (1991).

38. F. Diederich and R.L. Whetten, *Acc. Chem. Res.* **25**, 119 (1992).

39. F. Diederich, R. Ettl, Y. Rubin, R.L. Whetten, R. Beck, M.M. Alvarez, S. Ang, D. Sensharma, F. Wudl, K.C. Khemani, and A. Koch, *Science,* **252**, 548 (1991).

40. K. Kikuchi, Y. Nakahara, S. Suzuki, K. Saito, I. Ikemoto, and Y. Achiba, *Proceedings of the Third C₆₀ Symposium,* July 14–15, Tokyo, 1992, p. 23.

41. T. Wakabayashi, K. Kikuchi, Y. Nakajima, H. Shiromaru, S. Suzuki, and Y. Achiba, *Proceedings of the Fourth C₆₀ Symposium,* January 26–27, Toyohashi, 1993, p. 144.

42. J.F. Anacleto and M.A. Quilliam, *Anal. Chem.* **65**, 2236 (1993).

43. K. Jinno, H. Matsui, H. Ohta, Y. Saito, K. Nakagawa, H. Nagashima and K. Itoh, *Chromatographia,* **41**, 353 (1995).

44. A. Dupont, J.-P. Gisselbrecht, E. Leize, L. Wagner, and A. van Dorsselaer, *Tetrahedron Lett.* **35**, 6083 (1994).

45. K. Kikuchi, private communication.

46. K. Jinno, J. Wu, M. Ichikawa, and I. Takata, *Chromatographia*, **37**, 627 (1993).

47. K. Jinno, *J. Chromatogr. Sci.* **27**, 729 (1989).

48. P. Shah, T.B. Tsu, and L.R. Rogers, *J. Chromatogr.* **396**, 31 (1987).

49. K.R. Sentell, *J. Chromatogr.* **656**, 231 (1993).

50. H. Ohta, Y. Saito, K. Jinno, J.J. Pesek, M.T. Matyska, Y.-L. Chen, J.C. Fetzer and W.R. Biggs, *Chromatographia*, **40**, 507 (1995).

51. Y. Saito, H. Ohta, H. Nagashima, K. Itoh, K. Jinno, M. Okamoto, Y.-L. Chen, G. Luehr, and J. Archer, *J. Liquid Chromatogr.* **18**, 1897 (1995).

52. R.C. Zeigler and G.E. Maciel, *J. Phys. Chem.* **95**, 7345 (1991).

53. K.Albert, M. Pursch, and S. Strohschein, Abstract of the Symposium on Chemically Modified Surfaces, June 19–21, 1995, San Jose, CA.

54. H. Ohta, K. Jinno, Y. Saito, J.C. Fetzer, W.R. Biggs, J.J. Pesek, M.T. Matyska, and Y.-L. Chen, *Chromatographia*, **42**, 56 (1996).

55. K. Jinno, Y. Saito, Y.-L. Chen, G. Luehr, J. Archer, J.C. Fetzer, and W.R. Biggs, *J. Microcolumn Sep.* **5**, 135 (1993).

56. K. Jinno, H. Ohta, Y. Saito, T. Uemura, H. Nagashima, K. Itoh, Y.-L. Chen, G. Luehr, J. Archer, J.C. Fetzer, and W.R. Biggs, *J. Chromatogr.* **648**, 71 (1993).

57. Y. Saito, H. Ohta, H. Nagashima, K. Itoh, K. Jinno, M. Okamoto, Y.-L. Chen, G. Luehr, and J. Archer, *J. Liquid Chromatogr.* **17**, 2359 (1995).

58. J.J. Pesek and E.J. Williamsen, *Trends Anal. Chem.* **11**, 259 (1992).

59. Y. Saito, K. Jinno, J.J. Pesek, Y.-L. Chen, G. Luehr, J. Archer, J.C. Fetzer, and W.R. Biggs, *Chromatographia*, **38**, 295 (1994).

60. K. Jinno, H. Mae, Y. Saito, J.J. Pesek, J.C. Fetzer, and W.R. Biggs, *J. Microcolumn Sep.* **3**, 417 (1991).

61. Y. Saito, H. Ohta, H. Nagashima, K. Itoh, K. Jinno, and J.J. Pesek, *J. Microcolumn Sep.* **7**, 41 (1995).

62. R.S. Ruoff, D.S. Tse, R. Malhotra, and D.C. Lorents, *J. Phys. Chem.* **97**, 3379 (1993).

63. K. Jinno, K. Nakagawa, Y. Saito, H. Ohta, H. Nagashima, K. Itoh, J. Archer and Y.-L. Chen, *J. Chromatogr. A*, **691**, 91 (1995).

64. C.E. Kibbey, M.R. Savina, B.K. Parseghian, A.H. Francis and M.E. Meyerhoff, *Anal. Chem.* **65**, 3717 (1993).

65. K. Jinno, T. Uemura, J. Haginaka and Y. Saito, *J. Microcol. Sep.* **4**, 325 (1992).

66. Y. Saito, M. Mifune, J. Oda, Y. Otsuki, M. Mitsuhashi, Y. Mori, A.H. Gassim and J. Haginaka, *Anal. Sci.* **7**, 805 (1991).

67. K. Jinno, C. Kohrikawa, Y. Saito, H. Ohta, J. Haginaka and Y. Saito, *J. Microcol. Sep.* **8**, 13 (1996).

68. Y. Saito, H. Ohta, H. Terasaki, H. Nagashima, K. Jinno and K. Itoh, *J. HRC. CC.* **18**, 569 (1995).

CHAPTER
4

Chromatographic Enantiomer Separation on Chiral Polymers

Yoshio Okamoto and Eiji Yashima

4.1 Introduction

Optically pure isomers may be obtained by the resolution of enantiomers, and many chromatographic methods, particularly high performance liquid chromatography (HPLC), have been developed for purposes of analytical and preparative-scale resolution. The key to effective separation by this method is a chiral stationary phase (CSP) capable of chiral recognition. In the past 10 years, many CSPs for HPLC have been prepared and nearly 100 of them have been commercialized [1,2]. The CSPs are classified into two categories: one consists of optically active small molecules with chiral recognition ability, usually bonded to silica gel; the other is based on chiral polymers [3]. This chapter focuses on the resolution of enantiomers by polymeric CSPs.

Figure 4.1 shows typical CSPs prepared from optically active polymers [3]. Elucidation of the chiral recognition mechanism in polymeric CSP systems is seldom easy because the determination of the precise steric structure of a polymer is an extremely laborious task. Nevertheless, polymeric CSPs are attractive and interesting because their chiral recognition depends on the higher order structure of the polymer, which may result in unexpected high chiral recognition ability. Most polymers have been used as CSPs, and the process includes immobilizing the polymers on silica gel to attain high efficiency.

This chapter discusses some of the characteristic features arising when the chiral polymers illustrated in Figure 4.1 are used as CSPs for HPLC. The polymers are classified into the following categories:

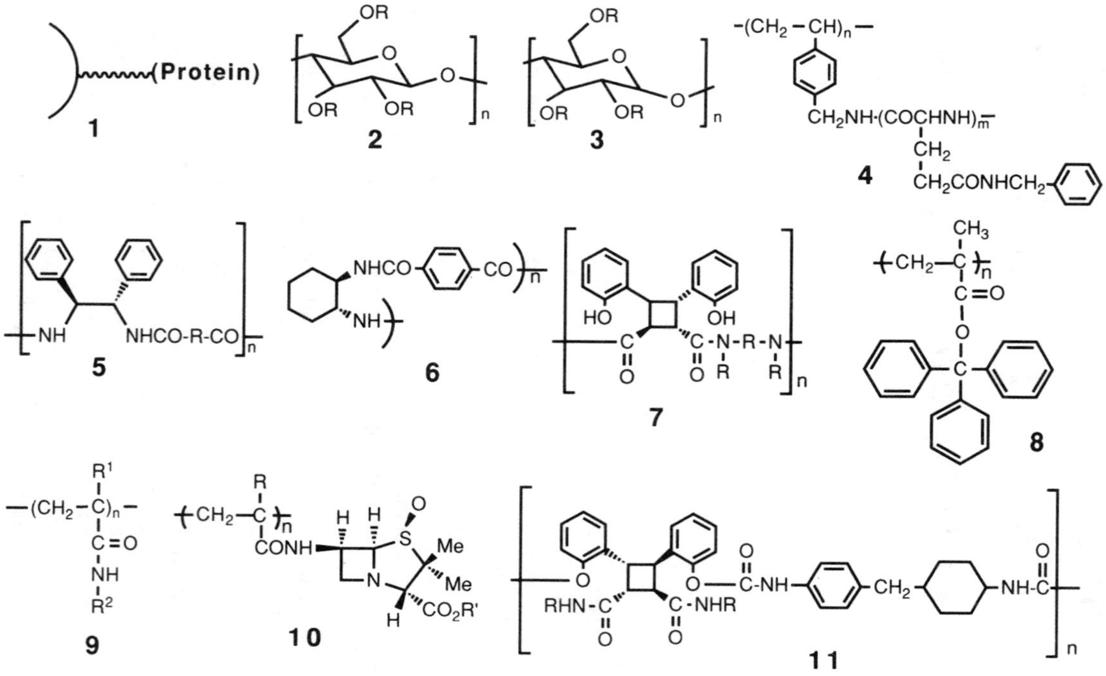

Figure 4.1 The principal chiral polymer stationary phases for HPLC discussed in this chapter.

Proteins (**1**)
Polysaccharides and their derivatives (**2,3**)
Polyamides (**4–7**)
Polymethacrylates (**8**)
Polyacrylamides and polymethacrylamides (**9,10**)
Polyurethanes (**11**)
Synthetic polymers with chiral cavities
Other chiral polymers

4.2 Protein

Proteins such as silk and wool show low chiral recognition as CSPs. However, enzyme proteins are known to possess high chiral recognition abilities and to act as catalysts for asymmetric synthesis in vitro as well as in vivo. It was expected, therefore, that some proteins, after being immobilized on silica gel, would show high chiral recognition as CSPs for HPLC. Several protein-immobilized silica gels have been prepared for optical resolution, and some of them have been commercialized.

4.2.1 Bovine Serum Albumin (BSA) (Resolvosil)

BSA is a globular protein consisting of 581 amino acid residues (molecular weight 66,210); it is relatively acidic, with an isoelectric point of 4.7. Numerous organic compounds are known to bind to BSA, indicating its hydrophobic character. In 1983 Allenmark and Bomgrem [4] used an aqueous buffer as an eluent to demonstrate that BSA covalently immobilized to silica gel is a useful CSP for the optical resolution of a variety of racemates including amine and carboxylic acid derivatives [5]. Racemic barbiturates (**12**) [6] used as sedatives and hypnotics, and two anti-thrombogenic coumarine derivatives, phenprocoumon (**13**) and warfarin (**14**) [7], are well resolved into enantiomers on Resolvosil columns.

The Resolvosil column is based on a spherical silica (7 or 10 μm particle, 30 nm pore diameter) to which BSA is immobilized. The details of the method for the immobilization of BSA are not reported. In a modified method for immobilization [8], BSA was covalently immobilized to silica gel using, for example, glutaraldehyde as a cross-linking agent.

A fragment (MW 38,000) of BSA degradated by enzyme-catalyzed hydrolysis was immobilized to silica gel and the chiral recognition ability was evaluated [9]. This fragment showed much lower recognition than that of intact BSA.

Human serum albumin (HSA) was also used as a CSP for the resolution of enantiomers of 2-arylpropionic acid nonsterodial anti-inflammatory drugs [10]. This column has also been commercialized.

4.2.2 α_1-Acid Glycoprotein (EnantioPac and Chiral AGP)

Hermannson found that α_1-acid glycoprotein (orosomucoid; AGP), present in human plasma, is also useful for the optical resolution of racemic compounds when immobilized to silica gel [11].

AGP (MW 44,100) has 14 residues of sialic acid units in the sugar parts showing an isoelectric point of 2.7. AGP was first covalently immobilized to silica gel through amino groups of AGP, followed by cross-linking. However, the CSP (EnantioPac) [11] thus obtained showed low stability. Further improvements were necessary to overcome the unstability (Chiral AGP) [12].

The Chiral AGP column shows effective resolving power for many racemic drugs (Figure 4.2).

4.2.3 Ovomucoid (Ultron ES-OVM)

A glycoprotein ovomucoid (MW 28,000; isoelectric point 3.9–4.5) isolated from egg white has been used as a practical CSP after immobilization to silica gel [13]. Sugar units of ovomucoid are essential for chiral recognition [14], and hydrophobic

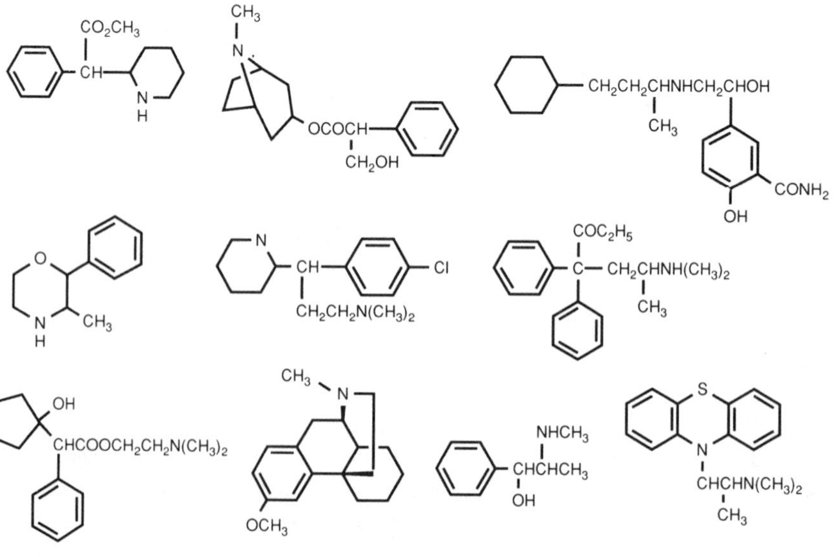

Figure 4.2 Compounds resolved on the Chiral AGP column.

racemic compounds with aromatic groups are well resolved on the ovomucoid column (Figure 4.3). The chiral recognition ability of the ovomucoid column is somewhat complementary to that of the Chiral AGP column [15].

Other proteins, cellulase [16], α-chymotrypsin [17], and avidin [18], were chemically bonded to silica gel for optical resolution. The cellulase column effectively separates β-adrenergic blockers into enantiomers.

These proteins should be handled with special care, since some of them lack conformational stability and the degree of chiral recognition may change with time.

The CSPs based on proteins possess effective chiral adsorbing sites on only a part of the macromolecules. Thus the sample loading capacity is low, and therefore the columns are not suitable for preparative-scale separation.

4.3 Polysaccharides and Derivatives

Polysaccharides, such as cellulose and amylose, are the most readily available polymers having with optical activity, and although their chiral resolving ability is not especially high, they are known to perform optical resolution as CSPs. On the other hand, the esters and carbamates of polysaccharides have been found to resolve successfully a variety of racemic compounds as CSPs for HPLC [19,20], and many CSPs consisting of polysaccharide derivatives are sold and extensively used for both analytical and preparative separation of enantiomers. Some characteristics of these CSPs are briefly described in Sections 4.3.1 through 4.3.5, together with those of underivatized polysaccharides themselves.

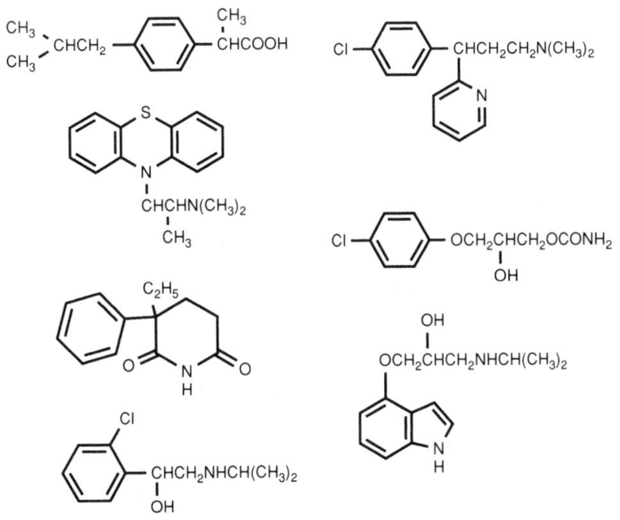

Figure 4.3 Compounds resolved on the ovomucoid column.

4.3.1 Cellulose

More cellulose—linear poly(β-D-1,4-glucoside) (**2** in Figure 4.1, R = H)—is produced than any other polymer on earth. The first utilization of the chirality of cellulose for optical resolution was demonstrated by Kotake et al. [21], who separated a racemic amino acid derivative into two spots of enantiomers by paper chromatography. More recently, it has been shown that thoroughly purified cellulose with a high degree of crystallinity (microcrystalline cellulose) is able to completely resolve amino acids subjected to liquid chromatography [22–24]. Although these materials themselves do not offer practically useful CSPs because of low optical resolving abilities and difficulty in handling, cellulose has been easily converted to a variety of derivatives (e.g., triacetate, tribenzoate, trisphenylcarbamate) by reaction on active hydroxy groups with corresponding reagents.

4.3.1.1 Cellulose Ester

Microcrystalline cellulose triacetate (MCTA), which is obtainable by means of the heterogeneous acetylation of native microcrystalline cellulose in benzene [25,26], shows interesting chiral resolving properties when used as a CSP for liquid chromatography, although the chiral resolution ability of a partially acetylated cellulose is very low [27]. Hesse and Hagel pointed out that the microcrystallinity of MCTA is essential for chiral recognition, since optical resolving ability is substantially reduced once the triacetate has been dissolved in a solvent [25,26]. The MCTA column has been employed for the resolution of various racemic compounds, especially aromatic compounds, using an ethanol/water mixture as the eluent (Figure 4.4) [27–32]. The mechanism for chiral recognition is not yet satisfactorily elucidated, but inclusion seems to be a reasonable explanation by which aromatic compounds may be adsorbed into the chiral cavities of MCTA matrix.

When this MCTA was supported on silica gel from a solution, it also gave a useful CSP (Chiralcel OA) having a chiral recognition ability completely different

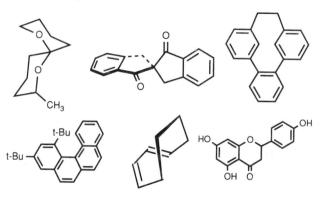

Figure 4.4 Compounds resolved on the MCTA column.

from that of MCTA [33,34], suggesting in turn that microcrystallinity is not essential for chiral recognition. The chromatograms of the resolution of Tröger base on the two CSPs are shown in Figure 4.5. Enantiomers of Tröger base were eluted in reversed order on the two triacetate columns. This result is ascribed to the different higher order structures of the two triacetates. Therefore, the chiral recognition of a CSP depends highly on the conditions of its preparation [35].

Cellulose tribenzoate (CTB: **2** in Figure 4.1, R = C_6H_5CO, Chiralcel OB) and cellulose tricinnamate (**2**, R = $C_6H_5CH{=}CHCO$, Chiralcel OK) are also useful CSPs when coated on silica gel [33,34]. The introduction of substituents on the phenyl groups of CTB was systematically studied [36]. Alkyl, halogen, trifluoromethyl, and methoxy groups were selected as the substituents. The inductive effect of the substituents greatly influenced their optical resolution ability. The benzoate derivatives having electron-donating substituents such as methyl groups showed better chiral recognition ability than those having electron-withdrawing substituents such as halogen groups. However, the most electron-donating methoxy group was not suitable because of the high polarity of the substituent itself. Among the benzo-

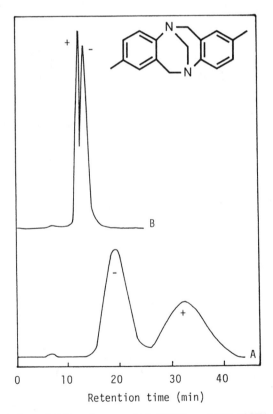

Figure 4.5 Resolution of Tröger base on (A) MCTA and (B) MCTA-coated silica gel columns (25 cm × 0.46 i.d. cm; eluent, H_2O/methanol (30:70); flow rate, 0.5 mL/min.

ates, cellulose tris(4-methylbenzoate) (Chiralcel OJ) exhibited chiral recognition for many racemic compounds (Figure 4.6) and a practically useful CSP [37,38]. Porous beads of cellulose 3- and 4-methylbenzoates were also prepared as CSPs [39].

The main chiral adsorbing sites are considered to be the polar carbonyl groups of esters, which can interact with racemic compounds through hydrogen bonding and dipole–dipole interaction for chiral discrimination [40,41]. Chiral recognition ability was greatly dependent on the conditions of preparation of CSP [35], particularly on the solvent used to dissolve cellulose tribenzoate derivatives in the coating process, as observed in the case of cellulose triacetate.

Among these cellulose triester, triacetate, tribenzoate, tris(4-methylbenzoate), and tricinnamate have been commercialized as chiral columns.

4.3.1.2 Phenylcarbamates of Cellulose

Cellulose trisphenylcarbamate derivatives, which are prepared by the reaction of microcrystalline cellulose with substituted phenylisocyanates, also show a very high optical resolving power when coated on silica gel. The chiral recognition abilities of a series of cellulose phenylcarbamate derivatives have been evaluated (Figure 4.7)

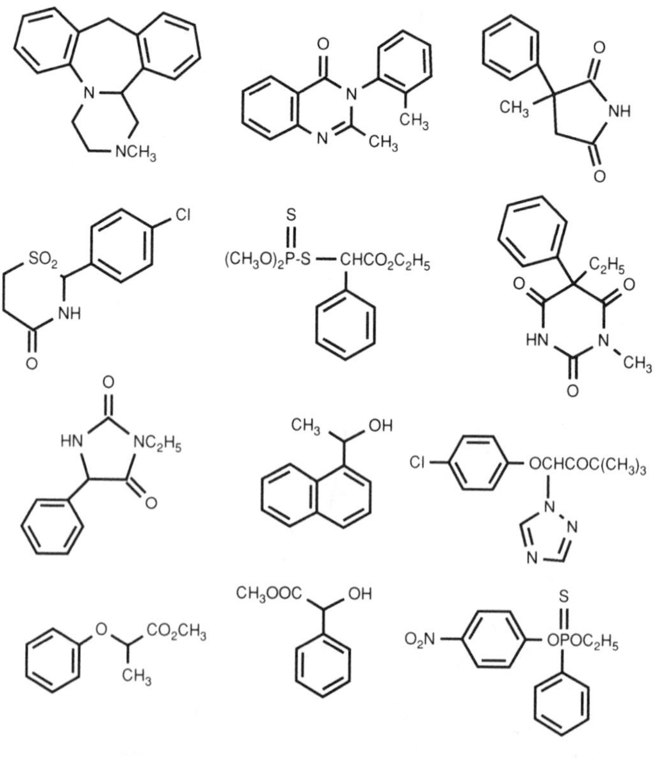

Figure 4.6 Compounds resolved on cellulose tris(4-methylbenzoate).

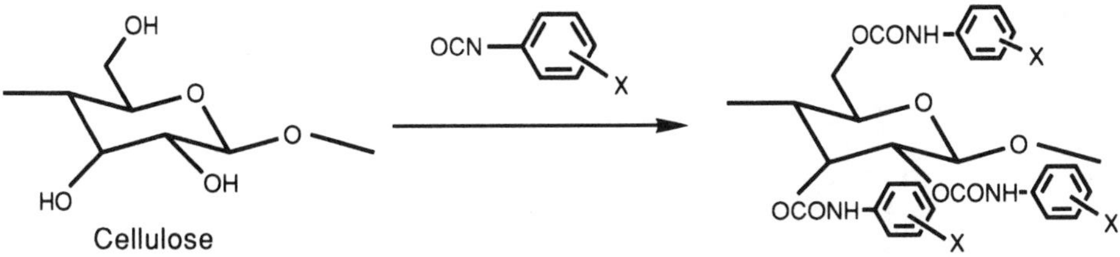

Cellulose

Cellulose tris(phenylcarbamate) derivatives

X =

a: 4-CH_3O	h: 4-Cl	o: 3,5-$(CH_3)_2$
b: 4-CH_3	i: 4-Br	p: 2,6-$(CH_3)_2$
c: 4-CH_3CH_2	j: 4-CF_3	q: 3,4,5-$(CH_3)_2$
d: 4-$(CH_3)_2CH$	k: 4-NO_2	r: 3,5-Cl_2
e: 4-$(CH_3)_3C$	l: 3-CH_3	s: 3,5-F_2
f: H	m: 2-CH_3	t: 3,5-$(CF_3)_2$
g: 4-F	n: 3,4-$(CH_3)_2$	

Figure 4.7 Cellulose tris(phenylcarbamate) derivatives.

[42–45]. The chiral resolving power of CSPs was greatly influenced by the introduction of substituents on the phenyl groups. Substitution at the 3- or 4-position improved the optical resolution ability for many racemates, but 2-substituted derivatives showed low chiral recognition. The separation factors (α) for five racemates (**15–19**) and retention times for acetone T_a on 4-substituted derivatives are shown in Table 4.1, where the substituents on the phenyl groups (R) are arranged from left to right in order of increasing electron-withdrawing power of the substituents. Most CSPs resolve all the racemic compounds effectively except a few cases, though the separation factors depend greatly on the 4-substituents. The derivatives with heteroatom substituents such as methoxy and nitro groups show poor chiral recognition.

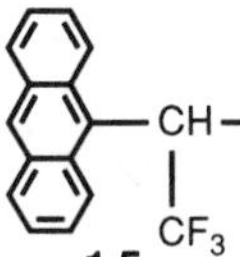

15

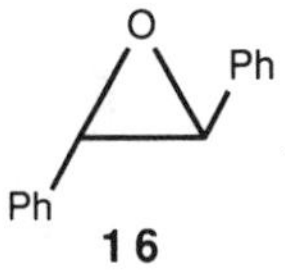

16

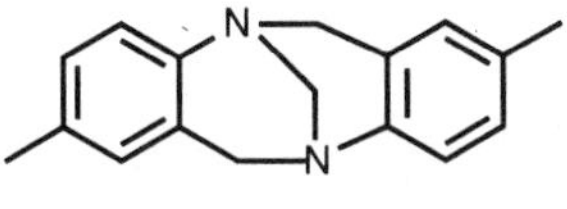

17

$Co(acac)_3$

18

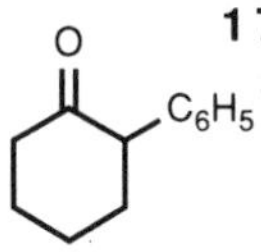

19

The retention times of acetone on these CSPs tend to increase as the electron-withdrawing power of the substituents increases. On the other hand, the retention times of the first eluted isomer of alcohol (**15**) decrease roughly as the electron-withdrawing power of the substituents increases [43]. These results indicate that the main chiral adsorbing sites are probably the polar carbamate groups, as shown in

Table 4.1 SEPARATION FACTOR IN THE OPTICAL RESOLUTION OF **15–19** AND RETENTION TIME OF ACETONE (T_a) ON CELLULOSE TRIS(4-SUBSTITUTED PHENYLCARBAMATE) DERIVATIVES[a]

	CH$_3$O—	Et	CH$_3$	(phenyl)	Br
15	1.35(−)	1.57(−)	1.52(−)	1.45(−)	1.29(−)
16	1.34(+)	1.55(+)	1.55(+)	1.46(+)	1.70(+)
17	~1(+)	1.11(+)	1.48(+)	1.37(+)	1.19(+)
18	~1(+)	1.76(+)	1.75(+)	1.24(+)	1.79(+)
19	1.13	1.19(−)	1.20(−)	1.17(−)	1.17(−)
T_a, min	10.0	9.4	10.1	10.9	11.2

	Cl	F	CF$_3$	NO$_2$
15	1.29(−)	1.26(−)	1.30(−)	~1(+)
16	1.68(+)	1.38(+)	1.61(+)	1.33(+)
17	1.16(+)	1.14(+)	1.23(+)	~1(−)
18	1.46(−)	1.53(+)	2.06(+)	~1(+)
19	1.16(−)	1.12(−)	1.18(−)	~1(−)
T_a, min	11.3	12.3	16.8	13.8

[a] Eluent, hexane-2-propanol (90 : 10).

Figure 4.8, and the groups are capable of interacting with a racemic compound via hydrogen bonding with NH and C=O groups and the dipole–dipole interaction on C=O. The acidity of the NH proton increases with an increase in the electron-withdrawing power of the substituents (X) on the phenyl groups. Therefore, acetone is more strongly adsorbed on the CSPs having the more acidic NH protons via hydrogen bonding interaction. This observation is associated with the downfield shift of the NH resonances as the electron-withdrawing power of the substituents (X) on the phenyl groups increases [43].

On the other hand, when X is the electron-donating substituent, such as a methyl group, the electron density of the carbonyl oxygen of the carbamates is expected to increase. Therefore, alcohols are more strongly adsorbed on the CSPs via hydrogen bonding interaction. When X itself is a polar group, such as a nitro or methoxy group, racemic compounds can interact with X substituents. Since X is far from a chiral glucose residue, such an interaction on X should reduce the optical resolving power. The interaction of the carbamate groups of CSPs with racemic compounds seems to be most important for effective chiral resolution. Besides these polar interactions, the π-π interaction between the phenyl group of a CSP and an aromatic group of a solute may play some role in chiral recognition because several nonpolar aromatic compounds have also been resolved [44]. As an eluent, a hexane/2-propanol mixture appears to be suitable for the efficient separation of enantiomers. When an analyte has a basic amino group, the addition of a small amount of an amine such as diethylamine or isopropylamine for the foregoing eluent is recommended to avoid the tailing of peaks [45,46]. To separate acidic compounds, a small amount of a strong acid such as CF_3COOH can be added [47]. Aqueous eluents are also usable to resolve several drugs [48,49].

Most cellulose trisphenylcarbamate derivatives form a lyotropic liquid crystalline phase in a highly concentrated solution [43,50] and show very high crystallinity under a polarizing microscope when they are cast from a solution. This means that the carbamates coated on silica gel from a solution also have an ordered structure in which phenylcarbamate groups are regularly arranged. Such an ordered structure seems to be very important for efficient chiral recognition on CSPs derived from

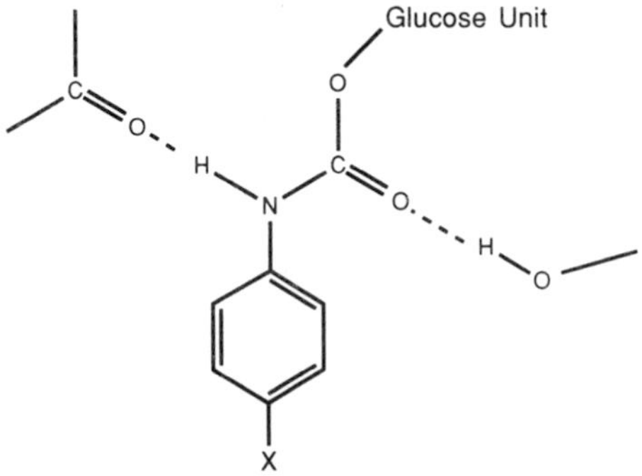

Figure 4.8   Sites for hydrogen bonding on substituted phenylcarbamates.

polymers. A few of the cellulose phenylcarbamate derivatives fail to form a liquid crystalline phase, showing low chiral resolving power.

Among many cellulose trisphenylcarbamate derivatives, 3,5-disubstituted derivatives such as 3,5-dimethyl- (Chiralcel OD) and 3,5-dichlorophenylcarbamates show particularly interesting and efficient optical resolving abilities for a variety of racemic compounds [43]. Table 4.2 summarizes the results of optical resolution studies of some racemic compounds (15–24) on these CSPs. The 3,5-dimethyl-phenylcarbamate resolves a variety of racemic compounds including aromatic hydrocarbons, amines, carboxylic acids [47], alcohols, amino acid derivatives [51], and many drugs [52] containing β-adrenergic blocking agents (β-blockers) [46].

Figure 4.9 shows some compounds resolved on the CSP. We tested 505 racemic compounds for optical resolution on the CSP, and 228 of them were almost completely resolved, with 86 partially resolved (i.e., showing two peaks). This means that about 62% of the 505 racemic compounds could be separated into enantiomers. The cellulose tris(3,5-dimethylphenylcarbamate) may be one of the most powerful CSPs.

Table 4.2   CAPACITY FACTORS AND SEPARATION FACTORS FOR 15–24 ON TRIS(3,5-DIMETHYLPHENYLCARBAMATE) AND TRIS(3,5-DICHLOROPHENYLCARBAMATE) OF CELLULOSE[a]

Racemate	3,5-Me$_2$		3,5-Cl$_2$	
	k'_1	α	k'_1	α
15	2.13(−)	2.59	0.28(−)	1.38
16	0.74(−)	1.68	0.56(+)	1.84
17	0.97(+)	1.32	0.87(+)	1.65
18	0.42(+)	~1	0.76(+)	1.82
19	1.17(−)	1.15	2.65(−)	1.26
20	1.37(+)	1.34	0.40(+)	1.29
21	2.36(−)	1.83	1.62(+)	1.11
22	1.47(−)	1.41	1.55(−)	1.20
23	2.43(+)	1.58	3.08(−)	1.21
24	0.83(+)	3.17	0.59(+)	1.41

[a]   Eluent, hexane 2-propanol (90 : 10); flow rate, 0.5 mL/min.

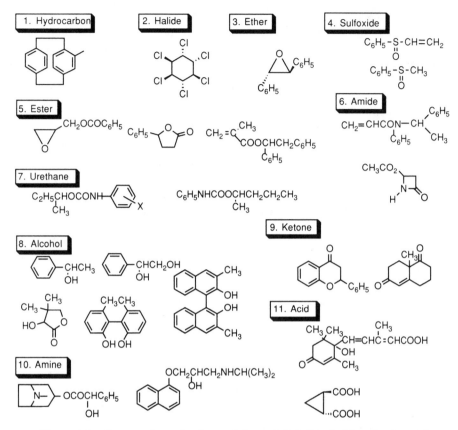

Figure 4.9 Compounds resolved on cellulose tris(3,5-dimethylphenylcarbamate).

Although cellulose tris(3,5-dichlorophenylcarbamate) shows unique chiral recognition ability, particularly with respect to bulky esters [43], this column has a defect; its solubility in hexane/2-propanol is exceptionally high. To mitigate the effects of this defect, the derivative was chemically bonded to silica gel, using diisocyanate as a spacer. However, the chiral recognition ability was greatly changed when hydroxy groups of cellulose amounting to several percent of the total were allowed to react with the spacer for immobilization on silica gel [53].

4.3.2 Phenylcarbamates of Amylose

Amylose (**3** in Figure 4.1; R = H), poly(α-D-1,4-glucoside), also can be converted to the carbamate derivatives as well as cellulose, and the chiral recognition capabilities of these compounds have been appraised in detail [42,45,54]. Again, amylose tris(3,5-dimethylphenylcarbamate) (Chiralpak AD) is the most efficient CSP among the various phenylcarbamate derivatives of amylose. The chiral recognition of the amylose derivatives differs from that of the corresponding cellulose derivatives. Some enantiomers elute in reversed order on the two CSPs.

We tested the optical resolution of 384 racemic compounds on the amylose tris(3,5-dimethylphenylcarbamate) column: 107 of them were almost completely resolved and 102 were partially resolved, with two overlapped peaks. Consequently, when two tris(3,5-dimethylphenylcarbamates) of cellulose and amylose were used for 505 racemic compounds, 185 racemates were separated only on the cellulose derivative, 82 only on the amylose derivative, and 129 on both columns. This means that 396 racemates, about 78% of the 505 racemic compounds, were separated into enantiomers. Many racemates have a chance to be resolved on the tris(3,5-dimethylphenylcarbamate) derivatives of either cellulose or amylose.

4.3.3 Phenylcarbamates of Oligosaccharides

3,5-Dimethylphenylcarbamates of oligosaccharides, cellooligosaccharide (**25**), maltooligosaccharide (**26**), and cyclodextrins (**27**) were prepared, and in an effort to understand the influence of the higher order structure of the polysaccharide derivatives, their chiral recognition abilities were compared with those of the corresponding polysaccharide derivatives [55].

Cellooligosaccharide derivative (n = 2, 4) **25**

Maltooligosaccharide derivative (n = 2 - 7) **26**

Cyclodextrin derivative (n = 6, 7, 8) **27**

$$R = \quad \text{(3,5-dimethylphenyl)}-NHCO-$$

The carbamates of cellooligosaccharides ($n = 2, 4$) were lower in chiral recognition ability than the cellulose derivative, indicating that cellooligosaccharide derivatives differ structurally from cellulose tris(3,5-dimethylphenylcarbamate). In case of the carbamates of linear maltooligosaccharides ($n = 2–7$), the 4- to 7-mers were similar to one another and not very different from amylose tris(3,5-dimethylphenylcarbamate) with respect to resolving power. These results were supported by conformational studies of the carbamates of the oligomers using circular dichroism (CD) spectroscopy.

β-Cyclodextrins, cyclic oligomers with the same glucose unit as amylose, can separate racemic compounds by inclusion [56], and the CSP consisting of cyclodextrin bonded silica gel has been commercialized. On the other hand, 3,5-dimethylphenylcarbamate of cyclodextrins showed chiral recognition quite different from that of amylose tris(3,5-dimethylphenylcarbamate). The amylose derivative must possess higher order structure different from those of the cyclodextrin derivatives.

Two different types of 3,5-dimethylphenylcarbamate of β-cyclodextrin immobilized on silica gel have been prepared; in one case immobilization probably proceeds via the 6-position [57] and in the other via the 2- and/or 3-position [58] of the respective glucose units. The chiral recognition abilities of these CSPs are rather different, and the latter seems to show higher chiral recognition ability [59].

4.3.4 Other Phenylcarbamates of Polysaccharides

Phenylcarbamates of other polysaccharides such as chitosan (**28**), xylan (**29**), curdlan (**30**), dextran (**31**), and inulin (**32**) were also prepared to evaluate their chiral recognition abilities [42]. Some of the CSPs showed unique and interesting chiral properties, although none resolved as many racemic compounds as did cellulose trisphenylcarbamate.

4.3.5 Aralkylcarbamates of Cellulose and Amylose

Trisalkylcarbamates such as methylcarbamate and cyclohexylcarbamate of cellulose showed very low chiral recognition abilities. However, some of aralkylcarbamates of cellulose and amylose (Figure 4.10) were found to show characteristic chiral recognition for many racemic compounds. Among several aralkylcarbamates of cellulose and amylose including (RS)-, (R)-, and (S)-1-phenylethylcarbamates and (RS)-1-(1-naphthyl)ethylcarbamate, cellulose tris[(R)- or (RS)-1-phenylethylcarbamate] and amylose tris[(S)- or (RS)-1-phenylethylcarbamate] exhibit high optical resolving abilities, and some of racemic compounds shown in Figure 4.11 are better

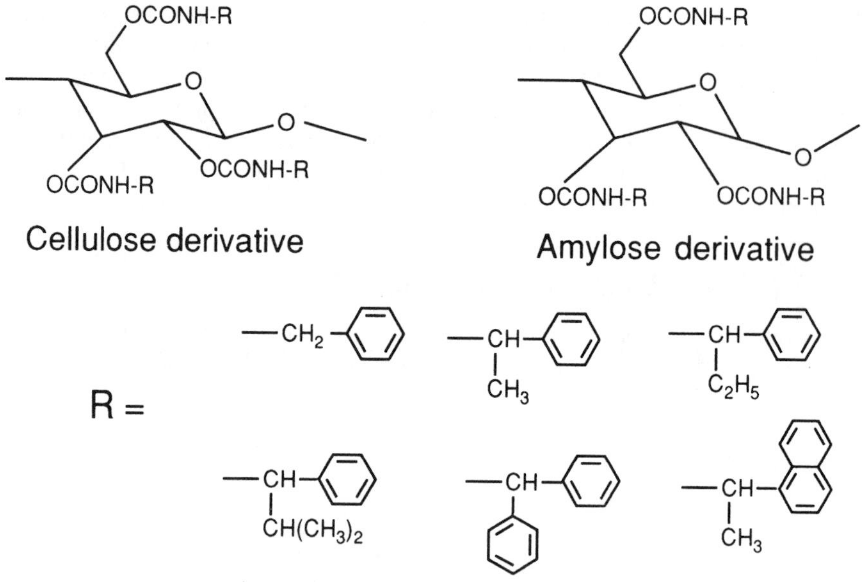

Figure 4.10 Aralkylcarbamates of cellulose and amylose.

resolved on amylose tris[(S)-1-phenylethylcarbamate] (Chiralpak AS) than other polysaccharides carbamates including phenylcarbamate derivatives [60–63].

4.4 Polyamides

Poly(α-amino) acids (**4** in Figure 4.1) have been extensively studied as model protein compounds, and a few of them have been evaluated as CSPs. For example, cross-linked porous polystyrene resin beads incorporating poly(N-benzyl-L-glutamine) or poly(γ-benyzl-L-glutamate) were prepared. The resin can separate racemic mandelic acid and hydantoin derivatives into enantiomers [64]. A CSP containing poly(L-leucine) or poly(L-phenylalanine) was also prepared, and some amino acid derivatives were resolved [65]. However, the resolving power of these CSPs for many racemates appears low compared with that of the protein-based CSPs discussed earlier. The optical resolving abilities of these compounds are greatly influenced by their higher order structures. The major binding force of these adsorbents to analytes is probably hydrogen bonding interacting. Therefore, nonpolar eluents such as toluene and hexane result in better separation of enantiomers, differing from the protein phases with respect to which water and buffer are usually used as an eluent; in addition, hydrophobic and ionic interactions may be the driving force for the effective separation of enantiomers.

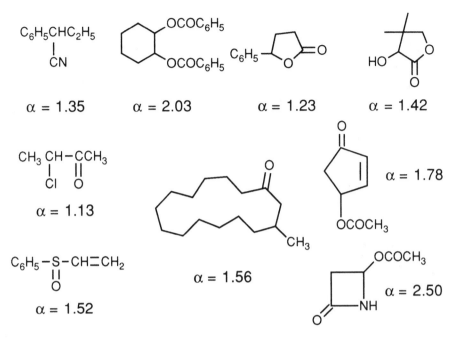

Figure 4.11 Compounds resolved on amylose tris[(S)-1-phenylethylcarbamate].

Fully synthetic chiral polyamides (**5** [66] and **6** [67] in Figure 4.1) prepared from chiral diamines and dicarboxylic acid derivatives showed characteristic resolving power for some racemic polar compounds with functional groups capable of hydrogen bonding.

The polyamide **7** derived from the (−)-anti head-to-head coumarine dimer **33** and diamines prepared by Saigo, Hasegawa, and co-workers [68] shows high chiral recognition ability as a CSP for HPLC depending on the number of the methylene units of the diamine components (Scheme 4.1). A distinct odd–even effect is observed in the chiral recognition on these polyamides. The polyamides with an even number of methylenes in the diamine residue show resolving ability, while those with an odd number show no chiral recognition ability [69]. This odd–even effect seems to be related to the crystallizability of the polyamides; only crystallizable polyamides with high order structures (even number of diamine residues) show chiral recognition ability. The hydroxy groups of **7** are easily carbamoylated with isocyanates. The carbamoylated polyamides showed high chiral recognition [70].

The chiral polyamides having axially dissymmetric binaphthalene moieties (**34**) were synthesized and their chiral recognition abilities were evaluated [71]. Several axially racemic biaryl compounds were resolved by HPLC with the polyamides coated on macroporous silica gel.

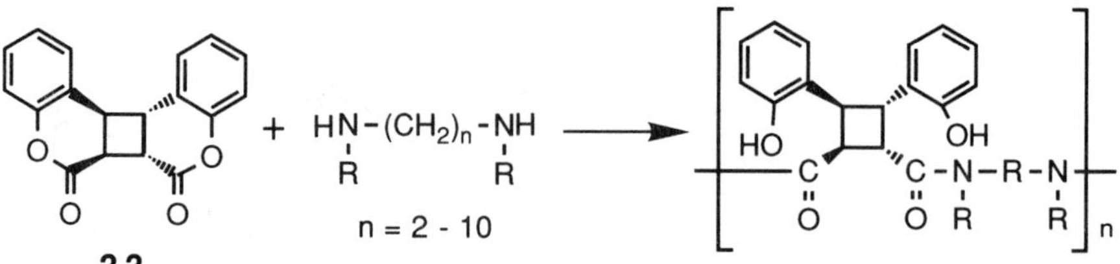

Scheme 4.1

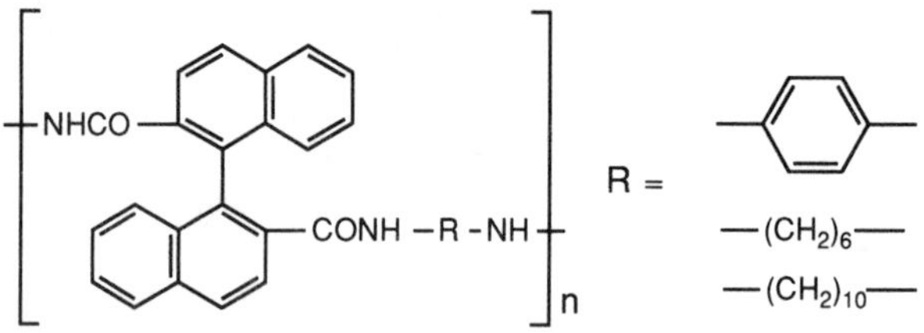

4.5 Polymethacrylates

Although many optically active polymethacrylates carrying chiral ester groups such as the (S)-1-phenylethyl group (**35**) have been prepared so far, these polymers show almost no separation for many enantiomers. However, the optically active polymer obtained by way of the polymerization of triphenylmethyl methacrylate (TrMA, **36**) [72] with chiral anionic initiators shows efficient optical resolving power and can be used as a CSP, especially when coated on silica gel.

4.5.1 Optically Active Poly(triphenylmethyl methacrylate)

Triphenylmethyl methacrylate is a unique monomer that affords only an isotactic polymer in both radical and anionic polymerization [73]. Anionic polymerization with butyllithium gives almost 100% isotactic polymer not only in toluene but also in tetrahydrofuran. This result has been ascribed to the bulky triphenylmethyl group, which forces the growing polymer chain to take a helical conformation. When TrMA is polymerized in toluene at −78°C with a chiral initiator, for example, (+)- or (−)-2,3-dimethoxy-1,4-bis(dimethylamino)butane (DDB)–lithium amide complex, it gives an optically active polymer (PTrMA) ($[\alpha]_D \approx 360°$), with chirality arising mainly from right- or left-handed helicity (Scheme 4.2) [74]. The polymer of high degree of polymerization (DP > 80) is insoluble in solvents because of high

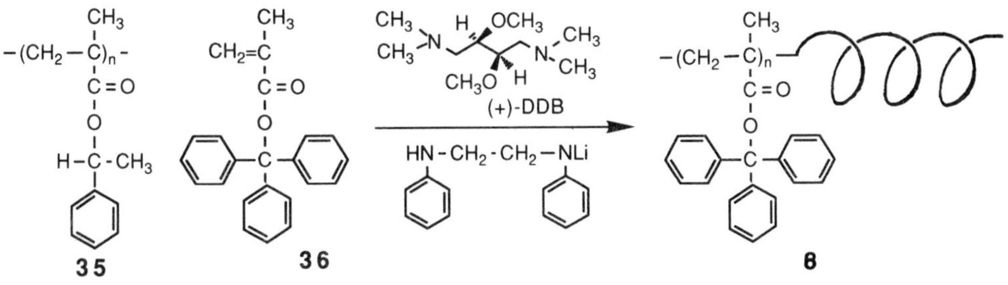

$$-(CH_2-\underset{\underset{H-\overset{\overset{\displaystyle CH_3}{|}}{C}-CH_3}{\overset{|}{O}}}{\overset{\overset{\displaystyle CH_3}{|}}{C})_n}-$$

35

$$CH_2=\underset{\overset{|}{O}}{\overset{\overset{\displaystyle CH_3}{|}}{C}}$$

36

$$\xrightarrow[\text{HN-CH}_2\text{-CH}_2\text{-NLi}]{\text{(+)-DDB}}$$

$$-(CH_2-\underset{\overset{|}{O}}{\overset{\overset{\displaystyle CH_3}{|}}{C})_n}-$$

8

Scheme 4.2

crystallinity, and the soluble polymer (DP ≈ 40) is conformationally stable enough to maintain its helical structure in solution at room temperature.

Three phenyl rings of an ester group are likely to have a propeller structure, which is also chiral owing to helicity. The polymer of purely one-handed helicity appears to show specific rotation ($[\alpha]_D$ 360–390°), depending on DP. We have succeeded in making soluble (+)- and (−)-PTrMA of almost 100% one-handed helicity by asymmetric polymerization [74]. PTrMA loses its optical activity when converted to poly(methacrylic acid) by solvolysis of ester groups in methanolic hydrochloric acid.

Insoluble (+)-PTrMA of high molecular weight shows chiral recognition [75] and can be used as a CSP for HPLC after being ground and sieved to small, uniform particles of 20–44 μm. Although when methanol served as an eluent the CSP resolved various racemic compounds having aromatic groups, the column was not practical because the brittleness of PTrMA reduces the lifetime of the column. This defect was overcome by using macroporous silica gel together with soluble PTrMA [76]. The soluble PTrMA was adsorbed by 20 wt % on silica gel (diameter, 10 μm; pore size, 100 or 400 nm) that had been treated with a silanizing agent such as dichlorodiphenylsilane. By this procedure, we could get easily a more effective CSP with higher resistance to compression and a longer lifetime relative to the ground, insoluble PTrMA.

Usually, better separation of enantiomers is attained on the (+)-PTrMA-coated silica gel than on the ground (+)-PTrMA (Figure 4.12). However, the two CSPs sometimes differed quite markedly in chiral recognition, as observed in the resolution of 1,1′-binaphthyl derivatives (Table 4.3) [77]. This discrepancy has been ascribed to the difference in the orientation of PTrMA molecules. So far more than 200 racemic compounds have been resolved on the PTrMA-coated silica gel [78]; examples are shown in Figure 4.13.

Trisacetylacetonates, Co(acac)₃ and Cr(acac)₃ [79], were completely resolved, and Al(acac)₃ [80] was for the first time separated into its optical active form on the (+)-PTrMA column, and Δ-isomers were eluted as the first fractions. It was possible to resolve racemic compounds having a phosphorus atom as a chiral center. The insecticides EPN (**37**) and salithion (**38**) were resolved [81]. Some halides were

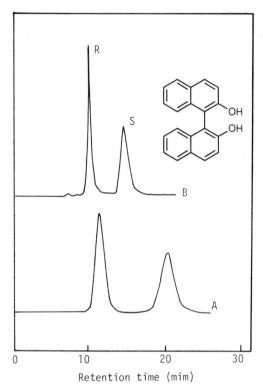

Figure 4.12   Resolution of binaphthol on (A) ground (+)-PTrMA and (B) (+)-PTrMA-coated silica gel columns (25 cm × 0.46 i.d. cm); eluent, methanol; flow rates, 0.72 mL/min (A) and 0.5 mL/min (B).

Table 4.3   RESOLUTION OF 2,2'-DISUBSTITUTED 1,1'-BINAPHTHYLS ON (+)-PTrMA COLUMNS[a]

Substituents	(+)-PTrMA-Coated Silica Gel Column				Ground (+)-PTrMA Column			
	k'_1	k'_2	α	R	k'_1	k'_2	α	R
HO—	0.50	1.16	2.32	3.83	1.50	3.20	2.13	2.39
CH$_3$O—	4.76	8.23	1.73	0.93	9.67	15.9	1.65	1.39
NH$_2$—	1.67	2.32	1.39	1.08	5.62	15.9	2.80	2.52
Br—	3.77	5.32	1.41	1.06	9.0	9.0	~1	~0
CH$_3$—	3.01	4.30	1.43	1.63	6.1	6.1	~1	~0

[a]   Columns, 25 cm × 0.46 i.d. cm; eluent, methanol; flow rate, 0.72 mL/min.

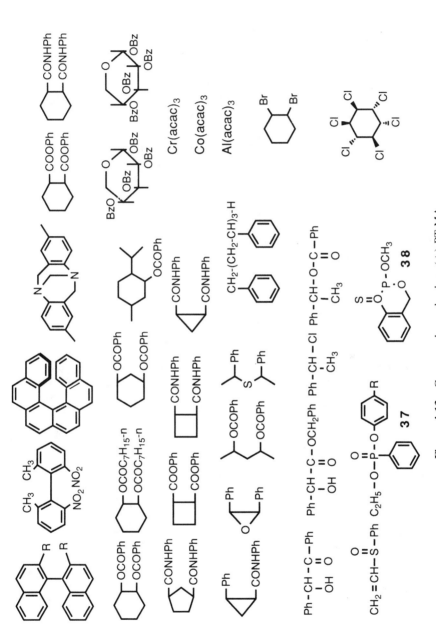

Figure 4.13 Compounds resolved on (+)-PTrMA.

also suitable compounds for resolution on (+)-PTrMA. The (+)-PTrMA column is particularly useful for the resolution of stereochemically interesting compounds (Figure 4.14) [82–101] that are difficult to resolve by other methods because of lack of functionalities. The enantiomers that racemize easily at room temperature may be separated by using the PTrMA column at low temperature. Tris(9-triptycyl)germanium chloride (56) was completely resolved on (−)-PTrMA at −30°C using methanol as an eluent [100].

The (+)-PTrMA columns tend to an elution order of enantiomers containing a C_2 axis and two aromatic groups (Figure 4.15). Usually, more highly retained enantiomers have P-helicity with respect to their aromatic groups if the molecule is inspected from the direction perpendicular to a C_2 axis [102]. This is also true to **42** and **50.** Therefore, the (+)-PTrMA molecule may be a right-handed helix, and an absolute configuration of compounds can be predicted from the results of the resolution if the compounds are similar in structure to those in Figure 4.15.

Generally, use of a polar eluent, such as methanol, rather than a nonpolar eluent gives better separation of enantiomers [103], although a hexane/2-propanol mixture sometimes is preferable. This suggests that nonpolar or hydrophobic interaction between PTrMA, probably the triphenylmethyl group with chiral propeller structure, and a nonpolar group of a solute may be important for achieving effective chiral recognition in the chromatography. Thus, 2-butanol and 1-phenylethanol

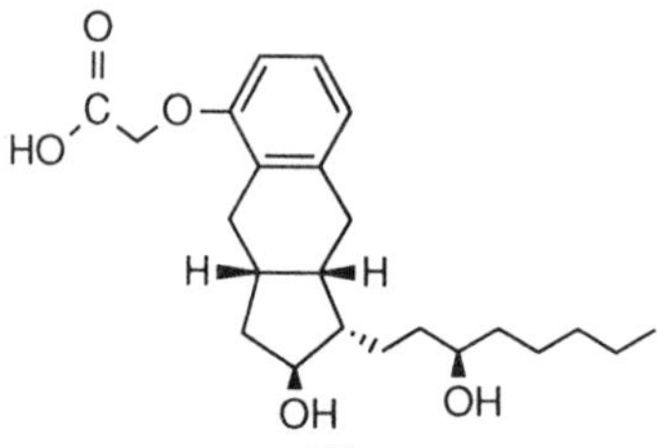

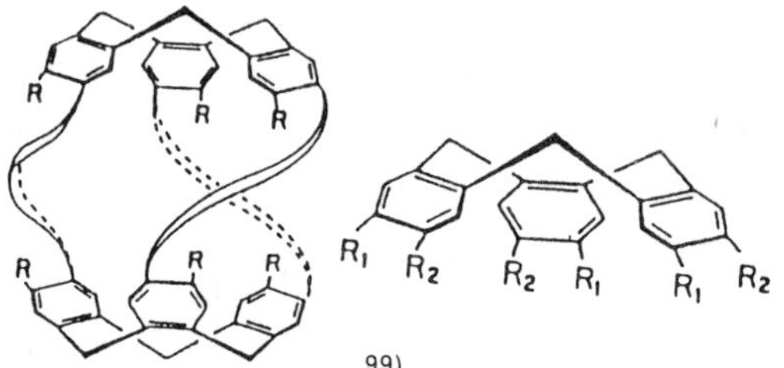

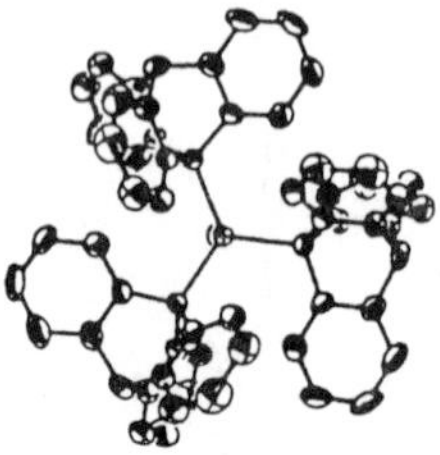

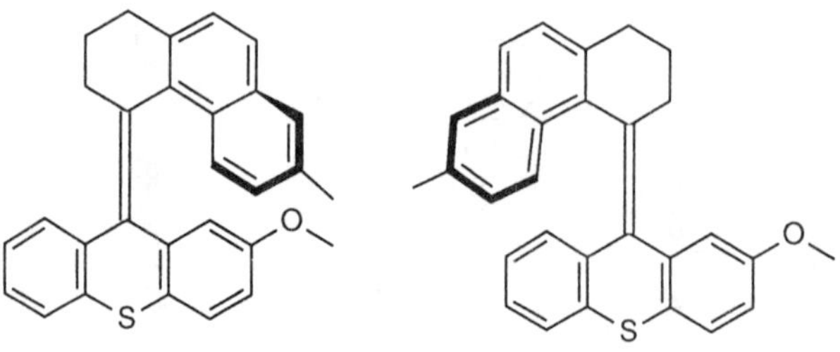

Figure 4.14 Compounds that lack functionalities, hence are difficult to resolve by many methods, resolved on (+)-PTrMA.

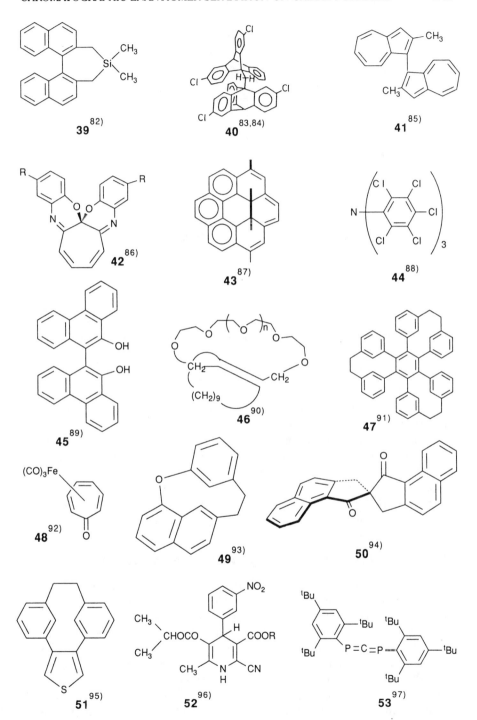

Figure 4.14   (*Continued*)

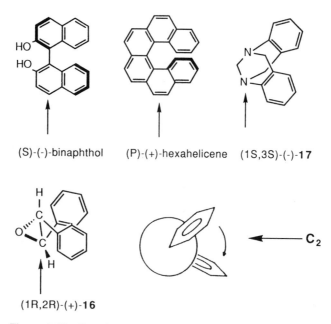

(S)-(-)-binaphthol   (P)-(+)-hexahelicene   (1S,3S)-(-)-**17**

(1R,2R)-(+)-**16**

Figure 4.15   Enantiomers adsorbed more strongly on (+)-PTrMA.

were not resolved on the column, but were resolved completely as their 3,5-dichlorobenzoate and benzoate, respectively [104].

The use of aromatic hydrocarbons, chloroform, and THF must be avoided because of the solubility of the PTrMA. However, these eluents can be used to resolve (±)-PTrMA and poly(diphenyl-2-pyridylmethyl methacrylate) into optical isomers on the (+)-PTrMA chemically bonded to silica gel column [105]. Chiral recognition of the (+)-PTrMA bonded silica gel was similar to that of the silica gel coated with PTrMA when methanol was used as an eluent.

4.5.2 Poly(diphenyl-2-pyridylmethyl methacrylate)

Diphenyl-2-pyridylmethyl methacrylate (DPyMA) also responds to asymmetric polymerization by yielding an optically active polymer (**58**) with highly one-handed helical structure [106]. The polymer supported on silica gel completely resolves many racemic compounds, showing a resolving ability slightly lower than that of PTrMA under the same chromatographic conditions with methanol as an eluent. However, when a nonpolar eluent, a hexane/2-propanol mixture, is used, the poly(DPyMA) exhibits higher resolving power for several alcohols than is obtained with PTrMA [107]. In this case, hydrogen bonding between a 2-pyridyl residue and alcohols seems to contribute to chiral discrimination. The poly(DPyMA) column has a longer lifetime than the PTrMA column because of slower solvolysis of ester group. These two columns are available commercially.

The chiral recognition ability of PTrMA is greatly affected by the introduction of substituents on the phenyl groups. The chiral recognition abilities of optically active poly[(*m*-chloro- or *m*-fluorophenyl)diphenylmethyl methacrylate] are slightly low compared with that of PTrMA, and poly[tris(*m*-chlorophenyl)methyl methacrylate] [poly(*m*-Cl$_3$TrMA) (**59**)] shows almost no resolving power, although the poly(*m*-Cl$_3$TrMA) possesses a one-handed helical structure. This set of properties has been ascribed to the different propeller structures of the triarylmethyl groups [108].

Optically active polymethacrylates bearing an axially dissymmetric 1,1'-binaphthalene moiety as a pendant group were synthesized by the radical polymerization of the corresponding methacrylates, and their chiral recognition abilities were evaluated. The CSPs resolved some racemic compounds such as 1-aryl-1-alkanols and 1,2-diols as the 3,5-dinitrophenylcarbamates [109]. The chiral recognition ability of the polymethacrylates was similar to that of the CSPs, which had been prepared by immobilization of the binaphthyl moiety onto silica. Thus it may be that the polymers lack a specific conformation that can contribute to chiral recognition.

4.6 Poly(meth)acrylamides

Blaschke et al. [110,111] performed extensive studies on the synthesis and application of chiral polymer gels of polyacrylamides and polymethacrylamides (**9** in Figure 4.1) bearing an optically active group on a side chain. The radical suspension copolymerization of chiral monomers, prepared from optically active amines or amino acid derivatives (**60, 61**) with a cross-linking agent such as ethylene glycol diacrylate, gives optically active polymer particle gels of the desired mean diameter. The gels are not suitable for HPLC work because of lack of resistance to compression, but they can be used in low pressure liquid chromatography.

$R_1 = C_6H_5, c\text{-}C_6H_{11},$
1-Naph, *p*-I-C$_6$H$_4$

$R_2 = H, CH_3$

$R_1 = CH_3, C_6H_5, C_6H_5CH_2,$
$c\text{-}C_6H_{11}CH_2, p\text{-}HOC_6H_4CH_2$
$p\text{-}CH_3CO_2C_6H_4CH_2$

$R_2 = C_2H_5, t\text{-}C_4H_9$

Blaschke et al. succeeded in resolving many drugs with these gels on a preparative scale and evaluated the differences in pharmacological behavior of the enantiomers [112]. It is noteworthy that these authors completely resolved ($\pm$)-thalidomide (**62**), which was used as a hypnotic drug, and found that the teratogenic action is caused by the (S)-isomer only; the (R)-isomer did not cause any teratogenic action, even at high doses [113]. This is one of key studies through which people recognized the importance of evaluating pharmacological behavior of both enantiomers of a given substance. The resolving power of the polymer gels is dependent on substituents (R_1, R_2) and the method of synthesis; for example, the racemic antiepileptic agent mephenytoin (**63**) is sufficiently resolved on **61** ($R_1 = C_6H_5CH_2$, $R_2 = C_2H_5$), but poorly on **60** ($R_1 = c\text{-}C_6H_{12}$, $R_2 = CH_3$). However, thalidomide (**62**) is hardly separated into enantiomers at all on the latter column but completely resolved on the former column [114].

Usually, these adsorbents are suitable for the resolution of racemic compounds containing functional groups capable of hydrogen bonding: amide, imide, carboxylic acid, or alcohol units. Mesuximide (**64**), with no amide hydrogen, is only very slightly resolved [110]. Therefore, nonpolar eluents such as benzene and toluene bring about better separation of enantiomers than polar eluents.

The optically active polymer gels can also be prepared by the reaction of optically active amines with cross-linked poly(acryloyl chloride). However, the gels thus obtained show lower resolving ability [110]. This result suggests that the resolving ability of the gel may not be ascribable simply to the chiral side group but may be partly due as well to the chiral higher order structure (e.g., a regular helical structure).

The acrylamide monomer **61** ($R_1 = C_6H_5CH_2$, $R_2 = C_2H_5$) was polymerized in the presence of silica gel. The polymer-bound silica gel obtained was useful for HPLC, and column efficiency was considerably improved. This CSP has been marketed [111]. Figure 4.16 shows some compounds resolved on the column. The column packed with poly[1-(naphthyl)ethyl methacrylamide] chemically bonded to silica gel also is available commercially [115]. The polymethacrylamides bearing (−)-menthylamine and (S)-amino acids as pendant groups were synthesized by radical polymerization of the corresponding methacrylamides (**65**), and their chiral recognition abilities were evaluated. The CSPs can effectively resolve some drugs [116].

62   **63**   **64**

65 R = CH_3, $CH_2CH(CH_3)_2$, (S)-$CH(CH_3)C_2H_5$

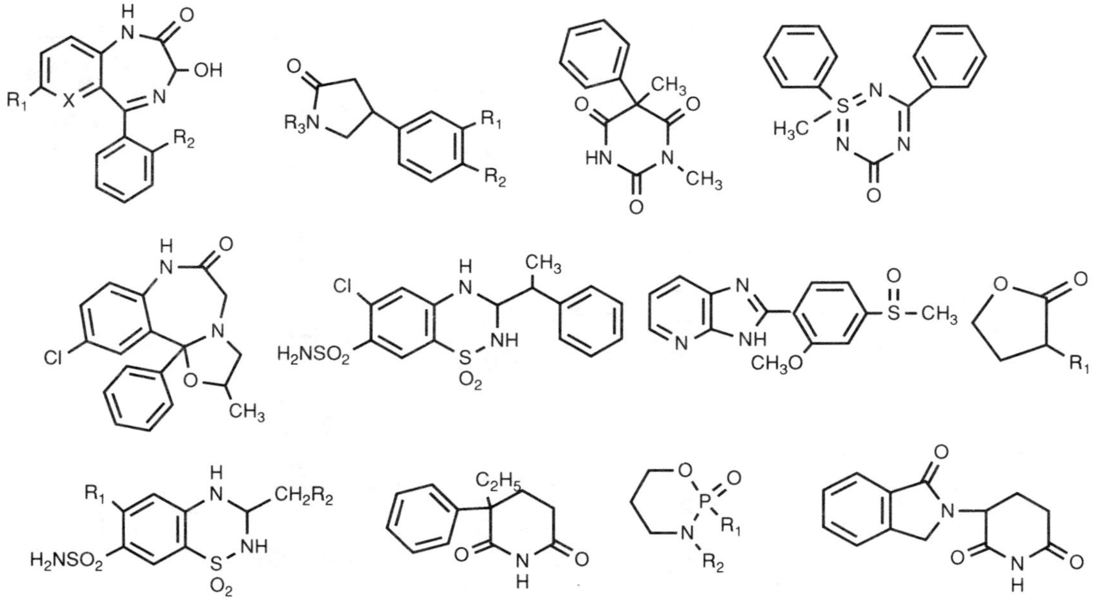

Figure 4.16 Compounds resolved on polyacrylamide.

Optically active poly(meth)acrylamides bearing penicillin sulfoxide skeleton (**10** in Figure 4.1) were prepared and coated on silica gel to use as CSPs. Benzylester derivative resolved some aromatic racemic compounds [117].

4.7 Polyurethanes

In 1992 optically active polyurethanes were prepared and used as a CSP (**11** in Figure 4.1) [118]. The polymer adsorbed on silica gel can separate several polar racemates. Polyurethane (**66**), which can separate 1,1'-binaphthyl compounds, has also been prepared [119].

$$\left[\begin{array}{c} R_1 \quad R_1 \\ \end{array} \right]_n$$

O-C-NH-R_2-NH-C-O

66

4.8 Synthetic Polymers with Chiral Cavities

A very interesting approach for obtaining chiral gels has been studied extensively by Wulff et al. [120,121]. A "molecular imprinting" technique, incorporating a chiral molecule as a template, was used to prepare cross-linked polymer gels containing chiral cavities, and the racemic templates were separated into enantiomers with the gels by liquid chromatography. The procedure is schematically shown in Figure

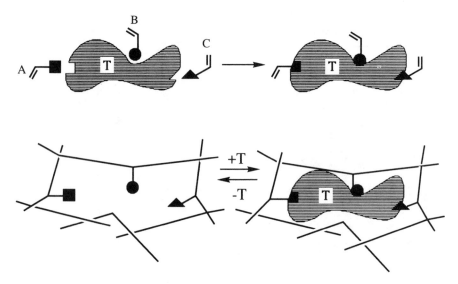

Figure 4.17 Procedure of "molecular imprinting" by means of a template T.

4.17. A template molecule, 4-nitrophenyl-α-D-mannopyranoside (**67**), was converted to the corresponding polymerizable boronic ester (**68**), which was copolymerized to form highly cross-linked polymers. Removal of the templates leaves the polymer gels with chiral cavities imprinting the template molecules. The column packed with the polymer gels adsorbed D-isomer preferentially over L-isomer, and baseline separation was achieved.

Besides sugar derivatives, amino acid derivates [122] and (S)-timolol [123] were used by Mosbach et al. as template molecules. A noncovalent interaction was used for fixation of the templates during polymerization, and amino acid derivatives and β-blockers other than racemic templates were separated into enantiomers.

Most polymer gels show relatively high chiral recognition abilities, but are not practical because of the difficulty of preparing an efficient CSP.

4.9 Other Chiral Polymers

Besides the polymers shown in Figure 4.1, the chiral polymers **69** and **70** have been utilized as CSPs. Several alcohols and amines with aromatic groups were resolved on **69** using hexane as an eluent [124], and amino acids were partially resolved on **70** [125].

References

1. S.G. Allenmark, *Chromatographic Enantioseparation: Methods and Application*. Wiley, New York, 1988.

2. S. Ahuja, in S. Ahuja, Ed., *Chiral Separation by Liquid Chromatography*, ACS Symposium Series 471. American Chemical Society, Washington, DC, 1991, Chapter 1.

3. Kikan Kagaku Sosetsu, No. 6, *Resolution of Optical Isomers*. Gakkai Shuppan Center, Tokyo, 1989.

4. S. Allenmark and B. Bomgrem, *J. Chromatogr.* **264**, 63 (1983).

5. S. Allenmark, *J. Liquid Chromatogr.* **9**, 425 (1986).

6. S. Allenmark, S. Andersson, and J. Bojarski, *J. Chromatogr.* **436**, 479 (1988).

7. Y.-Q. Chu and I. Wainer, *Pharm. Res.* **5**, 680 (1988).

8. R.A. Thompson, S. Andersson, and S. Allenmark, *J. Chromatogr.* **465**, 263 (1989). S. Andersson and S. Allenmark, *J. Chromatogr.* **591**, 65 (1992).

9. S. Andersson, S. Allenmark, and S. Nilsson, *J. Chromatogr.* **498**, 81 (1990).

10. E. Domenici, C. Bertucci, P. Solvadori, G. Felix, I. Cahagne, S. Motellier, and I.W. Wainer, *Chromatographia*, **29**, 170 (1990). T.A.G. Noctor, G. Felix, and I.W. Wainer, *Chromatographia*, **31**, 55 (1991).

11. J. Hermansson, *J. Chromatogr.* **269**, 71 (1983).

12. J. Hermansson and K. Strom, *Chromatographia*, **24**, 520 (1987).

13. T. Miwa, M. Ichikawa, M. Tsuno, T. Hattori, T. Miyazawa, M. Kayano, and Y. Miyake, *Chem. Pharm. Bull.* **35**, 682 (1987). T. Miwa, T. Miyazawa, M. Kayano, and Y. Miyake, *J. Chromatogr.* **408**, 316 (1987).

14. T. Miwa, H. Kuroda, S. Sakashita, N. Asakawa, and Y. Miyake, *J. Chromatogr.* **511**, 89 (1990).

15. K. M. Kirkland, K. L. Neilson, and D.M. McCombs, *J. Chromatogr.* **545**, 43 (1991).

16. P. Erlandsson, I. Mark, L. Hansson, R. Isaksson, C. Pettersson, and G. Pettersson, *J. Am. Chem. Soc.* **112**, 4573 (1990). S. Jönsson, A. Schön, R. Isaksson, C. Pettersson, and G. Pettersson, *Chirality*, **4**, 505 (1992).

17. I.W. Wainer, P. Jadaud, G.R. Schombaum, S.V. Kadadker, and M.P. Henry, *Chromatographia*, **25**, 903 (1988). P. Jadaud and I.W. Wainer, *J. Chromatogr.* **476**, 165 (1989).

18. T. Miwa and T. Miyakawa, *J. Chromatogr.* **457**, 227 (1988).

19. T. Shibata, K. Mori, and Y. Okamoto, in A.M. Krstrulovic, Ed., *Chiral Separation by HPLC*. Ellis Harwood, Chichester, 1989, p. 336.

20. Y. Okamoto and Y. Kaida, *J. Chromatogr. A*, **666**, 403 (1994).

21. M. Kotake, T. Sakan, N. Nakamura, and S. Senoh, *J. Am. Chem. Soc.* **73**, 2975 (1951).

22. S. Yuasa, A. Shimada, K. Kameyama, M. Yasui, and K. Adzuma, *J. Chromatogr. Sci.* **18**, 311 (1980).

23. T. Fukuhara, M. Isoyama, A. Shimada, M. Itoh, and S. Yuasa, *J. Chromatogr.* **387**, 562 (1987).

24. G. Gubitz, W. Jellenz, and D. Schonleber, *J. High Resolution Chromatogr., Chromatogr. Commun.* **3**, 31 (1980).

25. G. Hesse and R. Hagel, *Chromatographia*, **6**, 227 (1973).

26. G. Hesse and R. Hagel, *Liebigs Ann. Chem.* **1976,** 966.

27. A. Luttringhaus, V. Hess, and H.J. Rosenbaum, *Z. Naturforsch.* **B22,** 1296 (1967).

28. A. Mannschreck, H. Koller, and R. Wernicke, *Kontakte,* **1,** 40 (1985).

29. A. Hussenius, R. Isaksson, and O. Matsson, *J. Chromatogr.* **405,** 155 (1987).

30. E. Francotte and D. Lohman, *Helv. Chim. Acta,* **70,** 1569 (1987).

31. A.M. Rizzi, *J. Chromatogr.* **513,** 195 (1990).

32. M. Krause and R. Galensa, *Chromatographia,* **32,** 69 (1991).

33. A. Ichida, T. Shibata, I. Okamoto, Y. Yuki, H. Namikoshi, and Y. Toda, *Chromatographia,* **19,** 280 (1984). T. Shibata, I. Okamoto, and K. Ishii, *J. Liquid Chromatogr.* **9,** 313 (1986).

34. Y. Okamoto, M. Kawashima, K. Yamamoto, and K. Hatada, *Chem. Lett.* 739 (1984).

35. T. Shibata, T. Sei, H. Nishimura, and K. Deguchi, *Chromatographia,* **24,** 552 (1987).

36. Y. Okamoto, R. Aburatani, and K. Hatada, *J. Chromatogr.* **389,** 95 (1987).

37. F.A. Maris, R.J.M. Vervoort, and H. Hindriks, *J. Chromatogr.* **547,** 45 (1991).

38. E.J. Corey and D.-H. Lee, *J. Am. Chem. Soc.* **113,** 4026 (1991).

39. E. Francotte and R.M. Wolf, *Chirality,* **3,** 43 (1991). E. Francotte, R.W. Lang, and T. Winkler, *Chirality,* **3,** 177 (1991).

40. I.W. Wainer and M.C. Alembik, *J. Chromatogr.* **358,** 85 (1986). I.W. Wainer, M.C. Alembik, and E. Smith, *J. Chromatogr.* **388,** 65 (1987).

41. I.W. Wainer, R.M. Stiffin, and T. Shibata, *J. Chromatogr.* **411,** 139 (1987).

42. Y. Okamoto, M. Kawashima, and K. Hatada, *J. Am. Chem. Soc.* **106,** 5357 (1984).

43. Y. Okamoto, M. Kawashima, and K. Hatada, *J. Chromatogr.* **363,** 173 (1986).

44. Y. Okamoto, K. Hatano, R. Aburatani, and K. Hatada, *Chem. Lett.* **1989,** 715. H. Hopf, W. Grahn, D.G. Barrett, A. Gerdes, J. Hilmer, J. Hucker, Y. Okamoto, and Y. Kaida, *Chem. Ber.* **123,** 841 (1990).

45. Y. Okamoto, R. Aburatani, and K. Hatada, *Buil. Chem. Soc. Japan,* **63,** 955 (1990).

46. Y. Okamoto, M. Kawashima, R. Aburatani, K. Hatada, T. Nishiyama, and M. Masuda, *Chem. Lett.* 1237 (1986).

47. Y. Okamoto, R. Aburatani, Y. Kaida, and K. Hatada, *Chem. Lett.* 1125 (1988).

48. K. Ikeda, T. Hamasaki, H. Kohno, T. Ogawa, T. Matsumoto, and J. Sakai, *Chem. Lett.* 1089 (1989).

49. A. Ishikawa and T. Shibata, *J. Liquid Chromatogr.* **16,** 859 (1993).

50. U. Vogt and P. Zugenmaier, *Makromol. Chem. Rapid Commun.* **4,** 759 (1983).

51. Y. Okamoto, Y. Kaida, R. Aburatani, and K. Hatada, *J. Chromatogr.* **477,** 367 (1989).

52. Y. Okamoto, R. Aburatani, K. Hatano, and K. Hatada, *J. Liquid Chromatogr.* **11,** 2147 (1988).

53. Y. Okamoto, R. Aburatani, S. Miura, and K. Hatada, *J. Liquid Chromatogr.* **10,** 1613 (1987).

54. Y. Okamoto, R. Aburatani, T. Fukumoto, and K. Hatada, *Chem. Lett.* 1857 (1987).

55. R. Aburatani, Y. Okamoto, and K. Hatada, *Bull. Chem. Soc. Japan,* **63,** 3606 (1990).

56. D.W. Armstrong, *J. Liquid Chromatogr.* **7,** 353 (1984).

57. D.W. Armstrong, A.M. Stalcup, M.L. Hilton, J.D. Duncan, J.R. Faulkner, Jr., and S.-C. Chang, *Anal. Chem.* **62,** 1610 (1990).

58. T. Hargitai, Y. Kaida, and Y. Okamoto, *J. Chromatogr.* **628,** 11 (1993).

59. T. Hargitai and Y. Okamoto, *J. Liquid Chromatogr.* **16,** 843 (1993).

60. Y. Okamoto, Y. Kaida, H. Hayashida, and K. Hatada, *Chem. Lett.* 909 (1990).

61. Y. Okamoto, Y. Kaida, R. Aburatani, and K. Hatada, in S. Ahuja, Ed., *Chiral Separations by Liquid Chromatography,* ACS Symposium Series 471. American Chemical Society, Washington, DC, 1991, Chapter 5.

62. Y. Kaida and Y. Okamoto, *Chem. Lett.* 85 (1992).

63. Y. Kaida and Y. Okamoto, *Chirality,* **4,** 122 (1992).

64. H. Kiniwa, Y. Doi, T. Nishikaji, and N. Ogata, *Makromol. Chem.* **188,** 1841 (1987); *J. Chromatogr.* **396,** 395 (1987).

65. C. Hirayama, H. Ihara, and K. Tanaka, *J. Chromatogr.* **450,** 271 (1988).

66. K. Saigo, Y. Chen, N. Kubota, K. Tachibana, N. Yonezawa, and M. Hasegawa, *Chem. Lett.* 515 (1986).

67. Y. Okamoto, Y. Nagamura, T. Fukumoto, and K. Hatada, *Polym. J.* **23,** 1197 (1991).

68. K. Saigo, Y. Chen, N. Yonezawa, K. Tachibana, T. Kanoe, and M. Hasegawa, *Chem. Lett.* 1891 (1985). K. Saigo, N. Nakamura, Y. Adegawa, S. Noguchi, and M. Hasegawa, *Chem. Lett.* 337 (1989).

69. K. Saigo, Y. Chen, K. Fujioka, N. Yonezawa, and M. Hasegawa, *Chem. Lett.* 647 (1988). K. Saigo, T. Shiwaku, K. Hayashi, K. Fujioka, M. Sukegawa, Y. Chen, N. Yonezawa, M. Hasegawa, and T. Hashimoto, *Macromolecules,* **23,** 2830 (1990).

70. K. Saigo, M. Nakamura, Y. Adegawa, S. Noguchi, and M. Hasegawa, *Chem. Lett.* 337 (1989).

71. Y. Tamai, Y. Matsuzaka, S. Oi, and S. Miyano, *Bull. Chem. Soc. Japan,* **64,** 2260 (1991).

72. Y. Okamoto, K. Suzuki, K. Ohta, K. Hatada, and H. Yuki, *J. Am. Chem. Soc.* **101,** 4768 (1979).

73. H. Yuki, K. Hatada, T. Niinomi, and Y. Kikuchi, *Polym. J.* **1,** 130 (1970).

74. Y. Okamoto, H. Shohi, and H. Yuki, *J. Polym. Sci., Polym. Lett. Ed.* **19,** 451 (1981). T. Nakano, Y. Okamoto, and K. Hatada, *J. Am. Chem. Soc.* **114,** 1318 (1992).

75. H. Yuki, Y. Okamoto, and I. Okamoto, *J. Am. Chem. Soc.* **102,** 6356 (1980).

76. Y. Okamoto, I. Okamoto, H. Yuki, S. Murata, R. Noyori, and H. Takaya, *J. Am. Chem. Soc.* **103,** 6971 (1981).

77. Y. Okamoto and K. Hatada, *J. Liquid Chromatogr.* 9(2&3), 369 (1986).

78. Y. Okamoto, *CHEMTECH,* 176 (1987).

79. Y. Okamoto, S. Honda, E. Yashima, and H. Yuki, *Chem. Lett.* 1221 (1983).

80. Y. Okamoto, E. Yashima, and K. Hatada, *J. Chem. Soc., Chem. Commun.* 1051 (1984).

81. Y. Okamoto, S. Honda, K. Hatada, I. Okamoto, Y. Toga, and S. Kobayashi, *Bull. Chem. Soc. Japan,* **57,** 1681 (1984).

82. R. Noyori, N. Sano, S. Murata, Y. Okamoto, and H. Yuki, *Tetrahedron Lett.* **23,** 2969 (1982).

83. Y. Kawada, H. Iwamura, Y. Okamoto, and H. Yuki, *Tetrahedron Lett.* **24,** 791 (1983).

84. Y. Kawada, Y. Okamoto, and H. Iwamura, *Tetrahedron Lett.* **24,** 5359 (1983).

85. A. Tajiri, M. Fukuda, M. Hatano, T. Morita, and K. Takase, *Angew. Chem., Int. Ed. Engl.* **22**, 870 (1983).

86. Y. Okamoto, S. Honda, H. Yuki, H. Nakamura, Y. Iitake, and T. Nozoe, *Chem. Lett.* **1984**, 1149.

87. K. Yamamoto, T. Ueda, H. Yumioka, Y. Okamoto, and T. Yoshida, *Chem. Lett.* 1977 (1984).

88. Y. Okamoto, E. Yashima, K. Hatada, and K. Mislow, *J. Org. Chem.* **49**, 557 (1984).

89. K. Yamamoto, H. Fukushima, Y. Okamoto, K. Hatada, and M. Nakazaki, *J. Chem. Soc., Chem. Commun.* 1111 (1984).

90. K. Yamamoto, K. Noda, and Y. Okamoto, *J. Chem. Soc., Chem. Commun.* 1421 (1985).

91. W. Kissener and F. Vogtle, *Angew. Chem., Int. Ed. Engl.* **24**, 222 (1985).

92. A. Tajiri, N. Morita, T. Aso, and M. Hatano, *Angew. Chem., Int. Ed. Engl.* **24**, 329 (1985).

93. K.-H. Duchene and F. Vogtle, *Angew. Chem., Int. Ed. Engl.* **24**, 885 (1985).

94. N. Harada, J. Iwabuchi, Y. Yokota, H. Uda, Y. Okamoto, H. Yuki, and Y. Kawada, *J. Chem. Soc., Perkin Trans.* 1845 (1985).

95. K. Meurer, A. Aigner, and F. Vogtle, *J. Inclusion Phenom.* **3**, 51 (1985).

96. Y. Tokuma, T. Fujiwara, and H. Noguchi, *J. Pharm. Sci.* **76**, 310 (1987).

97. M. Yoshifuji, K. Toyota, T. Niitsu, N. Inamoto, Y. Okamoto, and R. Aburatani, *J. Chem. Soc., Chem. Commun.* 1550 (1986).

98. C.P. Clark, B.G. Snider, and P.B. Bowman, *J. Chromatogr.* **408**, 275 (1987).

99. A. Tambute, J. Canceill, and A. Collet, *Bull. Chem. Soc. Japan,* **62**, 1390 (1989).

100. J.M. Chance, J.H. Geiger, Y. Okamoto, R. Aburatani, and K. Mislow, *J. Am. Chem. Soc.* **112**, 3540 (1990).

101. B.L. Feringa, W.F. Jager, and B. de Lange, *J. Am. Chem. Soc.* **113**, 5468 (1991).

102. Y. Okamoto, I. Okamoto, and H. Yuki, *Chem. Lett.* 835 (1981).

103. Y. Okamoto, S. Honda, K. Hatada, and H. Yuki, *J. Chromatogr.* **350**, 127 (1985).

104. Y. Okamoto, S. Honda, K. Hatada, and H. Yuki, *Bull. Chem. Soc. Japan,* **58**, 3053 (1985).

105. Y. Okamoto, H. Mohri, M. Nakamura, and K. Hatada, *J. Chem. Soc., Japan,* 435 (1987).

106. Y. Okamoto, H. Mohri, and K. Hatada, *Chem. Lett.* 1897 (1988).

107. Y. Okamoto, H. Mohri, and K. Hatada, *Polym. J.* **21**, 439 (1989).

108. Y. Okamoto, E. Yashima, M. Ishikura, and K. Hatada, *Bull. Chem. Soc. Japan,* **61**, 255 (1988).

109. Y. Tamai, P. Qian, K. Matsunaga, and S. Miyano, *Bull. Chem. Soc. Japan,* **65**, 817 (1992).

110. G. Blaschke, *Angew. Chem, Int. Ed. Engl.* **19**, 13 (1980). G. Blaschke, in M. Zief and L.J. Crane, Eds., *Chromatographic Chiral Separation,* Chromatographic Science Series, Vol. 40. Dekker, New York, 1988, Chapter 7.

111. G. Blaschke, W. Bröeker, and W. Frankel, *Angew. Chem., Int. Ed. Engl.* **25**, 830 (1986).

112. G. Blaschke, *J. Liquid Chromatogr.* **9**, 341 (1986).

113. G. Blaschke, H.-P. Kraft, K. Fickentscher, and F. Köhler, *Arzneim.-Forsch.* **29**, 1640 (1979).

114. G. Blaschke, H.-P. Kraft, and H. Markgraf, *Chem. Ber.* **113**, 2318 (1980).

115. S. Hasegawa and Y. Sakurai, 52th National Meeting of the Chemical Society of Japan, August 1986, Abstract 2J36.

116. D. Arlt, B. Bömer, R. Grosser, and W. Lange, *Angew. Chem., Int. Ed. Engl.* **30**, 1662 (1991).

117. Y. Saotome, T. Miyazawa, and T. Endo, *Chromatographia,* **28**, 505, 509, 511 (1989).

118. Y. Chen and J.-J. Lin, *J. Polym. Sci.: Part A: Polym. Chem.* **30**, 2699 (1992).

119. T. Kobayashi, S. Watanabe, M. Kakimoto, and Y. Imai, *Polym. Prep. Japan,* **41**, 1875 (1992).

120. G. Wulff, W. Vesper, R. Grobe-Einsler, and A. Sarhan, *Makromol. Chem.* **178**, 2799 (1977). G. Wulff, H.-G. Poll, and M. Minarik, *J. Liquid Chromatogr.* **9**, 385 (1986).

121. M. Minarik and G. Wulff, *J. Liquid Chromatogr.* **13**, 2987 (1990). G. Wulff and G. Kirstein, *Agnew. Chem., Int. Ed. Engl.* **29**, 684 (1990).

122. L.I. Anderson and K. Mosbach, *J. Chromatogr.* **516**, 313 (1990).

123. L. Fischer, R. Muller, B. Ekberg, and K. Mosbach, *J. Am. Chem. Soc.* **113**, 9358 (1991).

124. N. Kunieda, H. Chakihara, and M. Kinoshita, *Chem. Lett.* 317 (1990).

125. T. Kakuchi, T. Takaoka, and K. Yokota, *Polym. J.* **22**, 199 (1990).

5

Combination of Biochromatography and Chemometrics: A Potential New Research Strategy in Molecular Pharmacology and Drug Design

Roman Kaliszan

Irving W. Wainer

5.1 Introduction

The fundamental processes of drug action have much in common with the processes that are the basis of chromatographic separations. Biological processes of absorption, distribution, excretion, and receptor activation are dynamic, as are the solute's distribution processes in chromatography. None of the essential pharmacological or chromatographic processes (excepting metabolism) implies the breaking of existing bonds in a drug (solute) molecule or the formation of new bonds. The same basic intermolecular interactions determine the behavior of chemical compounds in both biological and chromatographic environments [1]. However, the extreme stereospecificity of biological interactions has differentiated them from interactions determining regular chromatographic separations.

Modern techniques and procedures of high performance liquid chromatography (HPLC) and capillary electrophoresis (CE) allow for the inclusion of biomolecules as active components of chromatographic systems. This approach is now termed "biochromatography." The extreme complexity of biological systems limits rational design of an individual chromatographic system that would directly mimic a given biological system. On the other hand, chromatography is a unique method that can readily yield a great amount of diversified, precise, and reproducible data. It can be presumed that chemometric processing of appropriately designed and selected sets

273

of chromatographic data can reveal systematic information regarding the xenobiotics studied [2].

Chemometric processing of chromatographic data is synonymous with the analysis of quantitative structure–retention relationships (QSRR). QSRR is one of the most extensively studied manifestations of linear free energy relationships (LFER). It is assumed that demonstration of LFER suggests the presence of a real connection between certain correlated quantities and that the nature of this connection can be identified. One assumes that correlations among specific quantities are attributable to physicochemical relationships. The statistically derived correlations encourage attempts to identify the relationships behind them.

The following applications of QSRR studies have been described [1,3] (Figure 5.1):

1. Prediction of retention for a given solute.
2. Identification of the most informative (regarding properties) structural descriptors of solutes.

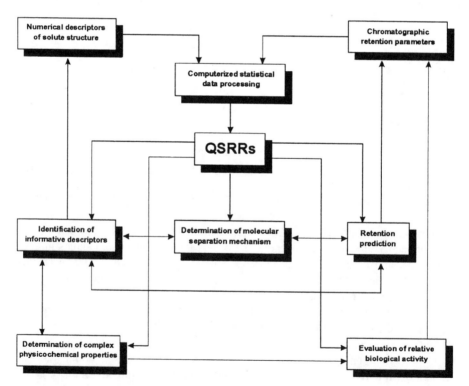

Figure 5.1   Methodology and applications of quantitative structure–retention relationships (QSRR).

3. Elucidation of molecular mechanism of separation operating in individual chromatographic systems.
4. Determination of hydrophobicity measures of drugs and other solutes.
5. Estimation of biological activities within a set of solute xenobiotic compounds.

QSRR studies require a set of quantitatively comparable retention data (dependent variable) for a sufficiently large set of solutes and a set of quantities (independent variables) reflecting structural features of these solutes (Figure 5.1). Through the use of chemometric computational techniques, retention parameters are characterized in terms of various combinations of solute descriptors or in terms of systematic knowledge extracted from these descriptors. When properly executed, QSRR can provide otherwise inaccessible information about the solutes and chromatographic systems investigated. This information may be of use in studies of other structure–property relationships, such as those characterizing specific pharmacological activities.

The development of statistically significant and physically meaningful QSRR requires reliable input data and stringent mathematical analysis. The ability of chromatography to readily produce a great amount of precise and reproducible data is an advantage of QSRR over quantitative structure–property relationship studies of other types. In a chromatographic process, all the experimental conditions can be kept constant and the solute structure becomes the single independent variable in the physicochemical system.

There are numerous reports regarding biochromatographic separations. Also, QSRR analysis of various chromatographic data has been the subject of many publications [1,4]. However, the combination of biochromatography and chemometrics is a research strategy very recently introduced and developed [2,5,6]. Such a combined approach (biochromatometrics?) is demonstrated here based on recent publications by the authors to provide information of direct relevance to molecular pharmacology and drug design.

5.2 Immobilized Artificial Membrane Stationary Phase in the Biochromatographic Determination of Drug Hydrophobicity

The hydrophobic effect is assumed to be one of the driving forces for both passive diffusion of xenobiotics through biological membranes and drug–receptor binding. For this reason, much effort has been devoted to the development of reliable and convenient systems for measuring hydrophobicity. All such systems yield quantities reflecting the degree of "phobia" of the solutes against aqueous phase and "philia" toward nonpolar species (lipophilicity). However, a complex property like hydrophobicity depends on solute environment and, thus, individual measuring systems produce data that highlight specifically different aspects of what is called (more or less intuitively) hydrophobicity [7].

The n-octanol/water partition system is the common reference system for determination of relative hydrophobicity [8]. Expressed as the logarithm of the partition coefficient, log P, this measure is the one most widely employed in medicinal chemistry and molecular pharmacology [8]. Large compilations of log P data have been developed using data acquired experimentally by the standard shake-flash method and from so-called CLOGP data, resulting from theoretical calculations based on the fragmental method [8–10].

Because of the tediousness and poor reproducibility of hydrophobicity parameters determined by the slow equilibrium methods, researchers turned toward reversed-phase high performance liquid chromatography (RPLC) as a tool for the convenient assessment of relative hydrophobicities within a series of solutes. The RPLC systems commonly used in general hydrophobicity determinations employ octanol-like or hydrocarbonaceous stationary phase materials [11].

The RPLC approach has been relatively successful in duplicating log P data derived by traditional techniques. However, if the reason to use the chromatographic system is to model processes in the biophase, the components of the chromatographic and the biological systems must be comparable. Thus, an RPLC system that is intended to model transport through biological membranes should be composed of an aqueous phase and an organized phospholipid layer.

Miyake et al. [12] derived LC hydrophobicity parameters employing a column of silica gel physically coated with dipalmitoyl phosphatidylcholine (DPPC). Leaving aside inconveniences regarding their preparation and stability, the systems with DPPC adsorbed onto silica most probably do not emulate the lipid dynamics of biological membranes, because the adsorbed lipids are not organized in a manner similar to natural (or artificial) membranes.

Recently, a new RPLC stationary phase material, the immobilized artificial membrane stationary phase (IAM), became available (Figure 5.2) [13]. This phase more closely models natural membranes that are composed of lipids with a polar head group and two nonpolar chains. In the IAM phase, a lipid molecule, lecithin, is covalently bound to propylamine silica, forming confluent monolayers of immobilized membrane lipids. In one form of the IAM, unreacted propylamine moieties are end-capped with methylglycolate. Only one of the alkyl chains is linked to the propylamine silica surface, and the immobilized lipid head groups protrude away from the stationary phase surface. These charged moieties are the first contact site between solutes and IAM.

Preliminary studies reported encouraging correlations between LC retention parameters determined on an IAM and human skin permeation data for a short series of alcohols and steroids [13,14]. The same biological data showed poor correlation with LC retention parameters determined on regular hydrocarbonaceous reversed-phase LC columns.

Correlations between the biochromatographic parameter of hydrophobicity determined on IAM and reference hydrophobicity data were studied in our laboratories

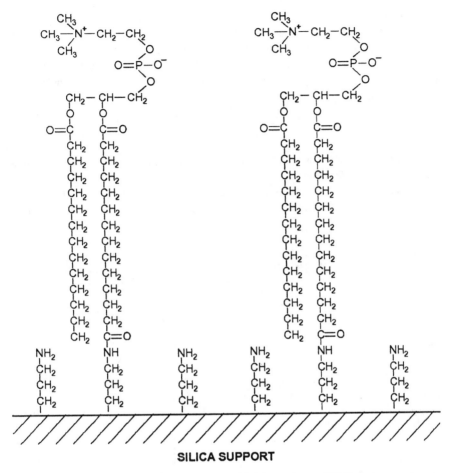

SILICA SUPPORT

Figure 5.2 Structure of the immobilized artificial membrane (IAM) stationary phase.

for groups of phenothiazine neuroleptics and antihistamine drugs [15], cardiovascular β-adrenolytics, α-adrenomimetics, and phenothiazine neuroleptics [16], as well as for a series of steroid hormones and phenol antiseptics [17]. Correlations varied depending on the class of agents studied.

The data considered in the case of β-adrenolytic drugs [16] are collected in Table 5.1. The relationship between the logarithm of the retention factor determined on the IAM with acetonitrile/0.1 M phosphate buffer pH 7.0 10:90% (v/v), log k'_{IAM}, and the n-octanol/water partition coefficient corrected for ionization at pH 7.0, log D, is described by the following equation:

$$\log D = 2.055(\pm0.206) \log k'_{IAM} - 1.560(\pm0.123) \qquad (5.1)$$
$$n = 10, R = 0.962, s = 0.32$$

where n denotes the number of solutes considered to derive the regression equation, R is the correlation coefficient, and s is the standard error of estimate; numbers in parentheses account for the standard error of the regression coefficients.

There is a lower correlation between log k'_{IAM} and the theoretically calculated from fragmental constants the n-octanol/water hydrophobicity parameter [10] corrected for ionization at pH 7.0, CLOGD:

$$CLOGD = 1.620(\pm 0.302) \log k'_{IAM} - 1.693(\pm 0.172) \qquad (5.2)$$
$$n = 11, R = 0.873, s = 0.47$$

Most probably the lower statistical quality of equation 5.2 with regard to equation 5.1 reflects the inadequacies of CLOGD.

For seven members of the set of psychotropic drugs (Table 5.2) [16] chromatographed on the IAM, both log P and pK_a data were found, and the log D values

Table 5.1 HYDROPHOBICITY PARAMETERS AND pK_a DATA FOR A SERIES β-ADRENOLYTIC DRUGS

Drug	pK_a[a]	Log P[b]	Log D[c]	CLOGP[d]	CLOGD[e]	Log k'_{IAM}[f]
Acebutolol	9.20	1.77	−0.43	1.61	−0.59	0.528
Alprenolol	9.65	3.10	0.45	2.59	−0.06	0.853
Atenolol	9.60	0.16	−2.44	−0.11	−2.71	−0.301
Betaxolol				2.17		0.929
Bisoprolol				1.69		0.574
Bopindolol				4.86		0.376
Carteolol		−0.46		1.17		0.301
Celiprolol				1.66		0.653
Cicloprolol				2.39		0.813
Dilevalol	9.45	3.09		2.18		1.211
Esmolol				1.53		0.574
Metoprolol	9.70	1.88	−0.82	1.20	−1.50	0.352
Nadolol	9.39	0.71	−1.68	0.23	−2.16	0.176
Nebivalol				3.50		
Nifenalol	8.78			1.16	−0.63	0.176
Oxprenolol	9.50	2.18	−0.32	1.62	−0.88	0.512
Pindolol	9.70	1.75	−0.95	1.65	−1.05	0.512
Practolol	9.50	0.79	−1.71	0.78	−1.72	−0.204
Propranolol	9.45	3.56	1.11	2.75	0.30	1.279
Sotalol	9.05	0.24	−1.81	0.23	−1.82	−0.301
Timolol						0.301

[a] Acidity constant.
[b] Logarithm of n-octanol/water partition coefficient for neutral forms determined experimentally by the shake-flask method.
[c] Logarithm of experimental n-octanol/water partition coefficient corrected for ionization at pH 7.0.
[d] Logarithm of theoretical n-octanol/water partition coefficient for neutral form of solute calculated by fragmental methods.
[e] Logarithm of theoretical n-octanol/uwater partition coefficient corrected for ionization at pH 7.0.
[f] Logarithm of LC retention factor determined on an immobilized artificial membrane column.

Source: Ref. 10 (data on pK_a, log P, and CLOGP from Craig) and Ref. 16 (data on log D, CLOGD, and Log k'_{IAM} calculated or determined by Kaliszan et al.).

Table 5.2 pK_a VALUES, HYDROPHOBICITY PARAMETERS AND ANTIHEMOLYTIC ACTIVITY OF A SERIES OF PHENOTHIAZINE NEUROLEPTICS

Drug	pK_a[a]	Log P[b]	Log D[c]	CLOGP[d]	CLOGD[e]	Log k'_{IAM}[f]	Log $1/C$[g]
Chlorpromazine	9.30	5.35	3.05	5.20	2.90	1.440	4.72
Fluphenazine	8.10	4.36	3.23	5.90	4.77	1.401	5.00
Perphenazine	7.80	4.20	3.34	5.57	4.71	1.373	4.82
Prochloperazine	8.10			6.15	5.02	1.782	5.22
Promazine	9.40	4.55	2.15	4.28	1.88	1.171	4.55
Promethazine	9.10			4.65	2.55	1.195	
Propiomazine	9.10			5.00	2.90	1.265	
Thioridazine	9.50	5.90	3.40	6.42	3.92	1.747	5.30
Trifluoperazine	8.10	5.03	3.90	6.48	5.35	1.754	5.22
Trifluopromazine	9.20	5.19	2.99	5.53	3.33	1.474	4.96

[a] Acidity constant.

[b] Logarithm of experimental n-octanol/water partition coefficient for neutral forms determined by the shake-flask method.

[c] Logarithm of experimental n-octanol/water partition coefficient corrected for ionization at pH 7.0.

[d] Logarithm of theoretical n-octanol/water partition coefficient for neutral form of solute calculated by fragmental methods.

[e] Logarithm of theoretical n-octanol/water partition coefficient corrected for ionization at pH 7.0.

[f] Logarithm of LC retention factor determined on an immobilized artificial membrane column.

[g] Logarithm of reciprocal of hemolytic concentration.

Source: Ref. 10 (data on pK_a, log P, and CLOGP from Craig) and Ref. 16 (data on log D, CLOGD, and Log k'_{IAM} calculated or determined by Kaliszan et al.), and Ref. 20 (data on log $1/C$ from Seeman).

could be calculated which corresponded to pH 7.0 of the LC eluent. The following equation related the two hydrophobicity parameters:

$$\log D = 2.142(\pm 0.621) \log k'_{IAM} - 0.019(\pm 0.927) \tag{5.3}$$
$$n = 7, R = 0.839, s = 0.32$$

Note that slopes are similar in equations 5.1 and 5.3.

In case of nonionized steroids, the correlation between $\log k'_{IAM}$ and the classical hydrophobicity parameter, $\log P$, was $R = 0.911$. On the other hand, there was no significant correlation between these quantities in the case of phenolic derivatives. Also Ong et al. [18] recently reported moderate correlations between $\log P$ and $\log k'_{IAM}$.

Certainly, $\log k'_{IAM}$ from IAM columns and the reference $\log P$ reflect different characteristics of hydrophobicity. In such a situation $\log k'_{IAM}$ is an interesting alternative to $\log P$ as a predictor of bioactivity, especially since, contrary to $\log P$, the $\log k'_{IAM}$ is derived in a simple, fast, and reproducible manner.

5.3 Prediction of Bioactivity Using Biochromatographically Determined Hydrophobicity

Retention parameters determined on an immobilized artificial membrane column, $\log k'_{IAM}$, were found [16] to correlate better with several pharmacokinetic parameters of β-adrenolytics than the logarithms of apparent n-octanol/buffer partition coefficient determined by Hinderling et al. [19] by the shake-flask method, $\log k'_{SF}$, and the logarithms of an LC retention factor determined in an n-octanol-like chromatographic system, $\log k'_{C_{18}}$. Table 5.3 gives the statistical parameters of the respective equations relating pharmacokinetic parameters to $\log k'_{IAM}$, $\log k'_{SF}$, and $\log k'_{C_{18}}$. In every case but one, the statistics of the regressions describing bioactivity in terms of $\log k'_{IAM}$ appear to be better than that provided by $\log k'_{SF}$ and by another LC hydrophobicity measure, $\log k'_{C_{18}}$.

In the same study [16], logarithms of the reciprocals of phenothiazine antihemolytic concentrations, $\log 1/C$, taken from Seeman [20], were related to (1) the standard n-octanol/water partition coefficients obtained from Craig [10], which were corrected for ionization, $\log D$, and (2) $\log k'_{IAM}$ (Table 5.2). The respective regression equations were:

$$\log \frac{1}{C} = 1.126(\pm 0.174) \log k'_{IAM} + 3.266(\pm 0.266) \tag{5.4}$$
$$n = 8, R = 0.935, s = 0.102$$

$$\log \frac{1}{C} = 0.405(\pm 0.130) \log D + 3.662(\pm 0.416) \tag{5.5}$$
$$n = 7, R = 0.812, s = 0.170$$

The $\log k'_{IAM}$ data for a series of steroid hormones [17] are given in Table 5.4, along with the reference n-octanol/water partition parameter $\log P$, taken from El

Table 5.3 STATISTICAL PARAMETERS OF REGRESSION EQUATIONS RELATING
PHARMACOKINETIC PARAMETERS TO EXPERIMENTAL
HYDROPHOBICITY PARAMETERS:
n, NUMBER OF DATA POINTS; R, CORRELATION COEFFICIENT;
s, STANDARD ERROR OF ESTIMATE

| | Hydrophobicity Descriptor[b] | | | | | | | | |
| | Log k'_{IAM} | | | Log k'_{SF} | | | Log $k'_{C_{18}}$ | | |
Pharmacokinetic Parameter[a]	n	R	s	n	R	s	n	R	s
Log t_T	9	0.83	0.44	11	0.84	0.46	9	0.74	0.55
Log f_b	13	0.83	0.37	14	0.80	0.40	10	0.76	0.47
Log r_A	6	0.95	0.15	6	0.90	0.21	6	0.88	0.23
Log K_{BC}	10	0.90	0.18	11	0.86	0.22	9	0.76	0.26
Log V_{uss}	9	0.86	0.23	11	0.84	0.30	9	0.73	0.31
Log K_p	10	0.87	0.45	13	0.90	0.43	9	0.85	0.46
Log r	9	0.82	0.58	11	0.93	0.38	9	0.95	0.32

[a] The following logarithms are listed: log t_T, fraction of drug bound and unbound to tissue; log f_b, traction of drug bound in plasma; log r_A, ratio of fraction of drug bound and unbound to albumin; log K_{BC}, true red cell partition coefficient; log V_{uss}, steady state volume of distribution referenced to the unbound drug in plasma; log K_p, partition coefficient of drug between plasma protein and plasma water; log r, ratio of the fraction of drug nonrenally and renally eliminated.
[b] The following logarithms are listed: log k'_{IAM}, LC retention factor determined on an immobilized artificial membrane column at pH 7.0; log k'_{SF}, apparent n-octanol/buffer pH 7.0 partition coefficient determined by the shake-flask method; log $k'_{C_{18}}$, LC retention factor determined in an octanol-like chromatographic system at pH 7.4.
Source: Ref. 19, except for data on log k'_{IAM}, which are from Ref. 16.

Tayar et al. [21], as well as the human skin/water partition coefficient log P_m and the human skin permeation coefficient log K_p, taken from Scheuplein et al. [22]. The relationship between log K_p and log k'_{IAM} for the steroids is given by the equation:

$$\log K_p = 1.77(\pm 0.22) \log k'_{IAM} - 10.19(\pm 0.37) \tag{5.6}$$
$$n = 10, R = 0.942, p < 10^{-5}$$

where p is the significance level of the regression equation. The extent of correlation provided by equation 5.6 is illustrated in Figure 5.3 (upper part).

The corresponding equation, which describes log K_p in terms of log P, has the form:

$$\log K_p = 1.18(\pm 0.20) \log P - 10.39(\pm 0.53) \tag{5.7}$$
$$n = 10, R = 0.900, p < 10^{-4}$$

For correlation between the observed log K_p data and the data calculated by equation 5.7, see Figure 5.3 (lower part).

The good statistical quality of equations 5.6 and 5.7 can be explained by reference to the evident intercorrelation ($R = 0.911$) of the two hydrophobicity parameters log k'_{IAM} and log P. However, Figure 5.3 shows log k'_{IAM} a better predictor of skin permeation than log P.

Table 5.4   HUMAN SKIN PERMEABILITY COEFFICIENTS (LOG K_p, log P_m) AND HYDROPHOBICITY PARAMETERS OF A GROUP OF STEROID HORMONES

Steroid Hormone	Log k_p (cm/s)[a]	Log P_m[a]	Log k'_{IAM}[b]	Log P[c]
Aldosterone	−9.08	0.83	0.649	1.08
Cortisone	−8.56	0.93	0.772	1.42
Deoxycorticosterone	−6.90	1.57	1.730	2.88
Estradiol	−7.08	1.66	2.179	2.69
Estriol	−7.95	1.36	1.368	2.47
Estrone	−6.00	1.66	2.001	2.76
Hydrocortisone	−9.08	0.85	0.843	1.53
Pregnenolone	−6.38	1.70	2.124	3.13
Progesterone	−6.38	2.01	2.199	3.70
Testosterone	−6.95	1.36	1.693	3.31

[a]   Data according to Schenplein et al. [22].
[b]   Data determined on an immobilized artificial membrane column by Nasal et al. [17].
[c]   Logarithms of n-octanol/water partition coefficient according to El Tayal et al. [21].

For the group of hormones analyzed, a highly significant correlation between the skin/water partition coefficient, log P_m, and hydrophobicity parameter determined on the IAM column, log k'_{IAM}, also was observed:

$$\log P_m = 0.64(\pm 0.06 \log k'_{IAM} + 0.40(\pm 0.10) \qquad (5.8)$$
$$n = 10, \; R = 0.966, \; p < 10^{-5}$$

The corresponding equation describing log P_m in terms of log P has the form:

$$\log P_m = 0.42(\pm 0.06) \log P + 0.33(\pm 0.17) \qquad (5.9)$$
$$n = 10, \; R = 0.917, \; p < 10^{-4}$$

In 1988 Raykar et al. [23] reported correlation between log P and log P_m. The high statistical quality of equations 5.8 and 5.9 proves that both log k'_{IAM} and log P can be used to evaluate the partitioning of steroids between water and the striatum corneum. Again, log k'_{IAM} seems to be more reliable for this purpose than log P.

5.4 Biochromatography in the Qualitative and Quantitative Assessment of Drug Binding to Serum Albumin

The reversible attachment of a drug to serum proteins often plays a significant role in the drug's pharmacokinetics and pharmacodynamics. A clear understanding of this process is fundamental to the safe and rational use of many therapeutic agents. Since human serum albumin (HSA) is a major component of plasma proteins, the mechanism and degree of drug–HSA binding have been extensively studied. The initial studies of drug–HSA binding described this process as nonspecific, comparing the mechanism to adsorbance by charcoal. However, it quickly became clear

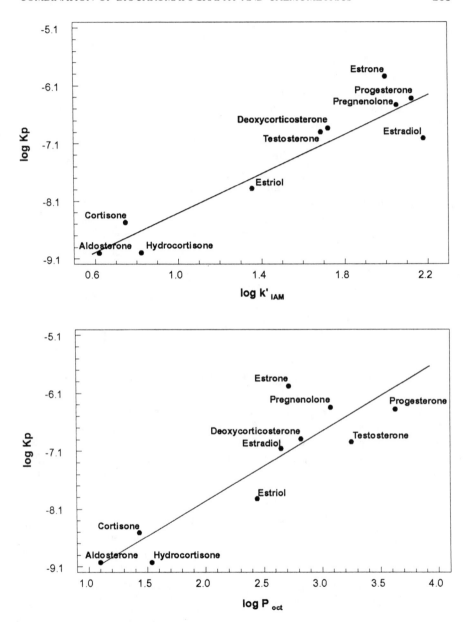

Figure 5.3 Relationships between the logarithm of the human skin permeation coefficient of steroids according to Scheuplein et al. [22] and the logarithm of the retention factor determined on an immobilized artificial membrane column and logarithm of n-octanol/water partition coefficient.

that the binding of a drug to HSA exhibits far greater structural and physicochemical selectivity. In the late 1950s, the three-dimensional structural selectivity (i.e., the ability of HSA to differentially bind the stereoisomers of a compound) was demonstrated for the enantiomers of several chiral compounds [24].

Since the extent to which a drug binds to plasma proteins can be an important factor in its pharmacological fate and effect, numerous methods for the determination of drug–protein binding have been described. Most involve equilibration of the drug with a solution of the isolated protein or with whole plasma, followed by the separation of free and bound drug using equilibrium dialysis, ultrafiltration, or ultracentrifugation.

Biochromatography offers an alternative experimental approach for the quantitative and qualitative determination of drug–protein binding. In this technique, the target protein is immobilized on an LC support, and the chromatographic retentions of the target solutes are determined on the resulting stationary phase.

In the past few years there has been rapid development of protein-based LC chiral stationary phases (CSPs) for the analytic and preparative separation of enentiomeric compounds [25]. These chromatographic supports include CSPs derived from α_1-acid glycoprotein (AGP-CSP) [26], ovomucoid (OVM-CSP) [27], bovine serum albumin (BSA-CSP) [28], and human serum albumin (HSA-CSP) [29].

Studies in these laboratories with the BSA-CSP [30] and HSA-CSP [29,31–36] have demonstrated that the stereochemical resolution achieved on these columns reflects the binding properties and stereoselectivity of the free protein. This description is typified by the use of biochromatography to detect and quantitate the drug–HSA binding of three series: 19 of the compounds were 1,4-benzodiazepines, 9 were coumarin derivatives, and 24 were structurally related triazole derivatives [35]. These compounds were chromatographed on an HSA-CSP and their k values determined. The percentage of binding of each compound to nonimmobilized HSA was also determined using standard ultrafiltration techniques. For each drug, the results from the ultrafiltration studies, expressed as percentage of drug bound to HSA, were correlated with chromatographic retention of the compound, expressed as $k'/(k' + 1)$.

The triazole derivatives studied showed very low amounts of binding to HSA (<40%) and were very poorly retained on the HSA-CSP; the correlation obtained between percentage of drug bound and $k'/(k' + 1)$ was poor, as well. For the more highly bound benzodiazepines (60–95%) and coumarins (80–98%), however, excellent correlations were obtained between chromatographic retention and extent of albumin binding, with correlation factors of 0.999 reported for both series (Figure 5.4).

These studies show that (1) the retention of a solute on the HSA-CSP is clearly related to its binding affinity for the free protein; (2) this relationship appears to be quantitative, at least when the albumin binding of the solute is 60% or higher; and (3) the HSA-CSP can be used as a rapid probe of drug–protein binding.

In addition to the determination of the extent of ligand binding to HSA, biochromatography can be used to probe interspecies differences in protein binding with respect, for example, to the magnitude and enantioselectivity of chiral ligand

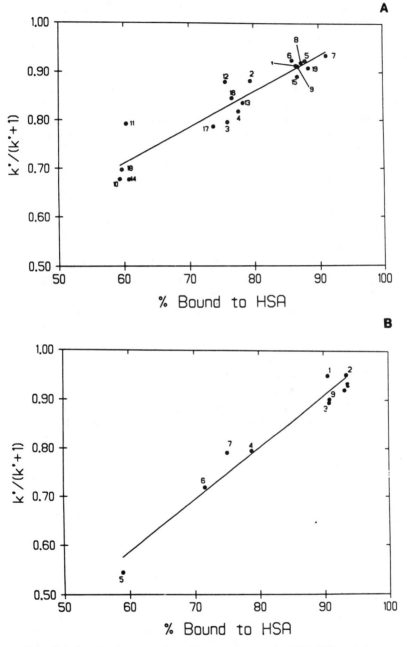

Figure 5.4   Relationships between the solute retention on the HSA-CSP and the extent of protein binding to free HSA: (A) relationship for a series of benzodiazepines and (B) relationship for a series of coumarins. For experimental details, see Noctor et al. [35].

binding to various animal proteins. This is illustrated by the results from the bio-chromatography of nonsteroidal anti-inflammatory drugs (NSAIDs), tryptophan, and warfarin on CSPs made from three different serum albumins: rat (RtSA-CSP), rabbit (RbSA-CSP), and human (HSA-CSP) [37].

The observed enantioselectivities and elution orders are presented in Table 5.5 [37]. When the chiral NSAIDs were chromatographed, there were no significant differences in the degree of enantioselectivity (expressed as enantioselectivity factor α) for fenoprofen, and flurbiprofen was higher on the HSA-CSP than on the other two, while for ibuprofen the order was RtSA > RbSA > HSA. For suprofen, no enantioselectivity was observed on the RtSA-CSP while moderate (α = 1.85) and high (α = 3.96) enantioselectivities were observed on the RbSA-CSP and HSA-CSP, respectively.

The relative enantioselectivities were reflected by the enantiomeric elution orders. On all the CSPs, the (S)-enantiomers eluted before the (R)-enantiomers, indicating that the proteins had a greater affinity for the (R)-form of the NSAIDs. These results were confirmed by ultrafiltration studies.

The chromatography of D- and L-tryptophan (TRP) on the CSPs resulted in the stereochemical separation of these enantiomers. In this case, the degree of enantioselectivity displayed by the RtSA-CSP was about 3.5-fold higher than the other two proteins, while the relative enantioselectivity, L-TRP bound to a greater extent than D-TRP, was the same for all three serum albumins. The relative enantioselectivity is consistent with findings reported earlier [24,38].

Table 5.5   ENANTIOSELECTIVITIES (α) AND ELUTION ORDERS OF SOLUTES CHROMATOGRAPHED ON CHIRAL STATIONARY PHASES (CSPs) BASED ON IMMOBILIZED RAT SERUM ALBUMIN (RtSA-CSP), RABBIT SERUM ALBUMIN (RbSA-CSP), OR HUMAN SERUM ALBUMIN (HSA-CSP)

Compounds	Eanantioselectivity (elution order)		
	RtSA-CSP	RbSA-CSP	HSA-CSP
Fenoprofen	1.24	1.22	1.51
	(R,S)	(R,S)	(R,S)
Flurbiprofen	1.43	1.75	2.09
	(R,S)	(R,S)	(R,S)
Ibuprofen	2.41	2.18	1.60
	(S,R)	(S,R)	(S,R)
Suprofen	1.00	1.85	3.96
		(S,R)	(S,R)
Tryptophan	7.40	2.14	2.56
	(D,L)	(D,L)	(D,L)
Warfarin	1.47	1.57	2.56
	(R,S)	(S,R)	(S,R)

Source: Data derived from Ref. 37.

When the enantioselective binding of (R)- and (S)-warfarin (WAR) was investigated, the magnitudes of the observed α factors were indistinguishable. However, on the RtSA-CSP and HSA-CSP, (R)-WAR eluted before (S)-WAR, whereas the opposite elution order was observed on the RbSA-CSP. These enantioselectivities are consistent with data reported from WAR–protein binding studies using HSA, RtSA, and RbSA [24,38,39].

5.5 Biochromatographic Determination of Drug–Drug Protein Binding Interactions

The displacement of a protein-bound drug from its binding sites results in increased circulating plasma concentration of the drug, thereby enhancing its pharmacological activity. This effect is particularly important in the case of highly protein-bound drugs with a small margin of safety [40]. An example of this phenomenon is the drug–drug interaction between warfarin and phenylbutazone, where the concomitant administration of the two drugs increases the circulating plasma level of warfarin, relative to the administration of warfarin alone, resulting in hypoprothrombinemia and bleeding [41].

There are a number of potential interactions between ligands that simultaneously bind to a protein [42]. The various possibilities are:

The ligands bind independently of each other and there is no interaction (*independent* binding).
The binding of one ligand facilitates the binding of the other (*cooperative* binding).
The binding of one of the ligands induces an allosteric change, which decreases the ability of the second compound to bind (*anticooperative* binding).
The two ligands bind competitively (*noncooperative* binding).

Drug–drug interactions at the protein binding level are often difficult to predict and measure, especially when standard protein binding techniques are used. These interactions can be readily observed and quantified using biochromatography, however, since the effect of one drug on the binding of the other is immediately reflected in a change in the chromatographic retention of the test solute.

The experimental technique used in the investigation of drug–drug protein binding interactions is as follows: (1) one of the drugs, the "solute," is injected onto the column containing the immobilized protein, and its retention k' is measured; (2) the other drug, the "displacer," is added to the mobile phase and its concentration is systematically increased during a series of experiments; and (3) the effect of the "displacer" concentrations on the k' of the "solute" is then measured. (Either drug can act as "displacer" or "solute." When the "solute" is chromatographed using a mobile phase that does not contain the "displacer," k' is directly proportional to its binding affinity for the immobilized protein (cf. Section 5.6). When the "displacer" is added to the mobile phase, the magnitude and direction of the resulting changes in k' can be used to determine the binding site of the ligand and to indicate whether cooperative, anticooperative, and noncooperative interactions occur between the "solute" and "displacer."

The relationship between the k' of the solute and the mobile phase concentration of the displacer is expressed by the following equation [36]:

$$\frac{1}{(k' - X)} = \frac{V_M K_2 [D]}{K_3 m_L} + \frac{V_M}{K_3 m_L} \tag{5.10}$$

where V_M is the void volume of the column; K_2 and K_3 are equilibrium constants for the binding of the displacer and solute, respectively; m_L represents moles of solute bound to the stationary phase; $[D]$ is the concentration of the displacer in the mobile phase; and X is residual k' resulting from binding at sites unaffected by the displacer.

The term X is a constant that represents the portion of k' resulting from the binding of the solute to sites at which the displacer does not compete. If both solute and displacer bind at only one identical site on the immobilized protein, then $X = 0$. Equation 5.10 predicts that when $X = 0$ a plot of $1/k'$ versus $[D]$ will produce a linear relationship with a slope of $V_M K_2 / K_3 m_L$ and an intercept of $V_M / K_3 m_L$. The value of K_2, the binding affinity constant for the displacer, can be determined directly by calculating the ratio of slope to intercept for this plot. The inverse of the slope gives m_L / V_m, which is the effective concentration of the binding sites in the column.

The application of this procedure to the study of drug–drug protein binding interactions is demonstrated by the biochromatographic study of the binding interactions on HSA between the enantiomers of ibuprofen (IBU) [43]. In this investigation, R- and S-IBU were chromatographed on an HSA-CSP using mobile phases that contained varying concentrations of R-IBU or S-IBU as the displacer; the results obtained when R-IBU was used as the displacer are presented in Figure 5.5. The data indicated that R- and S-IBU had one common binding site on the immobilized HSA and that S-IBU had at least one other major binding site. The association equilibrium constant for R-IBU with HSA was calculated to be 5.3×10^5 M^{-1}, and the constants for S-IBU at its two sites were 1.1×10^5 and 1.2×10^5 M^{-1}. These results are in good agreement with other reported association constants for the major binding site of IBU on HSA [44–46].

Similar studies have been carried out to investigate the effect of octanoic acid on the retention of NSAIDs [36,47]. In one study, anticooperative allosteric interactions were identified and attributed to conformational changes in the HSA induced by the binding of octanoic acid [36]. Displacement of the NSAID ketorolac was also observed in the second study, and the extent of this interaction was comparable to the effect seen in parallel ultrafiltration studies [47].

The biochromatographic approach to the study of ligand–protein binding does not have to be restricted to column liquid chromatography. Lloyd and co-workers [48,49] have demonstrated that capillary electrophoresis with HSA in the separation buffer can provide data on the extent and enantioselectivity drug–HSA binding.

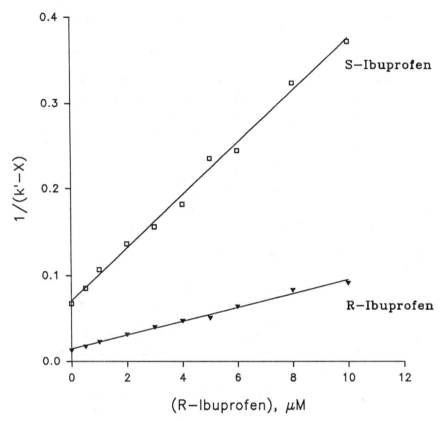

Figure 5.5 The effect of the mobile phase concentration of R-ibuprofen (R-IBU) on the chromatographic retention of R-IBU and S-IBU on the HSA-CSP. For experimental details, see Hage et al. [43].

Displacement techniques can also be used to probe binding sites and drug–drug binding interactions.

5.6 The Prediction and Description of Drug–Protein Binding from the Chemometric Analysis of Biochromatographic Retention Parameters

Chemometric analysis of enantiospecific biochromatographic data obtained on protein-based stationary phases can be used to relate solute structure and properties (topography) of binding site(s) on the immobilized protein. Biochromatography on an immobilized protein readily produces a substantial amount of precise and reliable binding-related data, which is necessary for chemometric analysis. An example of this approach is the biochromatographic measurement of the enantiospecific interactions of benzodiazepine (BDZ) derivatives with HSA [31].

A series of 22 BDZ derivatives (Table 5.6) was subjected to LC analysis [6,46]. There were nine achiral solutes, nine racemic mixtures, and four single enantiomers. The compounds were chromatographed on an HSA-CSP. The mobile phase was based on sodium dihydrogen phosphate/disodium hydrogen phosphate, pH 6.90, modified with 5% (v/v) propanol.

In the analysis of quantitative structure–enantiospecific retention relationships (QSERR) for chiral solutes, the logarithms of retention factors corresponding to both the first peak log k'_P and the second peak log k'_M were considered; retention parameters of the achiral agents, log k'_{AC}, were analyzed independently. The structure of the compounds was characterized by means of the following parameters (Figure 5.6) [6]: C_3, quantum chemically calculated electron excess charge on carbon C_3; P_{SM}, the excess charge difference between the hydrogen atom at C_3 and the most negatively charged atom in the other substituent at C_3 multiplied by the distance (in angstroms) between these two atoms; W, the width of the molecule along the phenyl substituent; f_{X+Y}, the sum of the hydrophobic constants of the substituent at position 7 in the fused benzene ring plus that of the substituent at position 2' of the phenyl substituent. Respective numerical data are collected in Table 5.7.

At first, a highly significant regression equation was obtained relating retention of the second eluting enantiomer, which is assumed to bind in the M conformation, log k'_M, and retention of the first eluting enantiomer, which is assumed to bind in the P conformation, log k'_P (Figure 5.7):

$$\log k'_M = 1.097(\pm 0.162) \log k'_P + 0.547(\pm 0.067) P_{SM} - 0.149 \qquad (5.11)$$
$$n = 8, R = 0.980, F = 61, p < 3 \times 10^{-4}$$

where F is the value of the F test of significance.

The retention of the first eluted enantiomer was described as follows:

$$\log k'_P = 0.183(\pm 0.079) f_{X+Y} - 0.278(\pm 0.055) W + 2.479 \qquad (5.12)$$
$$n = 13, R = 0.845, F = 12.5, p < 2 \times 10^{-3}$$

This equation can be interpreted such that retention (or binding) takes place at site that contains a hydrophobic pocket and some steric restrictions.

The retention of the second eluting enantiomer of benzodiazepines is described by the equation:

$$\log k'_M = 0.835(\pm 0.154) P_{SM} + 0.364(\pm 0.199) f_{X+T} \qquad (5.13)$$
$$- 2.690(\pm 0.932) C_3 + 0.556$$
$$n = 8, R = 0.938, F = 10, p < 0.03$$

This result indicates that binding occurs at a site containing both a hydrophobic pocket and a positively charged area. The cationic region of the binding site produces an attractive interaction with the submolecular dipole quantified by P_{SM} and a repulsive interaction with the excess positive charge at C_3.

Table 5.6 BENZODIAZEPINE TEST SOLUTES CHROMATOGRAPHED ON AN IMMOBILIZED HUMAN SERUM ALBUMIN COLUMN

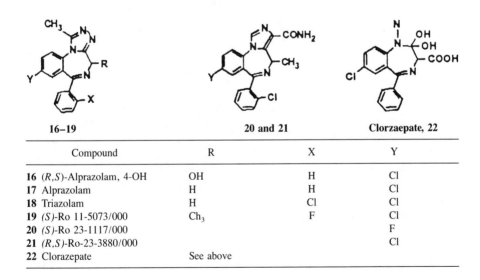

| 2–15 | | Chlordiazepozide, 1 | | |

Compound	R_1	R_2	X	Y
1 Chlordiazepoxide	See above			
2 (R,S)-Oxazepam H	H	OCO(CH₂)₂COO	H	Cl
3 Nitrazepam	H	H	H	NO₂
4 Flunitrazepam	CH₃	H	F	NO₂
5 Clonazepam	H	H	Cl	NO₂
6 Delorazepam	H	H	Cl	Cl
7 Desmethyldiazepam	H	H	H	Cl
8 Diazepam	CH₃	H	H	Cl
9 (R,S)-Lormetazepam	CH₃	OH	Cl	Cl
10 (R,S)-Lorazepam	H	OH	Cl	Cl
11 (R,S)-Oxazepam	H	OH	H	Cl
12 (R,S)-Temazepam	CH₃	OH	H	Cl
13 (S)-Ro 14-8935/000	CH₃	CH₃	Cl	NH₂
14 (S)-Ro 23-0983/001	H	CH₃	Cl	F
15 (R,S)-Ro 11-3128/002	H	CH₃	Cl	NO₂

| 16–19 | 20 and 21 | Clorzaepate, 22 |

Compound	R	X	Y
16 (R,S)-Alprazolam, 4-OH	OH	H	Cl
17 Alprazolam	H	H	Cl
18 Triazolam	H	Cl	Cl
19 (S)-Ro 11-5073/000	Ch₃	F	Cl
20 (S)-Ro 23-1117/000			F
21 (R,S)-Ro-23-3880/000			Cl
22 Clorazepate	See above		

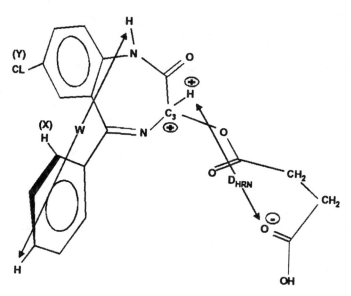

Figure 5.6   Structural descriptors of benzodiazepines used in quantitative structure–enantiospecific retention relationship (QSERR) equations. See text for explanation.

The retention of achiral benzodiazepines was described as:

$$\log k'_{AC} = 0.374(\pm 0.096)\, f_{X+Y} - 7.068(\pm 1.663)\, C_3 + 1.221 \qquad (5.14)$$
$$n = 9,\ R = 0.889,\ F = 11,\ p < 0.01$$

Since for achiral benzodiazepines the parameter P_{SM} is zero, equation 5.14 indicates that these compounds predominantly bind in the M conformation as reported in the literature [50].

Based on equations 5.11–5.14, a model was proposed [5] for the structural requirements of the two postulated modes of benzodiazepine binding to HSA (Figure 5.8). According to equation 5.12, benzodiazepines appear to bind within hydrophobic cavities, and substituents at N_1, C_2, and C_5 would then provide spatial orientation of the BDZ molecules within this cavity. Steric limitations suggest that the hydrophobic cavity has definite boundaries. In addition, the steric features at the stereogenic center, the third carbon atom, appear to play no role in this binding mode (Figure 5.8a).

The binding mode of the second eluting enantiomer, as described by equation 5.13, involves hydrophobic and electrostatic interactions. Thus, in addition to a hydrophobic cavity, there must be a cationic region in close proximity (Figure 5.8b). For the benzodiazepines in the M conformation, the electrostatic repulsion between the excess positive charge on the third carbon atom and the cationic site on the protein surface appears to be more than offset by the attraction of a negatively charged atom within the substituent at C_3 and the same area. In the case of the lesser

Table 5.7   LOGARITHMS OF CHROMATOGRAPHIC RETENTION FACTORS[a]
DETERMINED ON A HUMAN SERUM ALBUMIN LC COLUMN AND
STRUCTURAL PARAMETERS[b] FROM MOLECULAR MODELING
OF A SERIES OF BENZODIAZEPINE DERIVATIVES

Compound Number	Log k'_P	Log k'_M	Log k'_{AC}	P_{SM}	f_{X+Y}	C_3	W
1			0.8645	0.0849	1.05	0.1035	9.30
2	0.8512	1.8938		1.8635	1.05	0.2785	8.74
3			0.6243	0.0703	0.06	0.0960	8.54
4			0.4857	0.0609	0.20	0.0882	9.63
5			0.7679	0.0680	0.77	0.0966	8.67
6		1.0614		0.0635	1.76	0.0977	8.69
7		1.0969		0.0634	1.05	0.0979	8.59
8		1.1216		0.0578	1.05	0.0933	9.56
9	0.7672	0.9745		0.6120	1.76	0.2388	9.76
10	0.8068	0.9360		0.5953	1.76	0.2425	8.71
11	0.6561	1.0261		0.7049	1.05	0.2451	8.60
12	0.5224	1.1793		0.6113	1.05	0.2353	9.49
13	0.3892			0.0675	−0.18	0.0549	10.0
14	0.6628			0.0600	1.19	0.0624	8.64
15	0.7193	0.7193		0.0633	0.77	0.0651	8.70
16	0.2648	0.4533		0.5862	1.05	0.3169	10.5
17			0.4200	0.0784	1.05	0.1722	10.3
18			0.6243	0.0730	1.76	0.1725	10.3
19	0.3838			0.0484	1.19	0.1379	10.2
20	0.7404			0.0320	1.19	0.1309	9.20
21	1.0523	1.1156		0.0441	1.76	0.1184	9.22
22	0.9715	1.3992		1.0745	1.05	0.0916	8.14

[a] Retention parameters as follows: log k'_P, first eluting enantiomer; log k'_M, second eluting enantiomer; log k'_{AC}, achiral solutes.

[b] Structural parameters as follows: P_{SM}, submolecular polarity parameter (see Figure 5.6); f_{X+Y}, sum of hydrophobic constants of substituents X and Y (see Figure 5.6); C_3, electron excess charge on carbon atom C_3 of the diazepine system; W, molecular width (Å).

[c] Compounds are numbered as in Table 5.6.

Source: Ref. 5.

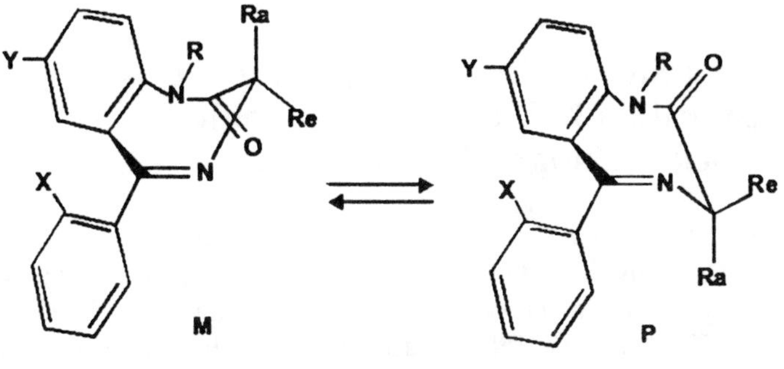

Figure 5.7   Conformations M and P of benzodiazepines.

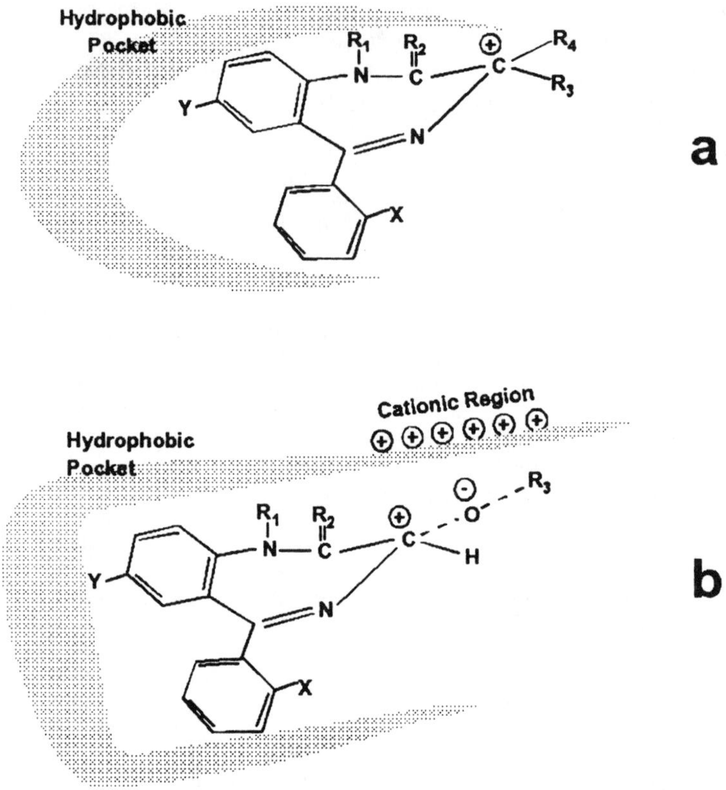

Figure 5.8   Two postulated modes of benzodiazepine binding to human serum albumin. See text for discussion.

retained enantiomer, binding at this site may be prohibited by the electrostatic repulsion between carbon C_3 and the cationic area and the steric hindrance due to the P conformation.

5.7 Binding Site for Basic Drugs on α_1-Acid Glycoprotein in View of QSRR Analysis of Biochromatographic Data

α_1-Acid glycoprotein (AGP) is a serum protein that mainly binds basic drugs [51]. The prevalent view is that AGP has only one common drug binding site, which binds drugs through hydrophobic and electrostatic interactions [52,53]. However, neither log P [52,54] nor pK_a [55] can account for binding differences within small sets of tested drugs.

Retention factors, log k'_{AGP}, were determined for 52 basic drugs of diverse chemical structure and pharmacological activity on an AGP-CSP [56–58]. The column was packed with human AGP chemically bound to 5 μm silica particles; the mobile phase was isopropanol/phosphate buffer [0.1 M, pH 6.5 (5:95 (v/v)]; eluent flow rate was 0.5 mL/min, and UV detection was at a wavelength of 215 nm. The retention factors k'_{AGP} were determined using the peak for the retention of sodium nitrate as the marker of column dead volume. The log k'_{IAM} parameters for the set of agents studied were also determined using an IAM column.

Molecular modeling was employed to determine the solute structural parameters that were important to describe their interactions with AGP. The following parameters were used: electron excess charge (N_{ch}) and the surface area (S_T) of a triangle having one vertex on the aliphatic nitrogen and the two remaining vertices on the extremely positioned atoms in the drug molecule (Figure 5.9) [58]. The structural parameters from molecular modeling and retention parameters determined on AGP and IAM columns are given in Table 5.8.

The QSRR equation relating retention on chemically immobilized AGP to a

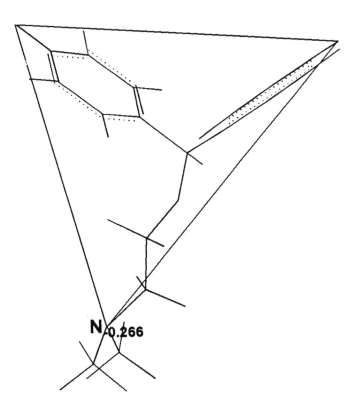

Figure 5.9 Structural descriptors of basic drugs binding to human α_1-acid glycoprotein (AGP) used in quantitative structure–retention relationships. See text for explanation.

Table 5.8 LOGARITHMS OF CHROMATOGRAPHIC RETENTION FACTORS DETERMINED ON AN α_1-ACID GLYCOPROTEIN LC COLUMN (log k'_{AGP}) AND ON AN IMMOBILIZED ARTIFICIAL MEMBRANE COLUMN (log k'_{IAM}) AND STRUCTURAL PARAMETERS[a] OF A SERIES OF BASIC DRUGS

Drug Name	Log k'_{AGP}	Log k'_{IAM}	CLOGP	N_{ch}	S_T
Antagonists of H_1 Histamine Receptor					
Antazoline	1.154	1.043	4.25	−0.285	25.23
Chlorpheniramine	1.202	1.055	2.73	−0.260	30.92
Chloropyramine	1.431	1.330	3.56	−0.268	31.11
Cinnarizine	2.148	2.250	6.14	−0.248	32.18
Dimethindene	1.382	1.194	3.42	−0.267	29.76
Diphenhydramine	1.140	1.006	3.36	−0.266	32.78
Isothipendyl	1.580	1.210	3.93	−0.260	20.41
Ketotifen	1.459	1.168	3.56	−0.245	25.54
Mepyramine	1.113	0.935		−0.272	32.81
Pheniramine	0.926	0.602	2.02	−0.260	27.37
Pizotifen	1.898	1.588		−0.247	29.13
Promethazine	1.833	1.508	4.65	−0.257	20.73
Tripelennamine	1.066	0.887	2.85	−0.264	30.37
Triprolidine	1.185	1.084	3.47	−0.240	24.02
Tymazoline	1.306	1.204		−0.287	17.07
Antagonists of H_2 Histamine Receptor					
Cimetidine	0.482	−0.271	0.21	−0.195	18.11
Famotidine	0.731	−0.271	−0.57	−0.161	27.44
Metiamide	0.517	−0.301	0.38	−0.193	19.19
Nizatidine	0.460	−0.368		−0.275	32.01
Ranitidine	0.600	−0.016	0.27	−0.262	31.86
Roxatidine	0.773	0.359	2.66	−0.279	28.99
Antagonists of β-Adrenoceptors					
Acebutolol	0.676	0.602	1.61	−0.300	49.35
Alprenolol	1.490	0.918	2.59	−0.299	25.99
Atenolol	0.499	−0.146	−0.11	−0.298	20.36
Betaxolol	0.838	0.994	2.17	−0.300	27.00
Bisoprolol	0.694	0.646	1.69	−0.297	45.07
Bopindolol	1.940	0.456	4.86	−0.301	38.39
Bupranolol	0.981	0.269		−0.291	21.76
Carteolol	0.706	−0.146	1.17	−0.289	24.88
Celiprolol	0.700	0.723	1.66	−0.304	37.28
Cicloprolol	0.735	1.012		−0.298	42.28
Dilevalol	1.106	1.272	2.18	−0.295	27.49
Esmolol	0.649	0.646	1.53	−0.299	30.30
Metoprolol	0.564	0.434	1.20	−0.301	22.66
Nadolol	0.606	0.269		−0.292	29.90
Nifenalol	0.639	0.269	1.16	−0.300	12.19
Oxprenolol	1.210	0.586	1.62	−0.299	28.60
Pindolol	0.870	0.586	1.65	−0.301	27.79
Practolol	0.509	−0.067	0.78	−0.301	22.74
Propranolol	1.612	1.340	2.75	−0.300	28.77
Sotalol	0.516	−0.146	0.23	−0.178	19.40
Timolol	0.696	0.385	1.63	−0.290	15.60

Table 5.8 (*Continued*)

Drug Name	Log k'_{AGP}	Log k'_{IAM}	CLOGP	N_{ch}	S_T
Agonists and Antagonists of α-Adrenoceptors					
Cirazoline	1.082	0.940	3.27	−0.288	16.78
Clonidine	0.847	0.410		−0.278	12.32
Doxazosin	1.798	1.983	3.77	−0.307	30.64
Moxonidine	0.528	−0.067		−0.286	13.35
Naphazoline	1.092	0.895	3.83	−0.295	17.61
Phentolamine	1.264	1.340	3.68	−0.289	24.30
Prazosin	1.390	1.594	2.16	−0.308	31.47
Tiamenidine	0.808	0.434		−0.284	9.96
Tramazoline	1.315	1.123	2.49	−0.285	21.22
UK-14,304	0.831	0.269		−0.276	15.74

a CLOGP, logarithm of *n*-octanol/water partition coefficient calculated by the fragmental method according to Craig [10]; N_{ch}, electron excess charge on aliphatic nitrogen; S_T, area of the size/shape triangle (see Figure 5.9).
Source: Ref. 58.

hydrophobicity measure log k'_{IAM}, electron excess charge on aliphatic nitrogen N_{ch}, and a size parameter S_T of drugs has the form:

$$\log k'_{AGP} = 0.6577(\pm 0.0402) \log k'_{IAM} + 3.342(\pm 0.841) N_{ch} \qquad (5.15)$$
$$- 0.0081(\pm 0.0030) S_T + 1.688(\pm 0.245)$$
$$n = 49, R = 0.929, s = 0.163, F = 92, p < 10^{-5}$$

Of 52 drugs studied, three (tripolidine, cimetidine, and bopindolol) did not fit equation 5.15. The reason may be the uncertainty in identifying the atom equivalent to the aliphatic nitrogen with regard to electrostatic interactions.

There are two aspects of QSRR equations: their predictive potency and their physical meaning. The predictive power of equation 5.15 is illustrated in Figure 5.10. Certainly, for some agents the predicted strength of chromatographically measured interactions with AGP differs from the observed one. However, no better means of evaluating the actual binding is known.

Equation 5.15 may be useful in drug design as a first approximation of relative binding of an agent to AGP without the need to perform biological experiments. It can also indicate relative differences in AGP binding among individual drugs and can help to predict the probability of pharmacokinetic interactions of two simultaneously administered drugs.

Equation 5.15 comprises a term (log k'_{IAM}) that cannot be obtained from computational chemistry and thus, to use it one needs some amount (however minute) of the existing substance. There is a hydrophobicity parameter that can be calculated from structural formula, namely, CLOGP [8–10]. For 38 drugs for which both log k'_{IAM} and CLOGP were available, the correlation between them was $R = 0.850$. Replacing log k'_{IAM} in equation 5.15 by CLOGP decreases correlation, but the resulting regression equation remains significant and thus informative. Most probably, the predictive power of CLOGP could be increased by introducing a correction

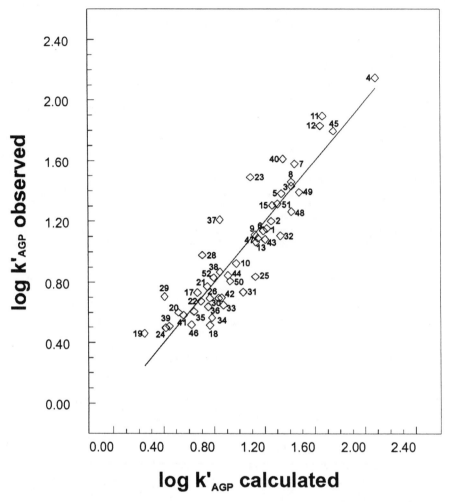

Figure 5.10   Logarithm of the retention factor determined on an α_1-acid glycoprotein (AGP) column against the value calculated by equation 5.15.

due to ionization at actual pH. The problem in individual cases is, however, to acquire the necessary pK_a values.

The QSRR equation derived here and the reported [51–55] qualitative characteristics of the mode of binding of xenobiotics by AGP allow for an indirect identification of structural features of the binding site of basic drugs (Figure 5.11). The site can be modeled by a conical pocket. Its wall (internal surface) contains lipophilic regions at the base of the cone. There is an anionic region close to the spike of the cone. Protonated aliphatic nitrogen guides drug molecules toward the anionic region. Hydrophobic hydrocarbon fragments of the interacting drugs provide anchoring in the lipophilic region(s) of the binding site. There is a steric restriction

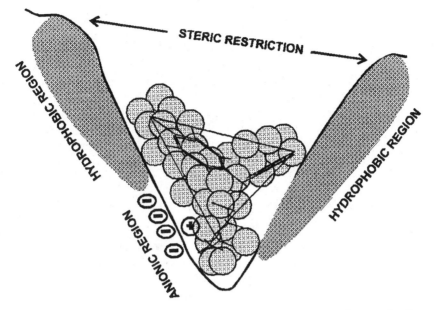

Figure 5.11 Mode of binding of basic drugs by α_1-acid glycoprotein. See text for discussion.

that prevents the molecule from plunging into the binding site. Asymmetric charge distribution accounts for the observed enantioselectivity of binding to AGP. Similarly, asymmetric distribution of charge (positive) in one of the two benzodiazepine binding sites on HSA accounted for the stereoselectivity of binding discussed earlier.

5.8 Concluding Remarks

Chemometric comparisons clearly demonstrate the advantages of biochromatographic measures of hydrophobicity determined by liquid chromatography on an immobilized artificial membrane stationary phase. The parameter $\log k'_{IAM}$ can conveniently be obtained and appears to be a better predictor of some kinds of pharmacological activity. It does not mean that $\log P$ and other chromatographic hydrophobicity parameters are useless. Hydrophobicity is a multidimensional property, and $\log k'_{IAM}$ simply increases our ability to quantify different features of this biologically important phenomenon.

QSRR equations 5.11–5.15 account in strict statistical terms for the properties qualitatively ascribed to the drug binding sites on HSA and AGP. Thus, pharmacologically relevant information can be assessed based on these indirectly derived models. Of course, it would be interesting to relate these working models to the structures of the drug–HSA and drug–AGP crystals once these have been resolved. It must be stressed, however, that while crystallographic analysis helps to visualize supposed active sites on biomacromolecules, the pictorial presentation is of limited use in predicting the binding of agents of a given chemical structure. Reliable predictions of activity and the determination of the required structural properties for the design of new drugs are facilitated by the derivation of good quantitative structure–pharmacological activity relationships. To derive such relationships, one needs larger sets of numerically expressed biological (biorelevant) data. Such are the retention data determined by means of the LC systems containing biomolecules. A combination of biochromatography and chemometrics appears to be a promising strategy in biological research.

Acknowledgment

This work was supported in part (R.K.) by the U.S.–Polish Maria Sklodowska-Curie Joint Fund II (project MZ/HHS-95-227).

References

1. R. Kaliszan, *Quantitative Structure–Chromatographic Retention Relationships.* Wiley, New York, 1987.

2. R. Kaliszan, *Chemometrics Intell. Lab. Syst.* **24,** 89 (1994).

3. R. Kaliszan, *Anal. Chem.* **64,** 619A (1992).

4. R. Kaliszan, *CRC Crit. Rev. Anal. Chem.* **16,** 323 (1986).

5. R. Kaliszan, T.A.G. Noctor, and I.W. Wainer, *Mol. Pharmacol.* **42,** 512 (1992).

6. I.W. Wainer, R. Kaliszan, and T.A.G. Noctor, *J. Pharm. Pharmacol.* **45**(Suppl. 1), 367 (1993).

7. R. Kaliszan, *Adv. Chromatogr.* **33,** 147 (1993).

8. C. Hansch and A. Leo, *Substituent Constants for Correlation Analysis in Chemistry and Biology.* Wiley, New York, 1979.

9. R.F. Rekker, *The Hydrophobic Fragmental Constant.* Elsevier, Amsterdam, 1977.

10. P.N. Craig, in C. Hansch, P.G. Sammes, and J.B. Taylor, Eds., *Comprehensive Medicinal Chemistry.* Vol. 6. Pergamon Press, Oxford, 1990, p. 237.

11. R. Kaliszan, *J. Chromatogr.* **656,** 417 (1993).

12. K. Miyake, F. Kitaura, N. Mizuno, and H. Terada, *J. Chromatogr.* **389,** 47 (1987).

13. H. Thurnhofer, J. Schnabel, M. Betz, G. Lipka, C. Pidgeon, and H. Hauser, *Biochim. Biophys. Acta,* **1064,** 275 (1991).

14. C. Pidgeon, C. Marcus, and F. Alvarez, in T.O. Baldwin and J.W. Kelly, Eds., *Applications of Enzyme Biotechnology.* Plenum Press, New York, 1992, p. 201.

15. R. Kaliszan, A. Kaliszan, and I.W. Wainer, *J. Pharm. Biomed. Anal.* **11,** 505 (1993).

16. R. Kaliszan, A. Nasal, and A. Bucinski, *Eur. J. Med. Chem.* **29,** 163 (1994).

17. A. Nasal, M. Sznitowska, A. Bucinski, and R. Kaliszan, *J. Chromatogr.* **692,** 83 (1995).

18. S. Ong, H. Liu, X. Qiu, G. Bhat, and C. Pidgeon, *Anal. Chem.* **67,** 755 (1995).

19. P.H. Hinderling, O. Schmidlin, and J.K. Seydel, *J. Pharmacokin. Biopharm.* **12,** 263 (1984).

20. P. Seeman, *Pharmacol. Rev.* **24,** 583 (1972).

21. N. El Tayar, R.-S. Tsai, B. Testa, P.-A. Carrupt, C. Hansch, and A. Leo, *J. Pharm. Sci.* **80,** 744 (1991).

22. R.J. Scheuplein, H. Blank, G.J. Brauner, and D. MacFarlane, *J. Invest. Dermatol.* **52,** 63 (1969).

23. P.V. Raykar, M.-C. Fung, and B.D. Anderson, *Pharm. Res.* **5,** 140 (1988).

24. W.E. Müller, in I.W. Wainer and D.E. Drayer, Eds., *Drug Stereochemistry: Analytical Method and Pharmacology.* Dekker, New York, 1988, p. 227.

25. I.W. Wainer, in I.W. Wainer, Ed., *Drug Stereochemistry: Analytical Methods and Pharmacology,* 2nd ed. Dekker, New York, 1993, p. 139.

26. J. Hermansson, *J. Chromatogr.* **269,** 71 (1983).

27. T. Miwa, M. Ichikawa, M. Tsuno, T. Hattori, T. Miyakawa, M. Kayano, and Y. Miyake, *Chem. Pharm. Bull.* **35,** 682 (1987).

28. S. Allenmark, *J. Liquid Chromatogr.* **9,** 425 (1986).

29. E. Domenici, C. Bertucci, P. Salvadori, G. Felix, I. Cahagne, S. Motellier, and I.W. Wainer, *Chromatographia,* **29,** 170 (1990).

30. I.W. Wainer and Y.-Q. Chu, *J. Chromatogr.* **455,** 316 (1988).

31. E. Domenici, C. Bertucci, P. Salvadori, G. Felix, I. Cahagne, S. Motellier, and I.W. Wainer, *Chirality,* **2,** 263 (1990).

32. E. Domenici, C. Bertucci, P. Salvadori, and I.W. Wainer, *J. Pharm. Sci.* **80,** 164 (1991).

33. T.A.G. Noctor, G. Felix, and I.W. Wainer, *Chromatographia,* **31,** 55 (1991).

34. T.A.G. Noctor and I.W. Wainer, *Pharm. Res.* **9,** 480 (1992).

35. T.A.G. Noctor, M.J. Diaz-Perez, and I.W. Wainer, *J. Pharm. Sci.* **82,** 675 (1993).

36. T.A.G. Noctor, I.W. Wainer, and D.S. Hage, *J. Chromatogr.* **577,** 305 (1992).

37. G. Massolini, A.-F. Aubry, A. McGann, and I.W. Wainer, *Biochem. Pharmacol.* **46,** 1285 (1993).

38. C. Langercrantz, T. Larsson, and I. Denfors, *Comp. Biochem. Physiol.* **69,** 375 (1981).

39. W. Schmidt and E. Jahnchen, *Experientia,* **34,** 1323 (1978).

40. T.A.G. Noctor, in I.W. Wainer, Ed., *Drug Stereochemistry: Analytical Methods and Pharmacology,* 2nd ed. Dekker, New York, 1993, p. 337.

41. E.M. Seller and J. Koch-Weser, *Pharmacol. Res. Commun.* **7,** 331 (1975).

42. B. Honore, *Pharmacol. Toxicol.* **66**(Suppl. II), 1 (1990).

43. D.S. Hage, T.A.G. Noctor, and I.W. Wainer, *J. Chromatogr. A,* **693,** 23 (1995).

44. N.P. Sollenne and G.E. Means, *Mol. Pharmacol.* **15,** 754 (1979).

45. A. Kober and I. Sjoholm, *Mol. Pharmacol.* **18,** 421 (1980).

46. T.A.G. Noctor, C.D. Pham, R. Kaliszan, and I.W. Wainer, *Mol. Pharmacol.* **42,** 506 (1992).

47. P.J. Hayball, J.W. Holman, and R.L. Nation, *J. Chromatogr. B,* **662,** 128 (1994).

48. D.K. Lloyd, S. Li, and P. Ryan, *Chirality,* **6**, 230 (1994).

49. D.K. Lloyd, S. Li, and P. Ryan, *J. Chromatogr. A,* **694**, 285 (1995).

50. T. Alebic-Kolbah, F. Kajifez, S. Rendic, V. Sunjic, A. Konowal, and G. Snatzke, *Biochem. Pharmacol.* **28**, 2457 (1979).

51. J.M.H. Kremer, J. Wilting, and L.H.M. Janssen, *Pharmacol. Rev.* **40**, 1 (1988).

52. J. Schley, *J. Pharm. Pharmacol.* **39**, 132 (1987).

53. L. Soltes, B. Sebille, and P. Szalay, *J. Pharm. Biomed. Anal.* **12**, 1295 (1994).

54. H. Glaser and J. Kriegelstein, *Naunyn-Schmiedebergs Arch. Pharmacol.* **265**, 321 (1970).

55. D.L. Goolkasian, R.L. Slaughter, D.J. Edwards, and D. Lalka, *Eur. J. Clin. Pharmacol.* **25**, 413 (1983).

56. A. Nasal, A. Radwanska, K. Osmialowski, A. Bucinski, R. Kaliszan, G.E. Barker, P. Sun, and R.A. Hartwick, *Biomed. Chromatogr.* **8**, 125 (1994).

57. R. Kaliszan, A. Nasal, and M. Turowski, *J. Chromatogr. A,* **722**, 25 (1996).

58. R,. Kaliszan, A. Nasal, and M. Turowski, *Biomed. Chromatogr.* **9**, 211, (1995).

Molecular Recognition Ability of Uniformly Sized, Polymer-Based Stationary Phases in HPLC

Ken Hosoya

6.1 Introduction

Polymer-based separation media have been utilized for a variety of separations in high performance liquid chromatography (HPLC) [1–5]. Among the interesting attributes of these separation media are the chemical stability and/or versatility of the related surface chemistry [6]. Because of slow mass transfer in micropores and solvation of the polymer chains, however, the chromatographic efficiency of polymer-based media in a reversed-phase mode is usually lower than that of silica-based packing materials in aqueous methanol, especially in the separation of low molecular weight compounds [7]. As a result, polymer-based separation media are mainly utilized when silica-based packing materials cannot be used because of their limited chemical stability or because of the risk of undesired secondary retention mechanisms associated with basic compounds that interact with residual silanol groups or metal impurities in the silica-based separation media [8–12].

Most spherical polymer-based separation media are prepared by a suspension polymerization method (SPM) [13]. That is, inert diluents, or porogens, of various kinds are utilized to produce porous structures [14]. Changes in the nature or amounts of the porogens used lead to different pore sizes and pore size distributions [15]. Even when polymerization conditions are strictly controlled, however, suspension polymerization produces particles having relatively broad size distributions [16]. Therefore, a size classification is necessary to improve the particle size distribution. Generally, wet sedimentation, counterflow settling, and counterflow centrifugation [17–19] are employed for this purpose, but these methods are relatively difficult to implement in a typical analytical laboratory and, even if the classification is done, some irregular particles remain in the final product. The presence of particles of irregular size may result in lower column efficiencies as well as poorer column stabilities [20]. Fine particles are particularly troublesome because they increase column pressure drop, which leads to shorter column lifetime, especially in a reversed-phase gradient mode, which is the usual chromatographic mode for the separation of proteins. Thus several disadvantages restrict the broad acceptance of polymer-based separation media in HPLC.

In 1990 polymer-based packing materials involving relatively wide pores (≈ 50 nm) were used to separate proteins in a reversed-phase gradient mode, and a better efficiency than silica-based packing materials with medium size pores (≈ 20 nm) was reported [8]. Moreover, a multistep swelling and polymerization (MSP) method was introduced for the preparation of size-monodisperse, polymer-based separation media, which provide excellent column efficiency [21,22] with very low column pressure drops. This good chromatographic performance was attributed to the size monodispersity of the beads, which contained no fine particles [23]. The low column pressure drop observed with these monodispersed beads is very important for column stability in view of the limited mechanical stability of most polymer-based packing materials.

6.2 Comparison of Properties of Uniformly Sized, Polymer-Based Stationary Phases with Those Prepared by Classical Suspension Polymerization

In view of the large amount of information available on the use of suspension polymerization to prepare porous structures and the benefits inherent in the MSP method that affords monodispersed beads, new techniques based on the concepts of both methods might lead to monodisperse particulate media of nearly ideal size.

The combination of the two techniques is complicated by the need to introduce a seed polymer in the swelling method. This seed acts as a shape template that retains its shape as it is swollen by monomers and solvents. Numerous approaches have been explored for the preparation of size-monodisperse seed particles [24]. Usually, polystyrene seed particles are prepared by an emulsifier-free emulsion polymerization [25]. These can then be used for MSP method. Classical suspension polymerization does not require seed polymer, but low molecular weight porogens are used to create the porous structures. Generally, a porogenic substance that is a good solvent for the growing polymer chains tends to produce only relatively small pores with a narrow pore size distribution as the result. In contrast, wide pores and a relatively broad pore size distribution are obtained through the use of a porogenic substance that is a poor solvent for the growing polymer chains and therefore leads to their faster precipitation [14]. When the porogenic solvent includes high molecular weight compounds (e.g., linear polymers), these substances act as porogens to create very large pores and a broad pore size distribution [26,27]. It is an interesting feature of the use of polymers as porogens that their effect is influenced greatly by both their molecular weight and their amount in the polymerizing mixture.

When swelling of the seed particle takes place, the magnitude of the enlargement in diameter of the particle can be calculated according to the following simple relationship:

$$\frac{(d_p)^3}{(d_{seed})3} = \frac{V_{monomers} + V_{seed}}{V_{seed}} \tag{6.1}$$

where d_p is the diameter of the final particle, d_{seed} is the diameter of the seed particles, $V_{monomers}$ is the volume of monomers used in the swelling, and V_{seed} is the volume of the seed particles. In most instances, the diameter of the seed particles is around 1 μm; thus when 5 μm particles are prepared, the seed particle occupies only 0.8% of the total volume of the final swollen particle. However, since seed polymer particles are prepared by an emulsifier-free emulsion polymerization method, their molecular weights are relatively high [25]. While the final beads are quite low in seed polymer content, their high molecular weight may modify the overall properties of the porogenic mixture in a way that affects the pore size and pore size distribution of the final beads. This change, in turn, may have a nonnegligible effect on the beads' chromatographic properties.

6.2.1 Particle Size Dispersity

Typical monomers and porogenic solvents were utilized to prepare size-monodisperse, polymer-based separation media by a multistep swelling and polymerization method. The objective was to investigate the influence of the seed polymer on the porous structures of the final beads through comparisons of their chromatographic properties with those of beads obtained by a classical suspension polymerization method involving the same reaction conditions, but without the seed polymer.

Poly(styrene–divinylbenzene) particles and poly(methyl methacrylate–ethylene dimethacrylate) particles were prepared by the MSP method, and by the corresponding SPM method with cyclohexanol or toluene as the porogenic solvent. The yields obtained in the preparation of eight different types of particle are summarized in Table 6.1.

The MSP methods generally gave very high yields, while suspension polymerization provided lower yields. Although size classification was not carried out, the very fine particles that could not been sedimented through the usual purification procedure were removed. As a result, the observed yields for beads obtained by SPM became even lower, as described in the next section. The yield advantage of the MSP method could be advantageous in the case of preparations involving expensive monomers (e.g., chiral media). Typical optical micrographs (Figure 6.1) of poly(methyl methacrylate–ethylene dimethacrylate) particles confirm that the particles prepared by the suspension polymerization method have various sizes, ranging from about 1 μm to about 20 μm. In contrast, the MSP method gives particles that have very good size monodispersity and are free of fines (some particles shown in Figure 6.1 probably were broken during sample preparation).

Table 6.1 CHEMICAL YIELDS AND SYMBOLS OF PREPARED PARTICLES

Entry	Method[a]	Monomers[b]	Porogen[c]	Yield (%)	Symbols
1	MSP	St-DVB	Toluene	91	St-tol-MSP
2	SPM	St-DVB	Toluene	74	St-tol-SPM
3	MSP	MMA-EDMA	c-Hexanol	95	Ma-chn-MSP
4	SPM	MMA-EDMA	c-Hexanol	75	Ma-chn-SPM
5	MSP	St-DVB	c-Hexanol	89	St-chn-MSP
6	SPM	St-DVB	c-Hexanol	76	St-chn-SPM
7	MSP	MMA-EDMA	Toluene	87	Ma-tol-MSP
8	SPM	MMA-EDMA	Toluene	65	Ma-tol-SPM

[a] Msp, multistep swelling and polymerization method; SPM; suspension polymerization method.
[b] St, styrene; DVB, commercial divinyl benzene; MMA, methyl methacrylate; EDMA, ethylene dimethacrylate.
[c] c-Hexanol, cyclohexanol.

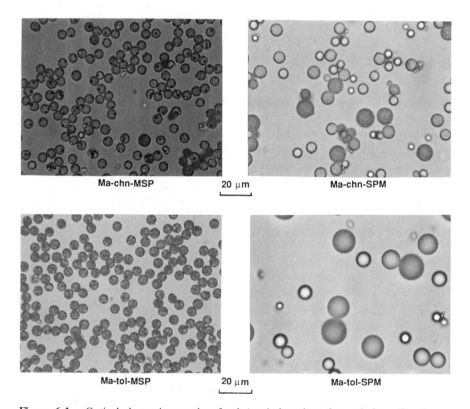

Figure 6.1. Optical photomicrographs of poly(methyl methacrylate–ethylene dimethacrylate) particles utilizing toluene (tol) or cyclohexanol (chn) as porogen. In this work Ma-tol-MSP was prepared by a multistep swelling and polymerization method, while Ma-tol-SPM was prepared by the corresponding suspension polymerization method.

The particle size distributions of typical poly(styrene–divinylbenzene) particles prepared by the MSP approach and by SPM are depicted in Figure 6.2, which also confirms the size monodispersity of the particles prepared by the MSP method, in sharp contrast to the broad distribution of the SPM beads. It must be emphasized here that since the same polymerization conditions were used for both methods, the reaction parameters were actually not optimized to obtain a narrower particle size distribution with the SPM beads [16]. Some optimization may afford narrower particle size distributions than reported herein, but, even if the reaction parameters are strictly controlled, it is clear that except in very special cases, size-monodisperse particles cannot be obtained by means of the suspension polymerization method [28].

6.2.2 Chemical Composition and Surface Chemistry of the Particles

Data obtained in elemental analyses of the particles are summarized in Table 6.2. Minor discrepancies were observed between the calculated and experimental values, but good agreement was achieved for all pairs of particles prepared from the same monomers and porogen by the two different polymerization methods. These findings confirm that the polystyrene seed particles utilized in the MSP method do not affect the polymerization itself, and the chemical composition of the final particles is almost identical in both cases. As mentioned earlier, despite the lower yields observed in the suspension polymerization method, the similar chemical compositions of the particles prepared by the two techniques strongly suggest that the lower yield in SPM is mainly due to the loss of fine particles during the purification steps. In addition, the FTIR spectra of the poly(methyl methacrylate–ethylene dimethacrylate) particles prepared by the two methods were essentially identical (Figure 6.3).

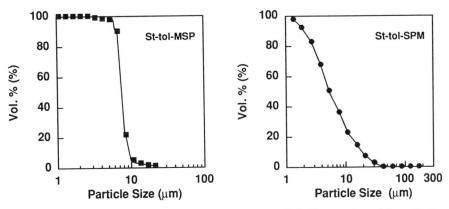

Figure 6.2. Particle size distribution of poly(styrene–divinylbenzene) particles utilizing toluene as porogen: MSP, preparation by a multistep swelling and polymerization method; SPM, preparation by the corresponding suspension polymerization.

Table 6.2   ELEMENTAL ANALYSIS DATA OF THE PARTICLES
FROM FIGURES 6.2 AND 6.3

Symbol	Amount Calculated			Amount Found (%)		
	C	H	O	C	H	O
St-tol-MSP	92.31	7.69	0	87.12	8.02	0
St-tol-SPM	92.31	7.69	0	88.07	8.17	0
Ma-chn-MSP	60.37	7.47	32.16	59.03	7.37	33.60
Ma-chn-MSP	60.37	7.47	32.16	59.30	7.38	33.32
St-chn-MSP	92.31	7.69	0	90.65	8.08	0
St-chn-SPM	92.31	7.69	0	90.38	8.10	0
Ma-tol-MSP	60.37	7.47	32.16	59.30	7.42	33.28
Ma-tol-SPM	60.37	7.47	32.16	59.31	7.49	33.20

Since FTIR measurements with diffuse reflectance attachment (DRIFT) tend to provide near surface chemistry, it may be concluded that both types of particle have very similar surface properties.

Since both the alkyl backbone and the pendant groups of vinyl polymers can play an important role in determining the retention and retention selectivity in reversed-phase liquid chromatography (RPLC), which is based on hydrophobicity, the surface chemistry of the particles as well as their microporosity should affect the retention of solutes in the RPLC mode [7]. Once again, DRIFT spectrometry shows that the sodium dodecyl sulfate (SDS) and polyvinyl alcohol used as stabilizers for the dispersion were completely removed during purification. In particular, no hydroxyl adsorption was seen in the infrared spectrum, while the removal of SDS was confirmed by the absence of sulfur in the elemental analysis of the beads. These findings suggest that purification methods involving repeated washings of the beads with solvents effectively remove impurities that could potentially affect the chromatographic properties of the cross-linked beads.

6.2.3 Pore Size and Pore Size Distribution of the Particles

Pore size and pore size distribution were determined by size exclusion chromatography (SEC) in tetrahydrofuran (THF). Other methods such as Equation made by *Brunauer*, *Emmett* and *Teller* isotherm or mercury intrusion porosimetry are also available for the evaluation of porous structures [29]. For this study, however, it was decided to take measurements under an actual set of separation conditions (SEC), since a view of pore size and pore size distribution can be obtained even if the material is affected by a swelling phenomenon [8]. Alkylbenzene as well as polystyrene standard samples were used for the measurements because micropores derived from cross-linked networks make an important contribution to the column efficiency and retention selectivity [7] of polymer-based packing materials.

The calibration curves of all the particles are shown in Figure 6.4. Poly(styrene–divinylbenzene) particles prepared by a multistep swelling and polymerization

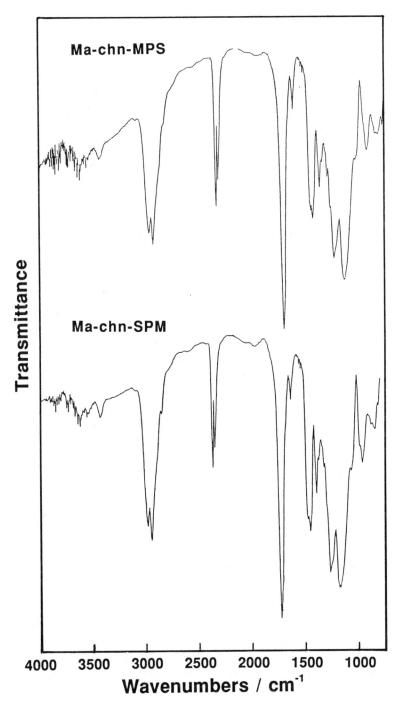

Figure 6.3. FTIR spectra of poly(methyl methacrylate–ethylene dimethacrylate) particles.

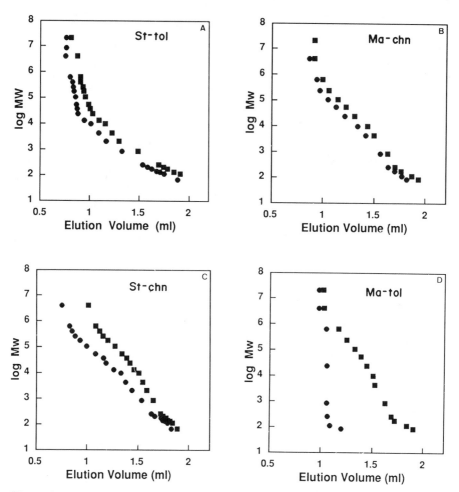

Figure 6.4. Calibration curves on prepared particles: (A) Poly(styrene–divinylbenzene) with toluene, (B) poly(methyl methacrylate–ethylene dimethacrylate) with cyclohexanol, (C) poly(styrene–divinylbenzene) with cyclohexanol, and (D) poly(methyl methacrylate–ethylene dimethacrylate) with toluene. Mobile phase, tetrahydrofuran; flow rate, 0.5 mL/min; UV detection, at 254 nm. Samples are polystyrene standards and alkylbenzenes: ■, particles prepared by a multistep swelling and polymerization method; ●, particles prepared by the corresponding suspension polymerization method.

method (St-tol-MSP) showed a larger contribution of macropore regions than those obtained by a suspension polymerization method (St-tol-SPM). The difference between the two calibration curves is small but highly reproducible. Since the two preparation methods differ only with respect to the use of a seed polystyrene in MSP, this small difference in calibration curves can be attributed to the influence of the seed polymer. In other words, since toluene is a good solvent for the polystyrene seed particles, it is the solution of the seed polymer that affects the pore structure (Figure 6.4A) [23]. It appears that St-tol-MSP may not display the clear exclusion limit that is observed with St-tol-SPM. In addition, St-tol-MSP was reported [23] to be able to resolve polystyrene standards with a molecular weight range of 10^7–10^2, while St-tol-SPM cannot separate above 10^5.

On the other hand, when methyl methacrylate, ethylene dimethacrylate, and cyclohexanol were employed, shape of the curve and volume of pores were the same for MSP and SPM (Figure 6.4B). This result may be due to the poor solubility of the polystyrene seed polymer in these monomers, solvents, and partly polymerized materials. As a result of a combination of lower solubility and lack of compatibility of the two polymers, the seed particle is excluded from the oil droplet containing absorbed monomers and solvents, as shown in Figure 6.5. Alternatively, the seed particle (perhaps partly solvated) may become segregated within, or at the outer edge of, the enlarged droplet because of its lack of solubility in the monomer-rich mixture.

Examination of the poly(methyl methacrylate–ethylene dimethacrylate) particles prepared by the multistep swelling and polymerization (Ma-chn-MSP) method by optical microscopy reveals the presence of a craterlike black shadow in all the particles. This anomaly may be a trace of the excluded polystyrene seed particle. A similar observation was made by Ugelstad et al. [30].

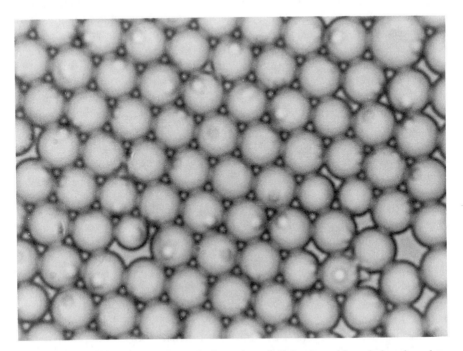

Figure 6.5.   Optical photomicrograph of an enlarged oil droplets with methyl methacrylate, ethylene dimethacrylate, and cyclohexanol.

For the MSP method to be successful, at least two critical conditions must be met. First, the seed particle must be soluble in the combination of monomers and porogenic solvent; otherwise the enlarged globules will not form. Second, the enlarged particles must not be allowed to coalesce—a condition that is easily met under the standard procedure used for the MSP method. While solubility is important for the establishment of the enlarged particles from the seed polymer and monomers/porogen, it need not be maintained once polymerization has started. As oligomeric and polymeric species form, thermodynamic factors may well combine to exclude the seed polymer; this exclusion will occur at different stages of the process depending on the combinations of monomers, porogens, and seeds selected for the experiment.

The foregoing characteristics are further illustrated by findings from other preparations with monomers and solvent of differing thermodynamic properties. When poly(styrene–divinylbenzene) particles were prepared by the two methods under similar conditions (MSP and SPM) utilizing cyclohexanol as the porogen, the calibration curves showed significant differences, mainly in the volume of pores (Figure 6.4C).

This difference in pore volume may be accounted for by subtle differences in the thermodynamic properties of the polymerizing mixtures, and the seeds play a role in changing the thermodynamic characteristics of the porogenic mixture in the case of the MSP method. In particular, in the MSP method a significant amount of phase-separated cyclohexanol was observed with the water phase after polymerization was complete, while no phase-separated cyclohexanol was observed in a parallel SPM experiment under conditions that were identical except for the absence of seed polymer. Clearly this loss of porogenic solvent is related to the subtle differences in porogens in the two polymerizations, with the small amount of dissolved seed polymer having an influence on the partition of the cyclohexanol. Although the pore volume was changed by about 20%, the calibration curves remained similar in shape. As will be described later, retention selectivity was essentially the same on both types of particle in reversed-phase mode. Overall, the role of the polymeric seed particles is limited with these methods, except for its influence on the exclusion of cyclohexanol porogen from the MSP system and the resulting smaller pore volume.

A much more drastic change in porous properties was observed when toluene was used as the porogen in the preparation of poly(methyl methacrylate–ethylene dimethacrylate) particles (Figure 6.4D). In the suspension polymerization process, particles containing almost no macro- or mesopores were obtained, as confirmed by size exclusion chromatography and microscopy, which revealed an almost smooth surface for the beads. On the other hand, porous particles with a large contribution of macropores were obtained in the MSP method. Similar particles free of macro- or mesopores were also reported when xylene or ethylbenzene was used as porogen. In the case of the MSP method, toluene is clearly a good solvent for the polystyrene seeds, and dissolution occurs with formation of the enlarged globules. The conditions of the MSP method, with a polystyrene solution within the porogenic mixture, are however sufficiently different from those of suspension polymerization to ensure that the integrity of the polymerizing droplets is maintained throughout the process. While some local separation does occur, it results only in the formation of numerous macropores, rather than in complete exclusion of the porogen. Obviously, the speculative arguments just presented will require further careful experimentation to provide a satisfactory explanation for these puzzling observations.

To gain a better understanding of the role of the seed, typical suspension polymerizations were carried out with the same amount of polymeric additives used in the case of MSP. A typical seed polystyrene prepared by an emulsifier-free emulsion polymerization has bimodal molecular weight distribution and molecular weight is usually higher than that prepared by a simple suspension polymerization method [25]. The measured number-average molecular weight (M_n) and weight-average molecular weight (M_w) for the seed polymer were 230,000 and 640,000, respectively. Calibration curves of Ma-tol-SPM prepared with three different polystyrene additives were measured. The seed polymer itself as well as two narrow dispersity polystyrene standards (M_n = 10,000 and 800) were used. As shown in Figure 6.6, porous particles with macro- and mesopores were obtained even with the suspension polymerization method. Moreover, pore size distribution was affected by changes in the molecular weight of the polymer additives: as the molecular weight increases, the polymer particles obtained have more macropores. These findings clearly suggest that the use of a porogen consisting of a polystyrene solution in toluene affords porous particles with a large contribution of macro- and

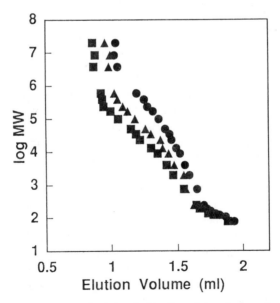

Figure 6.6. Calibration curves of poly(methyl methacrylate–ethylene dimethacrylate) particles prepared by suspension polymerization with polymeric additive: ●, with isolated seed polymer; ▲, with M_n = 10,000 (monodisperse); ■, with M_n = 800 (monodisperse). Chromatographic conditions are same as in Figure 6.4.

mesopores, whereas a corresponding suspension polymerization without the polymeric additive gives particles with micropores only.

Following the initial enlargement of the seed particles by dibutylphthalate, the second step of the MSP approach involves swelling of the slightly enlarged seeds with monomers (methyl methacrylate and ethylene dimethacrylate) and a porogenic solvent (toluene). This can be done as a single addition or through the sequential addition of the components. The solubilities of the various components are such that the polystyrene seed particles may be excluded from the monomers, while they are readily dissolved in the porogenic solvent. Therefore, three procedures were examined for the preparation of poly(methyl methacrylate–ethylene dimethacrylate) size-monodisperse particles using toluene as the porogen. These are represented as paths A–C in Figure 6.7.

Path A is the usual MSP method as described before, with simultaneous addition of the monomers and porogen. In path B, the swelling with toluene is carried out first; then the monomers are added to complete the swelling. Finally, in path C the monomers are added first, followed by the toluene. An interesting observation was made during the preparation of beads by path C. When the swelling with monomers (methyl methacrylate and ethylene dimethacrylate) was completed, a small sphere, which appeared to be the excluded seed particle, was observed at the surface on the size-monodisperse oil droplets (Figure 6.8a). The excluded seed particle disappeared again upon completion of the addition of toluene, but the swollen particles consisted of two parts (Figure 6.8b). This finding strongly suggests that the polystyrene seed particle, initially enlarged through solvation by dibutylphthalate, becomes excluded as the concentration of monomers grows in the vastly enlarged oil droplet because the seed is insoluble in the mixture of monomers. When toluene is added and migrates to the oil droplet, the polystyrene seed redissolves and therefore disappears again. Overall, the final monodispersed, enlarged oil particle is similar in composition to that of path A, the typical multistep swelling procedure. The particles produced by path B are also similar, since homogeneous swollen particles were observed after swelling with the toluene porogen, indicating that the seeds are dissolved. Subsequent addition of the monomers does not affect the macroscopic homogeneity of the swollen particle, and the enlarged oil droplets again have the same appearance (Figure 6.8c) and presumably the same macroscopic composition as those of path A.

The calibration curves obtained in the size exclusion chromatography mode for the three separation media obtained by paths A, B, and C are shown in Figure 6.9. Path A afforded macroporous poly(methyl methacrylate–ethylene dimethacrylate) particles with a calibration curve similar to that of Figure 6.4D. The particles obtained by paths B and C gave almost identical calibration curves (Figure 6.9). Thus despite obvious differences, all three paths outlined in Figure 6.7 eventually afford enlarged particles, similar in macroscopic composition, and subsequent polymerization gives particles nearly identical in porous structure.

At this stage it is difficult to paint a complete picture of the influence of the seed polymer. Clearly, the small amount involved (typically < 1%), exerts a very significant effect. This study reveals that the solubility of the polymer seeds in the porogen

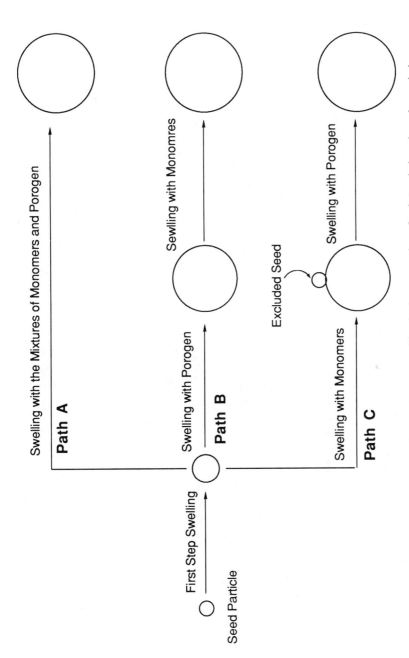

Figure 6.7. Swelling pathways of a step-by-step swelling in the preparation of poly(methyl methacrylate–ethylene glycol dimethacrylate) with toluene.

Path A

Swelling with the Mixtures of Monomers and Porogen

Path B

Swelling with Porogen

Sewlling with Monomres

Path C

Swelling with Monomers

Excluded Seed

Swelling with Porogen

First Step Swelling

Seed Particle

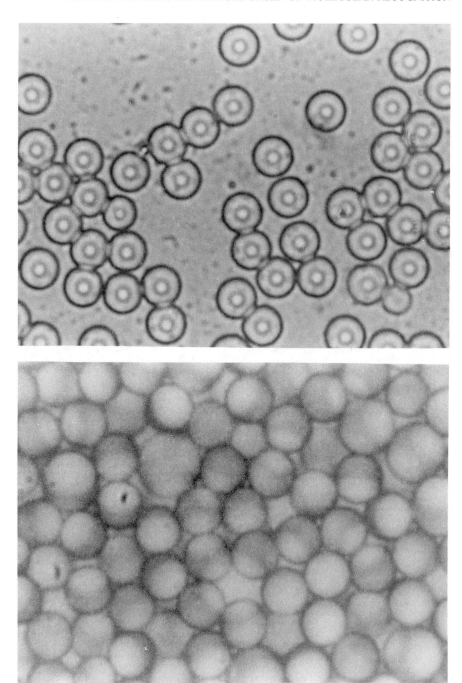

Figure 6.8. Optical photomicrographs of swollen oil droplets: (a) swollen particles with methyl methacrylate and ethylene dimethacrylate (path C in Figure 6.7), (b) further swelling is completed from (a) with toluene, and (c) swollen particles with methyl methacrylate, ethylene dimethacrylate, and toluene (path B in Figure 6.7).

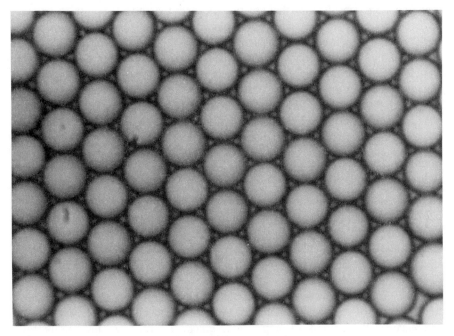

Figure 6.8 (*Continued*)

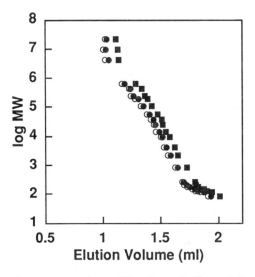

Figure 6.9. Calibration curves on the particles shown in Figure 6.8: mobile phase, tetra-hydrofuran; flow rate, 0.5 mL/min; UV detection at 254 nm. Samples are polystyrene standards and alkylbenzenes: ■, path A; ●, path B; ○, path C.

is an important consideration that greatly influences the ultimate porous structure of the beads. No influence of the seed polymer is seen when a poor solvent (of the seeds) is used as porogen. The influence of the seed polymer is not affected by the method of the swelling, which indicates that mixing of the reagents within a swollen particle is fast and that the swollen particle is almost homogeneous, at least on a macroscopic scale, unless the reagents are completely incompatible. These findings also suggest that even if the monomers and porogen have little affinity for each other, the addition of a seed polymer may actually improve the overall porogenic properties of the mixture and lead to a porous polymer.

6.2.4 Retention Selectivity in a Reversed-Phase Mode

The process of supercritical fluid extraction (SFE) is affected mainly by the pore size distribution of the separation medium, while retention selectivity in a reversed-phase mode is mainly affected by the surface chemistry of the separation medium. In some cases, separation media with specially designed pore shapes are able to selectively retain suitable solutes as a result of specific interactions analogous to those of a lock and a key.

However, since the various media reported in this study were made using exactly the same type of chemistry, differences in retention selectivity can be only due to differences in the methods of preparation, which, in turn, affect the porous structures. Figure 6.10 shows the retention selectivity of the various particles, except for Ma-tol-SPM, which is nonporous and provides very unstable retention times. The other three sets of beads show a linear relationship in the plot of their respective log k' (log k' for beads prepared by MSP vs. log k' for beads prepared by SPM), indicating that the beads prepared by both techniques have similar retention selectivities for the solutes tested.

The plot in Figure 6.10A is offset toward the St-tol-MSP axis, indicating preferred retention of the solutes by this medium, and lesser retention by the corresponding St-tol-SPM. This suggests different porous structures for the two types of bead, a result that correlates well with the findings in the SEC mode, where a small difference was observed in the macropore region.

In contrast, the plot for St-chn-MSP and St-chn-SPM (Figure 6.10C) is also linear but passes through the origin without offset, indicating that the two polymers have similar retention selectivities. Although these two separation media have different pore volumes as already described, their similar retention selectivities strongly suggest almost identical pore shapes and pore structures. Thus it can be concluded that the polystyrene seed particles have no effect on the pore size and pore size distribution of St-tol-MSP.

Similarly, the corresponding plot for Ma-chn-MSP and Ma-chn-SPM (Figure 6.10B) is also a straight line that passes through the origin, indicating that both types of particle are essentially identical in terms of chromatographic properties.

The multistep swelling and polymerization method requires calls for the use of a polymeric seed particle. Although this polystyrene seed particle represents less than 1 vol % in the swollen bead, it has an effect on the porous structure of the final size-

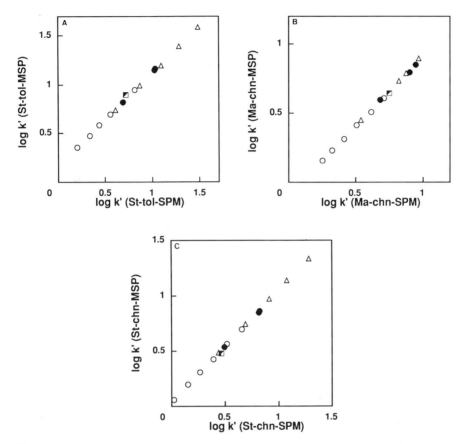

Figure 6.10. Retention selectivity in a reversed-phase mode: (A) poly(styrene–divinylbenzene) with toluene, (B) poly(methyl methacrylate–ethylene glycol dimethacrylate) with cyclohexanol, and (C) Poly(styrene–divinylbenzene) with cyclohexanol. Mobile phase, 80% aqueous acetonitrile for (A) and (C), 60% aqueous acetonitrile for (B); flow rate; 1 mL/min for (A) and (C), 0.8 mL/min for (B); UV detection at 254 nm. Samples: ○, alkylbenzenes; △, polycyclic aromatic hydrocarbons; ●, bulky aromatics; ◪, triptycene.

monodisperse beads if the porogenic solvent used in the swelling step is also a good solvent for the seed particle. Usually, the influence of the seed polymer is seen as a relatively modest but nevertheless significant change in porous structures with a shift toward larger pores. When poly(methyl methacrylate–ethylene dimethacrylate) particles are prepared with polystyrene seeds and toluene as the main porogen, porous beads are obtained, whereas nonporous particles result from a suspension polymerization carried out under similar conditions in the absence of the polystyrene seeds.

The changes in pore structures resulting from the presence of the seed polymers and of various porogens during polymerization result in different selectivities in reversed phase as well as in size exclusion mode. However, it is possible to use the MSP method to obtain porous particles that are monodispersed and have porous properties essentially identical to those of particles obtained by classical suspension polymerization. This is achieved through the selection of a seed polymer that becomes excluded from the swollen globules and therefore has no effect on the properties of the final beads. Such an approach allows the direct transfer to the preparation of monodispersed beads of the considerable amount of knowledge acquired in the preparation of macroporous beads by suspension polymerization.

6.3 Molecular Recognition of Uniformly Sized, Porous, Hydrophilic Polymer-Based Packing Materials

Restricted-access, reversed-phase, silica-based packing materials and internal surface reversed-phase, silica-based packing materials, as well as mixed functional silica supports, developed since the mid-1980s by means of various preparation techniques, have made great contributions to direct analyses of drug molecules existing in serum or plasma [31–44]. The preparation of those packing materials requires several reaction steps (e.g., introduction of hydrophobic ligand onto all the surfaces of porous silica gel and use of a special reaction technique to remove the ligand mainly from the exterior surface, followed by introduction of hydrophilic ligand on the exterior surface) [45]. Besides, mixed functional silica supports require at least two reaction steps to introduce hydrophilic and hydrophobic ligands, separately [46].

The other problem associated with the foregoing preparation methods for silica-based packing materials is that a hydrophilic functional group (e.g., a diol group) can hardly be introduced to silica support directly. In the case of silica-based, restricted-access, reversed-phase packing materials, diol groups can be introduced through a two-step reaction: namely, chemical introduction of a ligand, including oxirane-ring-to-silica supports, followed by hydrolysis of the oxirane ring to the diol group [45,46].

On the other hand, a polymer-based packing material, the hyper-cross-linked polystyrene called Styrosorb, was recently used as a restricted-access packing material in sample clean-up for HPLC [47]. In this material, the outer surface is Tris-modified and shows hydrophilic properties, while the unmodified internal surface remains hydrophobic. This material, which has micropores about 1.5 nm in diameter, prevents the access of matrix proteins and peptides into the pore system. Thus a potential restricted-access property has been conclusively demonstrated in a polymer-based packing material.

In this way, with polymer-based packing materials, micropores consisting of cross-linked polymeric structures play an important role in retention selectivity toward small molecules, while large molecules are retained or separated by macropores (real pores) [48]. Here, the cross-linked structures (i.e., the polymer backbone of polymer-based packing materials prepared by means of radical polymerization) always consists of hydrophobic carbon chains. Therefore, even polymer-based packing materials having identically hydrophilic outer and internal surfaces can be expected to find utility in the direct analysis of drugs in the presence of polypeptides, where small molecules can be retained by the polymer backbone and the hydrophilic properties of the packing material will serve to exclude the polypeptides. Indeed, a commercial polymer-based packing material for gel filtration chromatography, Asahipak GS-320, which was reported to be a polyvinyl alcohol derivative, also showed the desired chromatographic properties [49a]. When this packing material is used, however, additional (following) hydrolysis of ester groups of polyvinyl acetate particles is required to produce the hydrophilic monoalcohol groups, and the hydrophilicity is reported to be too high to achieve separation in reversed-phase mode, because the alcohol groups are introduced with relatively high density [49b].

In any event, known reactions entailing solid polymer particles (e.g., hydrolysis or introduction of hydrophilic groups) are essentially heterogeneous and may exert deleterious effects on reproducibility of chromatographic properties from batch to batch. Therefore, an easy one-step preparation method without any further reactions may present the advantage of improved reproducibility as well as ease of operation.

6.3.1 Chromatographic Properties

Uniformly sized, porous, hydrophilic polymer beads were prepared using glycerol monomethacrylate and glycerol dimethacrylate. Generally, the preparation of beaded particles utilizing water-miscible monomer occurs in conjunction with the use of the water-in-oil (W/O) suspension polymerization method, where aqueous monomer droplets are dispersed in appropriate organic solvent such as dichloroethane [13]. Although here the monomer, glycerol monomethacrylate, is water miscible, the yield of prepared particles is nearly 90% based on the total amount of monomers utilized by a normal oil-in-water (O/W) suspension system. Since the W/O suspension method is relatively complicated to operate, preparation in a normal O/W suspension system may be easier and more practical.

Although the prepared packing material involves both di- and mono-ol groups, in 40% aqueous acetonitrile, complete separations of series of hydrophobic small molecules such as alkylbenzenes and alkyl alcohols could be achieved on the prepared particles. Figure 6.11 indicates that the packing material has a typical reversed-phase characteristic that is attributable to a hydrophobic polymer backbone and also relatively hydrophobic ester groups. If retention selectivity of the prepared particle is compared with that on Asahipak GS-320 (Figure 6.12), the prepared particle is clearly the more hydrophobic: α_{CH_2} value of [k'(amylbenzene)/k'(butylbenzene)] of 1.19, versus 1.15 for Asahipak GS-320. Interestingly, the prepared particle has preferential retention toward aliphatic alkyl alcohols compared with alkylbenzenes.

Average pore diameter determined by a BET measurement is found to be 104.7 Å, as schematically represented in Figure 6.13. Thus cyclohexanol, the porogenic solvent employed in this report, is a suitable porogenic solvent for the preparation of a restricted-access packing material using this combination of monomers. Pore size

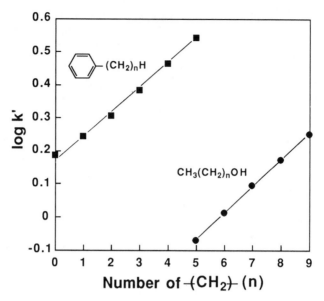

Figure 6.11. Plots of log k' versus carbon number of solutes: mobile phase, 40% aqueous acetonitrile; flow rate, 0.8 mL/min; UV detection at 254 nm and RI.

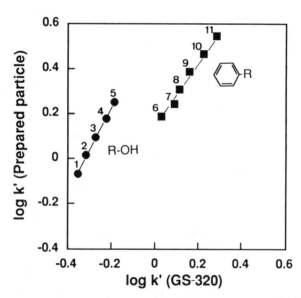

Figure 6.12. Retention selectivity of the prepared particle in a reversed-phase mode and on the chromatographic conditions given for in Figure 6.11. Samples: 1, hexanol; 2, heptanol; 3, octanol; 4, nonanol; 5, decanol; 6, benzene; 7, toluene; 8, ethylbenzene; 9, propylbenzene; 10, butylbenzene; 11, amylbenzene.

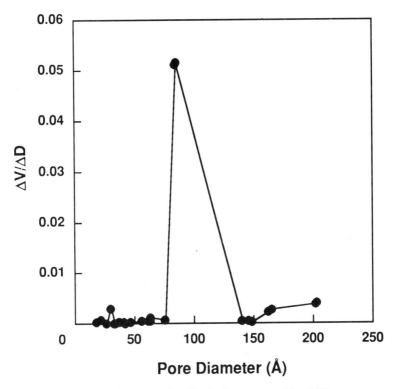

Figure 6.13. Pore size distribution measured by BET.

distributions are also depicted in Figure 6.14 and 6.15, which show calibration curves measured by size exclusion chromatography in tetrahydrofuran (THF) and water, respectively. Both calibration curves clearly indicate an exclusion limit around log MW = 4.8, where the difference in elution volume in THF and water is smaller than 15%. This finding means that the prepared particle has good solvent compatibility during the change of solvent systems. In fact, another advantage of this packing material is that a fast change of solvents from 100% water to 100% THF had no effect on column efficiency.

Figure 6.16 shows a chromatographic separation of hydrophilic drug molecules coexisting with bovine serum albumin (BSA). In 10% aqueous acetonitrile buffer as a mobile phase, BSA was excluded and eluted before the void volume of the column. The recovery of BSA obtained throughout at least 30 injections was higher than 98%, and any increment of column pressure drop was not observed during the operations. On the other hand, hydrophilic drug molecules were retained and well separated from each other as a result of hydrophobic interactions [45]. Retention selectivity toward selected drug molecules (Figure 6.17) indicates that the prepared particle shows retention factors larger than those on Asahipak GS-320 toward most of these drugs (Figure 6.18). In addition, retention selectivity is found to differ from

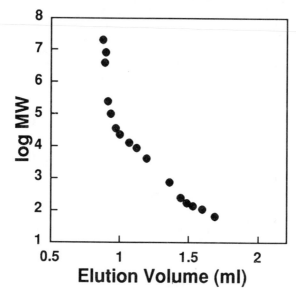

Figure 6.14. Calibration curve in THF: mobile phase, tetrahydrofuran; flow rate, 0.5 mL/min; UV detection at 254 nm; samples, polystyrene standards and alkylbenzenes.

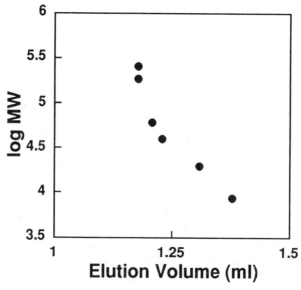

Figure 6.15. Calibration curve in water: mobile phase, water; flow rate, 0.5 mL/min; detection, RI; samples, dextrans.

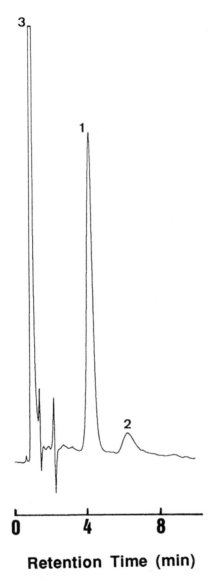

Retention Time (min)

Figure 6.16. Separation of hydrophilic drugs existing with BSA: packing material, poly(glycerol monomethacrylate-*co*-glycerol dimethacrylate); mobile phase, 10% aqueous acetonitrile 0.02 M phosphate buffer containing 0.1 M sodium sulfate (pH 7); flow rate, 1 mL/min; UV detection at 254 nm. Samples: 1, theophylline (0.3 μg); 2, barbital (1 μg); 3, BSA (60 μg).

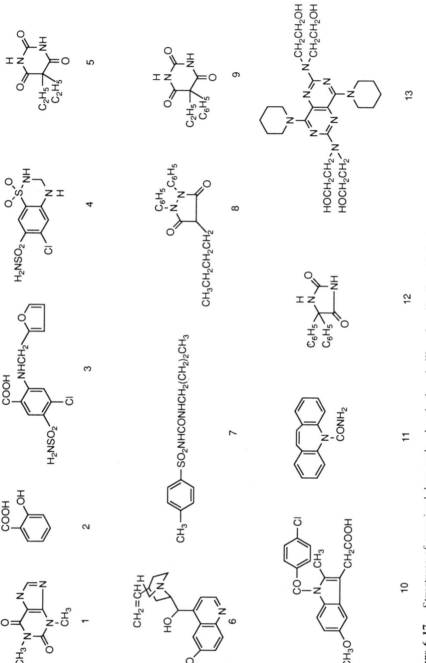

Figure 6.17. Structures of examined drug molecules: 1, theophylline; 2, salicylic acid; 3, furosemide; 4, hydrochlorothiazine; 5, barbital; 6, quinidine; 7, tolbutamide; 8, phenylbutazone; 9, phenobarbital; 10, indomethacin; 11, carbamazepine; 12, phenytoin; 13, dipyridamole.

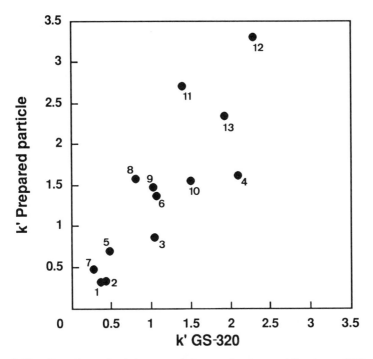

Figure 6.18. Retention selectivity toward drug molecules: mobile phase, 30% aqueous acetonitrile 0.02 M phosphate buffer containing 0.1 M sodium sulfate (pH 7); flow rate, 1 mL/min; UV detection at 254 nm; samples as listed for Figure 6.17.

drug to drug, which can be explained based on different functional groups of each packing material. The selected pair of hydrophilic drugs in Figure 6.16 cannot be separated on Asahipak GS-320; depending on the combinations of drugs to be separated, however, the observed differential retention selectivity (Figure 6.17) makes a choice of columns possible.

Compared with silica-based, restricted-access, reversed-phase packing material derived from the C_{18} stationary phase [45], hydrophobicity was lower, but introduction of another more hydrophobic monomer can control packing material hydrophobicity without losing complete recovery of the polypeptide. Actually, introduction of ethylene dimethacrylate instead of glycerol dimethacrylate increased the retention time of barbital with complete exclusion recovery of BSA (Figure 6.19).

The prepared packing material has not yet been optimized with respect to ratio of monomers, ratio of monomers and porogenic solvent, and reaction conditions. An easy one-step preparation by means of a seeded polymerization, however, might demonstrate the ability to produce a restricted-access, reversed-phase type of polymer-based packing material.

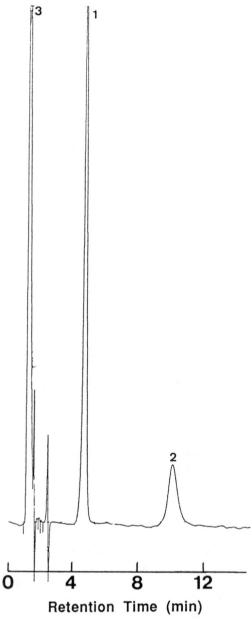

Figure 6.19. Separation of hydrophilic drugs existing with BSA: packing material, poly(glycerol monomethacrylate-*co*-ethylene dimethacrylate); mobile phase, 10% aqueous acetonitrile 0.02 M phosphate buffer containing 0.1 M sodium sulfate (pH 7); flow rate, 1 mL/min; UV detection at 254 nm. Samples: 1, theophylline (0.3 μg); 2, barbital (1 μg); 3, BSA (60 μg).

6.4 In Situ Surface-Selective Modification of Uniformly Sized, Macroporous Polymer Particles

6.4.1 Temperature-Responsive Polymeric Selector

The preparation and chromatographic use of macroporous glass beads modified with a temperature-responsive polymer, poly(N-isopropylacrylamide) (poly-NIPAM), was reported in 1992 [50]. The study cited took advantage of the temperature responsiveness of poly-NIPAM [51,52] to control pore size in high performance liquid chromatography through the changes in column temperature. This application is rather interesting because separation selectivity in HPLC had previously been controlled mostly by the changes in the mobile phase and/or the stationary phase, while this early report [50] suggested that temperature could be also used to control separation selectivity in HPLC. However, the preparation method was so complicated that its applicability to the preparation of stationary phases for HPLC was limited.

The surface modification of ordinary support materials with a variety of reactive substances is a very useful technique for the preparation of effectively functionalized materials because the reactive substances can be incorporated in such a way that they are located only at the surface of the support materials. For these modifications, the reactive functional groups located at the surface of the support materials are treated with the reactive substances to afford the surface-modified support materials. Because of the relatively slow mass transfer and steric hindrance within the support materials, however, these modification reactions are difficult and do not run to completion. This problem is especially acute with highly cross-linked macroporous polymeric support materials [53], for which accessibility to the reactive groups is severely limited.

Therefore, a copolymerization technique that directly incorporates functional monomers within the polymeric support materials has been used extensively. However, this method wastes a significant part of the functional monomers, which are incorporated at sites (e.g., the inner framework of the beads [53]) that are inaccessible to the substances that diffuse through the pores of the medium during the separation process.

A useful novel method achieves the in situ surface-selective modification of uniformly sized macroporous polymeric materials by means of graft-type copolymerization of the temperature-responsive polymer, poly-NIPAM. This new, widely applicable modification method improves the current modification and preparation techniques of surface-functionalized beads (Figure 6.20).

6.4.2 Concept of Surface-Selective Modification [54]

Because NIPAM monomer is water soluble and a water-soluble radical initiator is present, the polymerization of NIPAM itself is initiated within the aqueous phase. While this polymerization proceeds, the polymerization of ethylene dimethacrylate

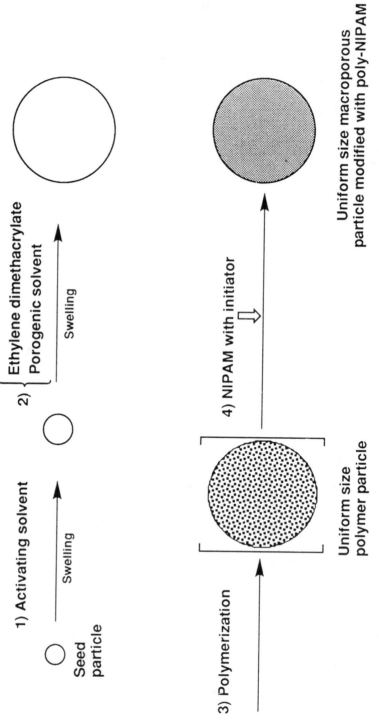

Figure 6.20. Preparation procedure of uniformly sized macroporous polymer particles modified with poly-NIPAM.

also continues independently in the adjacent organic phase that makes up the uniformly sized oil droplets as shown in Figure 6.20.

Once the growing NIPAM polymer chains have reached a certain size, they become water insoluble as a result of their temperature responsiveness at the high polymerization temperature of 80°C. This change leads to precipitation of growing poly-NIPAM in the aqueous phase. At this moment, the partly polymerized droplets containing the porogen and the residual ethylene dimethacrylate monomer are still dispersed in the aqueous phase; consequently, the precipitating poly-NIPAM should distribute itself between aqueous and organic phases according to its partition coefficient.

Table 6.3 shows the solubility parameters [55] and partition properties of the poly-NIPAM determined separately by simulated experiments. Cyclohexanol completely dissolves poly-NIPAM, as would be expected from the relatively similar solubility parameters of these compounds. On the other hand, poly-NIPAM is completely insoluble in the toluene; therefore, in experiments involving toluene as the porogen, a triphase system consisting of the aqueous phase, the organic phase, and the poly-NIPAM dispersed into the aqueous phase was observed.

According to these results, if cyclohexanol is used as the porogenic solvent, the growing poly-NIPAM presumably distributes itself into the cyclohexanol phase, filling the pores of the polymerizing porous particles, modifying their internal surfaces through a graft-type copolymerization. In contrast, if toluene is used as the porogenic solvent, poly-NIPAM, being insoluble in the porogen, cannot penetrate the inner pores of the beads that are filled with toluene, and therefore the internal surface of the beads remains unmodified. In this case, poly-NIPAM can only graft itself onto the beads' external surface, which is in contact with the aqueous phase.

6.4.3 Characteristics of the Prepared Particles

Figure 6.21 compares scanning electron micrographs of modified particles using the porogens mentioned in Section 6.4.2 (i.e., cyclohexanol and toluene) with their unmodified counterparts. The external appearance of the modified particles prepared with toluene as the porogen (Figure 6.21d) is drastically different from that of the unmodified particles (Figure 6.21c). In contrast, similar external appearances

Table 6.3 PARTITION OF NIPAM IN VARIOUS SOLVENTS[a]

Initiator	Cyclohexanol, $\delta = 11.4_{ppm}$[b]	Water	Toluene, $\delta = 8.91_{ppm}$[b]	Water
No	0.0	100.0[c]	0.0	100.0[c]
PPS[d]	99.0[e]	1.0	1.0[e]	99.0[e]

[a] Phase ratios corresponding to those of the actual polymerization mixture, Poly-NIPAM, $\delta = 11.18_{ppm}$.
[b] Ref. 55.
[c] NIPAM monomer was recovered.
[d] Potassium peroxodisulfate.
[e] Poly-NIPAM was yielded.

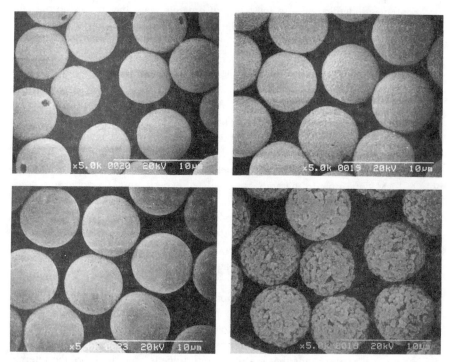

Figure 6.21. Scanning electron micrographs of (a) unmodified particles using cyclohex-anol as porogen, (b) modified particles using cyclohexanol as porogen, (c) unmodified particles using toluene as porogen, and (d) modified particles using toluene as porogen.

are observed for the NIPAM-modified particles utilizing cyclohexanol as the po-rogen (Figure 6.21b) and the unmodified particles (Figure 6.21a). These differences in external appearances demonstrate that surface-selective modification [54] with poly-NIPAM is achieved through the selection of the porogenic solvents according to the concept described in Section 6.4.2.

According to elemental analyses, 80% of the added NIPAM is incorporated into the base particles if cyclohexanol is used as the porogen, while only 60% of the added NIPAM is incorporated when toluene was used as the porogen. These data confirm that the poly-NIPAM is incorporated more extensively with cyclohexanol than with toluene because with toluene the added NIPAM, once polymerized, can reach only the external surface of the base particles. Therefore its effect is felt on a much smaller surface area than would be the case if it could also reach the much larger internal surface area of the beads.

Since one of the most persuasive proofs of the temperature-responsive physical transformation of poly-NIPAM is the reversible transformation of the apparent volume of the polymer in the aqueous phase, pore size and pore size distribution should change in a temperature-dependent manner if poly-NIPAM is effectively incorporated on the internal surface of pores. As a result of the temperature responsiveness of poly-NIPAM in the helix and random coil forms, a smaller pore size was reported below the critical temperature (32°C) than above it [50]. To verify that this transition occurs with the beads modified in surface-selective fashion, the author used size exclusion chromatography with water as the mobile phase to obtain information on the pore size and pore size distribution of macroporous particles in HPLC.

As shown in Figure 6.22, the calibration curves observed with poly-NIPAM particles prepared using cyclohexanol as the porogen affords the expected temperature-dependent pore size change. In contrast, the poly-NIPAM modified particles prepared using toluene as the porogen clearly show the opposite behavior that is also observed with the unmodified particles. These findings strongly suggest that the internal surface of the base particles was indeed selectively modified with poly-NIPAM when cyclohexanol was used as the porogen, while the external surface only was modified with poly-NIPAM when toluene was used as the porogen.

The in situ surface-selective modification of monodispersed macroporous beads by temperature-responsive poly-NIPAM has been accomplished, offering a novel type of separation medium. The porosity of this medium is readily adjusted by a change in temperature, a feature that may prove useful in a number of chromatographic applications. The ability afforded by the combination of upper critical solution temperature phenomenon and selection of porogen to control the site of modification within or at the surface of macroporous beads is unmatched in conventional techniques for bead preparation. While any conclusions are still somewhat speculative, this type of in situ modification method is extremely promising, and its extension to other systems is under active investigation.

6.5 Temperature-Controlled, High Performance Liquid Chromatography Using a Uniformly Sized, Temperature-Responsive, Polymer-Based Packing Material

Control of chromatographic separation selectivity in HPLC by applying external physical stimuli such as light [56,57], electric or magnetic fields [58,59], or temperature [60,61] is particularly interesting, because separation selectivity or resolution can be controlled without changing stationary phases and/or mobile phases. Indeed, photocontrolled chromatography has been examined utilizing photoresponsive cis–trans transformation or noninoncation transformation of functional polymeric chromatographic selectors [62–64].

Although the reported photocontrolled chromatography represents an interesting attempt to control separation selectivity by applying external physical stimuli, a phototransmittable tube (e.g., quartz) should be used as column material. Moreover, light can hardly pass homogeneously through a column having a large diameter. Similarly, special apparatus would be required to apply an electric or magnetic

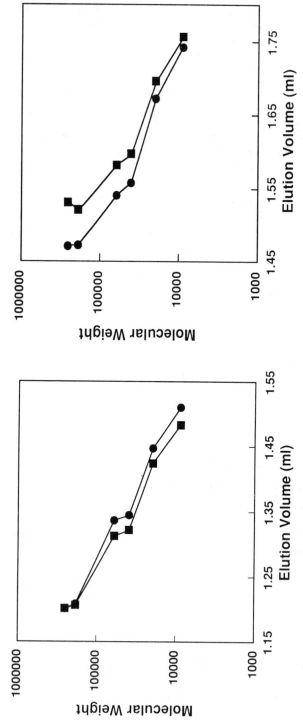

Figure 6.22. Temperature dependence of calibration curves for poly-NIPAM observed in size exclusion chromatography: mobile phase, water; dextran standard samples at 30°C (■) and 50°C (●).

field to the column. In contrast, temperature can be applied much more easily by the use of a thermostated column oven of the kind often utilized in GC or LC. On the grounds of facile realization, then, temperature will be the easiest and the most widely applicable physical stimulus for HPLC.

The polymer of *N*-Isopropylacrylamide is one of the reported temperature-responsive polymers [51,52] and has been utilized in temperature-responsive hydrogels for novel drug delivery systems [61]. Apparently the temperature-responsive phase transition of poly-NIPAM [51,52] causes the polymer to dissolve in water below its phase transition temperature (32°C), while the polymer becomes water insoluble above the phase transition temperature. This drastic change of the water solubility of the polymer can be explained on the basis of the conformational transformation of poly-NIPAM chains from the relatively hydrophilic helix form and to hydrophobic random coil form.

As mentioned before, porous glass-based packing material and silica gel modified with poly-NIPAM have been applied in temperature-dependent pore size control in size exclusion chromatography [50] as well as in procedures designed to achieve separation selectivity in reversed-phase liquid chromatography (RPLC) [65]. The preparation methods reported for the stationary phases were rather complicated, however [50].

Recently, the author and others invented a very simple method of preparing a uniformly sized, polymer-based packing material modified with poly-NIPAM as described in Section 6.4 [66]. The modification was achieved surface-selectively [54], and poly-NIPAM was combined selectively onto either the external or the internal surface of the macroporous, uniformly sized base packing material. We turn now to a description of the temperature-responsive control of chromatographic separation selectivity in reversed-phase liquid chromatography using a prepared, temperature-responsive, polymer-based packing material.

6.5.1 Separation Selectivity in Reversed-Phase Mode

According to the reported temperature-responsive conformational transition of poly-NIPAM chains [50], hydrated amide groups of the polymer chain are exposed as apparent hydrophilic external functional groups of the helical chain below the phase transition temperature of poly-NIPAM (32°C), which allows it to dissolve in water; in contrast, above the phase transition temperature, relatively hydrophobic isopropyl substituents and/or the polymer backbone begins to dominate the hydrophobic characteristics of the random coil polymer chain, which changes the hydrophilic water-soluble polymer into a water-insoluble polymer.

This phase transition of the poly-NIPAM chain between hydrophilic helix and hydrophobic random coil is accompanied by a change in the hydrodynamic volume of the polymer chain. Therefore, in the work described earlier [66], the authors proved that surface-selective modification had been achieved with poly-NIPAM by demonstrating temperature-responsive control of pore size of the prepared uniformly sized, polymer-based packing materials modified with poly-NIPAM on the internal or external surface. In fact, the packing material having a modified internal surface tended to show expansion of the pore size above the phase transition temperature due to shrinkage of the poly-NIPAM chain, while reduction of the pore size was found with the packing material having the modified external surface.

In addition to the pore size control that was found, this phase transition of the combined polymer chain will presumably lead to a change of separation selectivity of the stationary phase toward low molecular weight solutes as well as large molecules, because the temperature-responsive transformation of the polymer is due to the change of its apparent exposed functional groups lying between hydrophilic amide group and hydrophobic alkyl substituents as mentioned earlier. If this is found to be the case, upon raising the reaction temperature above the phase transition temperature, a stationary phase of the hydrophilic amide type should become a relatively hydrophobic alkyl-type stationary phase that necessarily displays a different separation selectivity.

Figure 6.23 compares separation selectivity of 10 solutes in an aqueous acetonitrile mobile phase by means of the changes in column temperature. Since all the plots with the unmodified packing material are placed on the straight dashed line, where retention factors (k') are shortened at higher temperature, no change in the separation selectivity is obtained with the unmodified packing material. On the other hand, the packing material having the modified internal surface shows affinity to hydrophilic solutes at the lower temperature (30°C) than those at 50°C, as well as different separation selectivity from that with the unmodified base packing material (see phenol, solute 8, in Figure 6.23). This change of separation selectivity can be explained based on the above-mentioned temperature-responsive transformation of the characteristics of apparent exposed functional groups of the poly-NIPAM selector, because below the phase transition temperature, amide groups are exposed as the apparent functional groups of the polymer chain that has preferential retention toward relatively hydrophilic solutes such as acetanilide, acetophenone, methyl benzoate, or cyanobenzene through interactions between hydrophilic functional groups (e.g., amide–amide interactions). These observations suggest that the separation selectivity of the modified packing material changes above and below the phase transition temperature.

6.5.2 Separation Selectivity in Drug Separation

Figure 6.24 presents another example of change in separation selectivity with the packing material modified on the internal surface: the three drugs employed were well separated with the unmodified base packing material at 30°C, while at 50°C, barbital and tolbutamide could not be separated because of peak overlaps due to the shorter retention time at 50°C. On the other hand, with the packing material having the modified internal surface, separation selectivity was found to be different, and Figure 6.25 shows that barbital and furosemide could not be separated at 30°C. Upon raising the column temperature to 50°C, however, good separations of the three drugs were obtained within 7 minutes—a shorter analysis time than that with the unmodified base packing material. The faster analysis observed is very important because it results in savings of time and mobile phase.

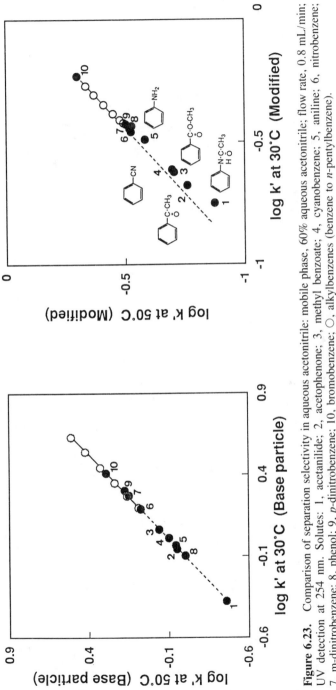

Figure 6.23. Comparison of separation selectivity in aqueous acetonitrile: mobile phase, 60% aqueous acetonitrile; flow rate, 0.8 mL/min; UV detection at 254 nm. Solutes: 1, acetanilide; 2, acetophenone; 3, methyl benzoate; 4, cyanobenzene; 5, aniline; 6, nitrobenzene; 7, m-dinitrobenzene; 8, phenol; 9, p-dinitrobenzene; 10, bromobenzene; ○, alkylbenzenes (benzene to n-pentylbenzene).

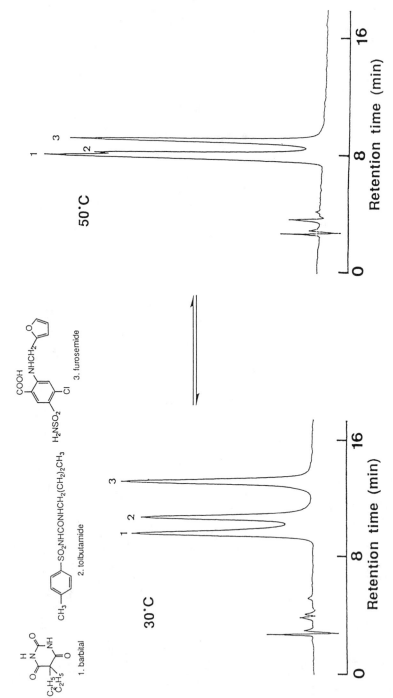

Figure 6.24. Separation of drugs with the unmodified base packing material: mobile phase. 20% aqueous acetonitrile/0.02 M phosphate buffer containing 0.1 M sodium sulfate (pH 7): flow rate, 0.6 mL/min; detection at 254 nm.

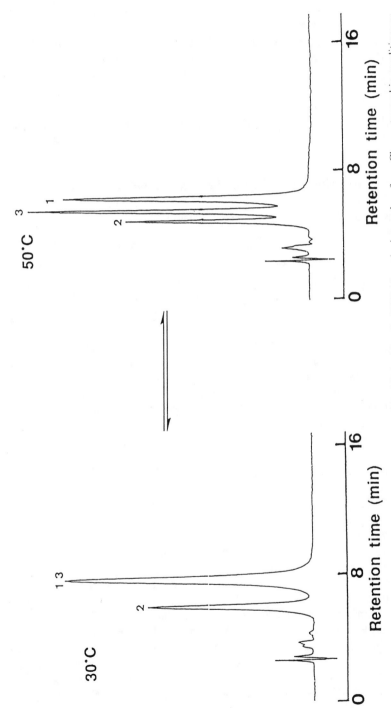

Figure 6.25. Separation of drugs with the packing material modified with poly-NIPAM on the internal surface. Chromatographic conditions as given for Figure 6.24.

The foregoing findings strongly suggest that the packing material modified with poly-NIPAM on the internal surface can be used as a stationary phase of temperature-controlled chromatography. In contrast, the packing material modified with poly-NIPAM on the external surface did not show any difference in separation selectivity toward low molecular weight solutes tested from the unmodified packing material. A similar result was also obtained for temperature-responsive pore size control [66]. Packing materials having a modified external surface may turn out to be applicable, however, in the case of separation of drugs coexisting with a polypeptide such as BSA.

6.5.3 Properties of the Modified External Surface

With both unmodified and modified packing material on the internal surface, large molecular weight polypeptide solutes such as BSA were adsorbed as a result of the hydrophobicity of the external surface of the packing materials and could not elute before the void volume of the column. On the other hand, BSA was completely excluded when the packing material was modified on the external surface with poly-NIPAM and eluted before the void volume of the column used at 30°C, as shown in Figure 6.26. Since the unmodified base packing material cannot exclude the polypeptide completely, these findings support the contention that the poly-NIPAM selector below the phase transition temperature is sufficiently hydrophilic to elute the polypeptide without hydrophobic adsorption. From those observations, this packing material modified with poly-NIPAM selector on the external surface is applicable as an internal surface packing material of the reversed-phase type [31].

Interestingly, the amount of the excluded polypeptide is drastically decreased when the column temperature is raised to 50°C, as a result of the temperature-responsive transition of hydrophobicity of the external poly-NIPAM layer. Thermal rearrangement of BSA is negligible under the chromatographic conditions employed, and since the observed change of polypeptide absorption is completely reversible [67], the column can be reused. The observed phenomenon means that with the packing material modified on the external surface, control of column temperature can shift the polypeptide's elution region, which is usually very large and sometime overlaps with the peaks of drug solutes having short retention times. In this case, two drugs used were well separated based on hydrophobic interaction with the unmodified internal surface.

Uniformly sized packing materials surface-selectively modified with poly-NIPAM are easily prepared and afford temperature-responsive control of chromatographic separation selectivity toward low molecular weight molecules involving drugs as well as large molecules such as polypeptides. The observed thermal control of separation selectivity is very useful because we just change column temperature to get a different separation selectivity, instead of resorting to the time-consuming process of changing columns and mobile phase. By choosing the packing material having modified internal and external surfaces, we can select different types of

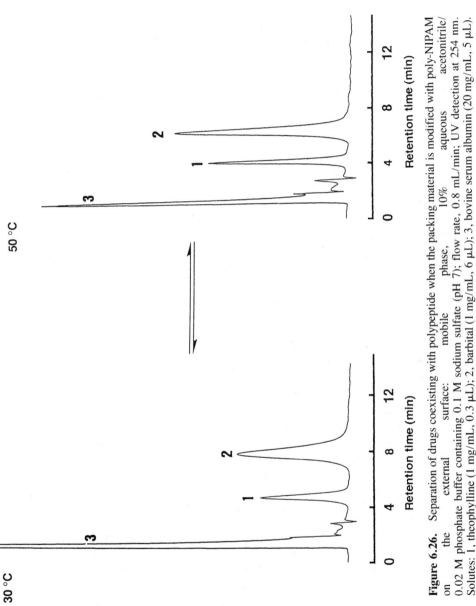

Figure 6.26. Separation of drugs coexisting with polypeptide when the packing material is modified with poly-NIPAM on the external surface: mobile phase, 10% aqueous acetonitrile/0.02 M phosphate buffer containing 0.1 M sodium sulfate (pH 7); flow rate, 0.8 mL/min; UV detection at 254 nm. Solutes: 1, theophylline (1 mg/mL, 0.3 μL); 2, barbital (1 mg/mL, 6 μL); 3, bovine serum albumin (20 mg/mL, 5 μL).

temperature-controlled chromatography. Thus the method reported offers a route to the control of separation selectivity by applying external physical stimuli.

6.6 Uniformly Sized, Hydrophobic, Polymer-Based Separation Media Selectively Modified with Hydrophilic External Polymeric Layer

Hydrophobic separation media have been utilized as packing materials for reversed-phase liquid chromatography (RPLC), and the silica-based C_{18} phase has been one of the most popular stationary phases for this mode of analysis. The silica-based hydrophobic stationary phases can be utilized in a wide range of water–organic solvent mixtures in high performance liquid chromatography, but the particles themselves, being highly hydrophobic, are not wettable by water, and thus it has been impossible to use the hydrophobic stationary phases as packing materials for open-column liquid chromatography. To eliminate the impasse due to these irreconcilable characteristics, we invented a method to convert hydrophobic silica-based stationary phases such as C_{18} phase into water-wettable stationary phases by a simple aqueous hydrolysis of ligand existing on the external surface of the hydrophobic stationary phase followed by the introduction of diol groups [39]. The modified stationary phases can be utilized as the packing materials for open-column liquid chromatography using water as mobile phase; in addition, these stationary phases can realize direct analysis of drugs in serum because the external and internal surfaces are hydrophilic and hydrophobic, respectively [45]. However, the original hydrophobicity of the unmodified stationary phases was always lowered by the hydrolysis and the diol groups that were introduced [45].

Polymer-based hydrophobic separation media such as poly(styrene–divinylbenzene) beads have also served as stationary phases in RPLC. Recent works suggested that the polymer-based separation media involved relatively preferential retention toward halogenated organic compounds as well as aromatic compounds, which was an advantage over silica-based separation media when it came to concentration of these toxic pollutants from aqueous environmental media [68]. In general, however, polymer-based packing materials are mechanically insecure; poly(styrene–divinylbenzene) particles, especially, tend to shrink in a water-rich mobile phase, resulting in the disintegration of the packed column. This disadvantage can be overcome to some extent by copolymerization of some hydrophilic monomer and/or modification of the surface with hydrophilic compounds; however, the hydrophobicity may be reduced by the introduction of hydrophilic portion, leading to the loss of the original separation selectivity and hydrophobicity of the base particles. Since copolymerization also affects pore size and pore size distribution of the original particles [53], a site-selective modification method that does not affect the separation selectivity of the original separation media should be selected.

Sections 6.6.1 through 6.6.3 describe a simple method for selectively modifying the external surface of hydrophobic polymer-based separation media by taking advantage of a hydrophilic polymer layer that does not affect the original separation selectivity and hydrophobicity of the base separation media.

6.6.1 Another Concept of Surface-Selective Modification

Polymerization of styrene and divinylbenzene under the polymerization conditions selected affords nearly quantitative conversion after 4–5 hours [53]. Therefore when hydrophilic monomers are added (4 hours after the start of the polymerization) to the aqueous polymerization medium of base particles consisting styrene and divinylbenzene, the base particles have almost been polymerized, and phase separation between cross-linked polymeric parts and porogenic solvent filling pores of the base particles is the result [68]. While the hydrophilic monomers are added directly into the polymerization medium, these should be distributed between the water phase and the porogenic solvent according to their distribution properties.

To estimate the distribution properties of glycerol monomethacrylate, glycerol dimethacrylate (Figure 6.27), and poly(glycerol monomethacrylate), experiments were carried out separately using a water/toluene two-phase system, which is a hypothetical polymerization mixture. The results obtained are summarized in Table 6.4. These observations ideally mean that the glycerol monomethacrylate monomer and its polymer are distributed dominantly into the water phase (aqueous polymerization medium), while the glycerol dimethacrylate distributes completely to the toluene phase [porogen filling the constructed pores of poly(styrene–divinylbenzene) particles]. Therefore, if each monomer is added separately, glycerol dimethacrylate can be polymerized to modify the internal surface of pores, while glycerol monomethacrylate tends to modify the external surface of the macroporous particles. In both cases, the polymerization of the added monomer can be initiated by radical species originating from those on the base particles, as discussed later.

Once polymerization has occurred in the pores (i.e., upon distribution of the added monomer into the toluene phase), the pore volume of the modified particles should be reduced in comparison with that of the base particles. Table 6.5 gives pore volumes of particles prepared by various modification conditions.

As expected from the estimated distribution properties, when 2 mL of glycerol

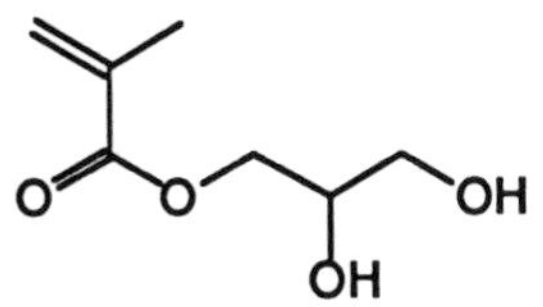

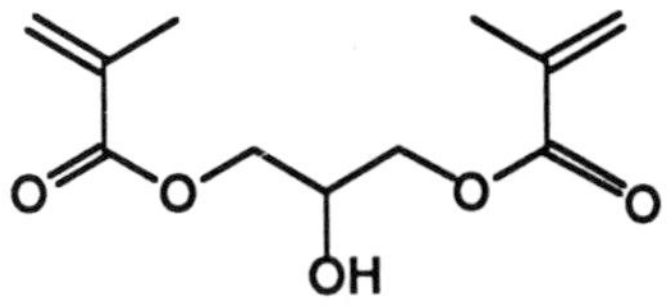

Figure 6.27. Structures of two hydrophilic monomers: **I**, glycerol monomethacrylate and **II**, glycerol dimethacrylate.

Table 6.4   DISTRIBUTION EXPERIMENTS
WITH ADDITIVES[a]

Additive	Toluene (wt %)	Water (wt %)
Glycerol monomethacrylate	2	98
Glycerol dimethacrylate	100	0
Poly(glycerol monomethacrylate)	8	92

[a]   The phase ratio was set up according to that of actual polymerization mixture.

dimethacrylate was added, pore volume was reduced by about 23 vol % compared with that of the base particles, in which the added glycerol dimethacrylate is 20 wt % of toluene. On the other hand, the addition of glycerol monomethacrylate did not affect pore volume in comparison with the addition of the unmodified original base particles, while quantitative incorporation of glycerol monomethacrylate, which was verified by elemental analysis.   As shown in Figure 6.28a and 6.28b, the added glycerol monomethacrylate does not corrupt the size uniformity of the base particles, but it does change their external appearance, which suggests that polymerization of the added glycerol monomethacrylate occurs mainly on the external surface of the base particles. These findings clearly reflect the results given in the distribution experiment (Table 6.4).

Interestingly, the addition of both glycerol monomethacrylate and glycerol dimethacrylate gave a reduction of pore volume by 16 vol % without water soluble initiator; the addition of the water-soluble initiator (potassium peroxodisulfate) with the monomers, in contrast, produced no observable change of pore volume. These findings imply that polymerization of the added monomers initiated by the added water-soluble initiator in the water phase prevents polymerization of glycerol dimethacrylate into the toluene phase to keep the original pore volume of the base particles. In this case, the size uniformity is well maintained, as shown in Figure 6.28c, and the external appearance is similar to that of Figure 6.28b.

Table 6.5   PORE VOLUME OF THE PREPARED PARTICLES

Monomer	Initiator[a]	Pore Volume (mL/g)[b]
No (base particle)	No	0.94
GMMA (1 mL)[c]	No	0.90
GDMA (2 mL)[d]	No	0.72
GMMA (1 mL) + GDMA (1 mL)	NO	0.78
GMMA (1 mL) + GDMA (1 mL)	PPS[e]	0.92

[a]   Added water-soluble radical initiator.
[b]   Determined by size exclusion chromatography.
[c]   Glycerol monomethacrylate.
[d]   Glycerol dimethacrylate.
[e]   Potassium peroxodisulfate (1 wt % of the added monomers).

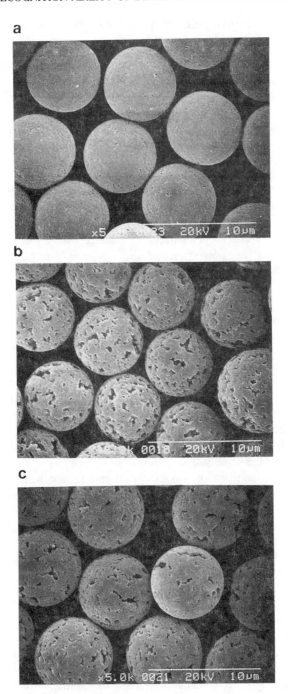

Figure 6.28. Scanning electron micrographs of prepared particles: (a) original base particles, (b) particles modified with glycerol monomethacrylate, and (c) particles modified with glycerol monomethacrylate and glycerol dimethacrylate.

6.6.2 Chromatographic Properties of the Prepared Particles

While the preceding findings suggest that glycerol monomethacrylate can modify the external surface of poly(styrene–divinylbenzene) base particles according to the properties of the distribution of glycerol monomethacrylate between water and toluene, the modified particles (with only glycerol monomethacrylate) were insecure in the water mobile phase, probably because there was swelling of the water-soluble poly(glycerol monomethacrylate) incorporated on the surface, which also gave very poor column efficiency. Modification with glycerol dimethacrylate affords secure particles in the water mobile phase, and earlier studies indicate that poly(glycerol dimethacrylate) particles are hydrophilic enough to avoid absorption of polypeptide [69]. In this modification, however, the distribution of glycerol dimethacrylate into toluene allows its polymerization into the pores of the poly(styrene–divinylbenzene) base particles, which is a selective modification of an internal surface. Even if glycerol monomethacrylate is added with glycerol dimethacrylate, the pore volume of the poly(styrene–divinylbenzene) base particles is reduced, which implies that the internal surface also is modified. Although it makes sense to modify the internal surface as a means of improving compatibility and/or wettability of the hydrophobic base particles in a water mobile phase, such a modification is likely to reduce the original hydrophobicity of the base particles, which is not the goal. On the basis of the results obtained, we selected a modification method in which glycerol monomethacrylate and glycerol dimethacrylate are added with the water-soluble radical initiator.

Figure 6.29 compares the separation selectivity of particles modified by the method just described with that on poly(styrene–divinylbenzene) base particles. In spite of quantitative introduction of the monomers, retention selectivity and hydrophobicity are found to be identical in both cases. This observation strongly suggests that modification occurs only on the surface and does not determine the retention selectivity toward the small solutes.

An effect of the modification was clearly observed when the particles were used in a water-rich mobile phase. Separation of two drugs existing with bovine serum albumin was carried out with the modified particles and with the base particles (Figure 6.30). Absorption onto the hydrophobic surface of the base particles prevented the elution of BSA before the void volume of the column; in addition, two drugs having different hydrophobic properties were eluted nearly at the void volume of the column with no separation, which means that the stationary phase does not work effectively in this mobile phase (Figure 6.30a). Actually, the packing bed of the base particles was found to shrink by 10% in the column. On the other hand, the modified particles afforded complete recovery of BSA before the void volume of the column and two drugs were well separated, reflecting the hydrophobicity of the solutes (Figure 6.30b). The complete recovery of BSA suggests that the external surface of the modified particle is hydrophilic enough to prevent absorption with the polypeptide, which is the typical characteristic of internal surface reversed-phase columns. Inasmuch as separation selectivity and hydrophobicity toward small molecules were identical for the modified particle and the base particle, it must be that

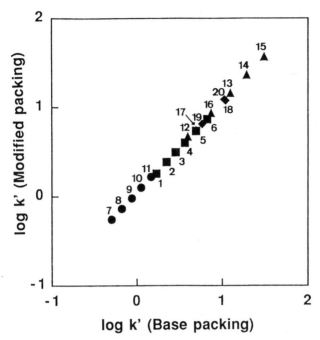

Figure 6.29. Separation selectivity of the prepared particles toward hydrophobic small particles in 80% aqueous acetonitrile: mobile phase, 80% aqueous acetonitrile; flow rate, 1 mL/min; UV detection at 254 nm and RI. Solutes: 1, benzene; 2, toluene; 3, ethylbenzene; 4, propylbenzene; 5, butylbenzene; 6, amylbenzene; 7, hexanol; 8, heptanol; 9, octanol; 10, nonanol; 11, decanol; 12, naphthalene; 13, anthracene; 14, pyrene; 15, triphenylene; 16, fluorene; 17, diphenylmethane; 18, triphenylmethane; 19, triptycene; 20, *o-ter*-phenyl.

the modification with the added glycerol monomethacrylate and glycerol dimethacrylate initiated by the water-soluble initiator occurs only on the external surface of the hydrophobic base particles.

This modification can be applied to another set of hydrophobic base particles—for example, poly(methyl methacrylate–ethylene dimethacrylate) particles. The modified particles were utilized in actual drug analysis of human serum, as shown in Figure 6.31. With the modified particles, polypeptides were completely recovered around the void volume of the column (a result that cannot be achieved with the base particles, and the hydrophilic drug theophylline could be retained to the point of detectability.

6.6.3 Summary

Surface modification of hydrophobic base particles with glycerol monomethacrylate was found to occur selectively on the external surface of the base particles according

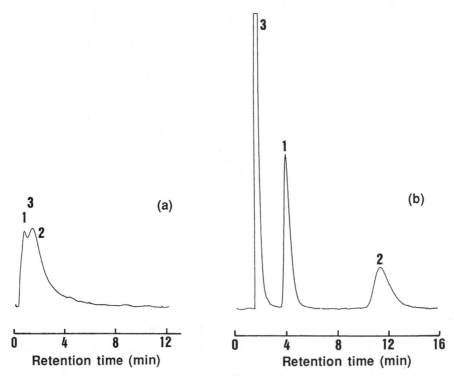

Figure 6.30. Separations of drugs coexisting with BSA (a) with base particles and (b) with modified particles: mobile phase; 10% aqueous acetonitrile 0.02 M phosphate buffer containing 0.1 M sodium sulfate (pH 7); flow rate, 0.8 mL/min; UV detection at 254 nm. Solutes: 1, theophylline (1 mg/mL, 0.3 μL); 2, barbital (1 mg/mL, 5 μL); 3, bovine serum albumin (20 mg/mL, 5 μL).

to the distribution property of the monomer. The swelling of the incorporated surface layer, however, made the modified particles insecure in the water-rich mobile phase. On the other hand, modification with glycerol monomethacrylate and glycerol dimethacrylate initiated by water-soluble initiator gave secure modified particles that retained the original separation selectivity and hydrophobicity of the base particles toward hydrophobic small solutes, while the external surface was converted to a substance that did not coabsorb with polypeptide. These characteristics could make possible the direct analysis of drugs in human serum.

Since polymer-based separation media involve preferential retention toward organohalides as well as aromatic hydrocarbons compared with silica-based hydrophobic stationary phases such as C_{18}, the former can be preferentially applied to concentration media in the solid phase extraction of aqueous environmental pollutants such as halogenated hydrocarbons or toxic dioxins. For this purpose, water

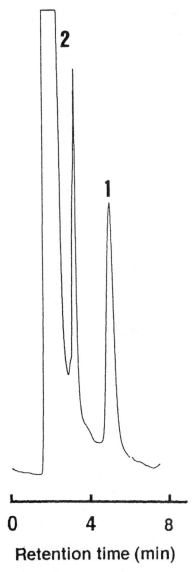

Figure 6.31.   Direct injection of serum sample: mobile phase, 30% aqueous acetonitrile/ 0.02 M phosphate buffer containing 0.1 M sodium sulfate (pH 7); flow rate, 0.8 mL/min; UV detection at 254 nm. Solutes: 1, theophylline; 2, polypeptides from serum.

wettability of the absorbent is very important to ensure reproducible results. From this point of view, the established modification method can improve the water wettability of a hydrophobic polymer-based absorbent without loss of the original retention property of the base absorbent.

6.7 One-Pot Preparation Method of Uniformly Sized, Polymer-Based, Chiral Stationary Phase for HPLC with Polymethacrylamide as a Chiral Selector

In 1979 G. Blaschke et al. reported the direct chromatographic resolution of some of racemic drugs [70] utilizing chiral polyacrylamide stationary phases [71,72]. The most famous example in this work was resolution of thalidomide, which made a great contribution to the elucidation of different medicinal actions of the enantiomers of this sedative.

Blaschke's chiral polyacrylamide stationary phases were prepared through a very simple procedure (Figure 6.32, path A). Chiral acrylamide or methacrylamide monomers were prepared by condensation reactions of acryloyl chloride or methacryloyl chloride with commercial chiral amines, while chiral polymer gels were prepared by copolymerization of these chiral acrylamide monomers with a cross-linking agent such as ethylene diacrylate [73]. The investigators used the prepared chiral polymer gels as the stationary phase in the chromatographic resolution of racemic drugs. Since, however, the polymer gels were soft gels as a result of low cross-linking (e.g., a 10:1 ratio of chiral monomer to cross-linking agent [73]), the polymer gels could not be used as stationary phases for HPLC.

The authors addressed this problem by preparing a corresponding stationary phase by the condensation reaction, in which a ready-made, cross-linked polymer gel including acryloyl chlorides was further modified with the corresponding chiral amines (Figure 6.32, path B). This stationary phase, however, showed lower resolution than that obtained by the copolymerization of the chiral monomer with cross-linking agent described earlier [74]. Apparently, then, polymerization of the chiral acrylamide is an important step in the preparation of an effective chiral stationary phase.

Thus we can hardly expect chiral recognition by a polymeric chiral selector on the basis of the chemical structure of the corresponding monomer unit, for the three-dimensional structure of chiral polymers was reported to play an important role in chiral recognition, whereas small differences in monomer unit and/or polymer structure made unexpectedly big differences in chiral recognition [75]. On the basis of observation, we can suggest the possibility that immobilization by covalent bonding of a polymeric chiral selector onto an achiral support material lowers or changes the chiral resolution of the selector because the effective three-dimensional structure of the polymeric chiral selector is somehow disturbed by the covalent bonding [76]. Moreover, since in this immobilization reaction between a solid (the support material) and a chiral polymer, it is most likely that steric hindrance will prevent the consumption for the immobilization reaction of all the reactive functional groups of both materials, the modified particle still will leave reactive functional groups, which probably will affect selectivity in HPLC as well as chemical stability of the stationary phase.

A problem in the preparation of highly cross-linked, polymer-based chiral stationary phases for HPLC by classical copolymerization methods is that a part of the

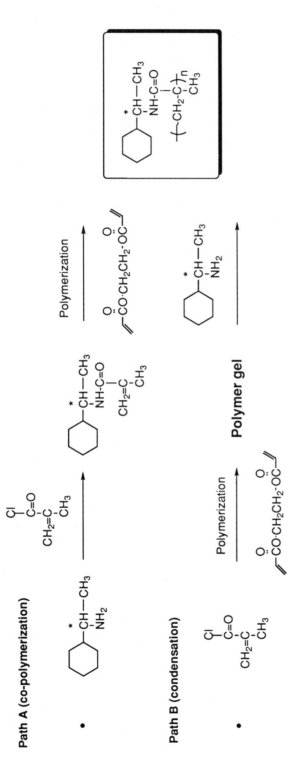

Figure 6.32. Preparation of chiral stationary phases including chiral acrylamides.

chiral monomer has to be incorporated into the chromatographically ineffective polymer structure, which means waste of this relatively expensive reagent.

Thus for the foregoing reasons, most stationary phases with polymeric chiral selectors for HPLC have been prepared by adsorption of the polymeric chiral selector onto a support material such as silica gel or poly(styrene–divinylbenzene) particles. Inability to use these stationary phases in the mobile phases that dissolve the polymeric chiral selectors remains a problem, however. In addition, physical exfoliation of the chiral selector cannot be avoided, either.

Later, Blaschke reported preparation of a silica-based, chiral polyacrylamide stationary phase by means of graft polymerization of the chiral acrylamides from commercial silica gel substituted with diol groups (LiChrosorb Diol) and further esterified with methacrylic acid [77]. These silica-based chiral stationary phases afford good resolution of the drugs in HPLC; since, however, the preparation procedure is relatively complicated, reaction conditions should be carefully controlled to keep good reproducibility. Moreover, a silica-based stationary phase is chemically unstable in both high pH and low pH environments.

We turn now to a simple, one-pot method of preparing a polymer-based chiral stationary phase (PCSP) of uniform size for HPLC. The phase contains chiral polymethacrylamide, which is incorporated by a graft-type copolymerization onto the surface of the base particles. In this method, solid chiral methacrylamide is also added directly into an aqueous polymerization medium of base particles of uniform size, prepared by a two-step swelling method (Figure 6.33). Molecular diffusion is expected to produce the adsorption of added methacrylamide and its polymerization by radical species on the surface of partly polymerized base particles. The chiral resolution of such base particles is compared with the ability in this respect of particles prepared by a traditional copolymerization method.

6.7.1 Chromatographic Properties

The quantity of the methacrylamide incorporated into the base polymer particles can be calculated on the basis of the percentage of nitrogen, found from elemental analysis, because the monomer used is somewhat hydrophobic. A relationship between theoretical content of PEA (Calcd. %) and the experimentally obtained content of PEA (Found %) is depicted in Figure 6.34. The experimentally obtained content agrees with the theoretical content of PEA nicely up to 20 wt %.

A separate attempt to determine experimentally the thermal polymerization of PEA dispersed in aqueous media without any radical initiator resulted in negligible yield of PEA polymer at the polymerization temperature. Therefore the initiation of polymerization should be induced by a radical species derived from the polymerizing base particle. In addition, since sodium nitrite, the added water-soluble radical inhibitor, prohibits chain transfer and migration of radial species from base particles to PEA monomer through the aqueous medium, we can conclude that the polymerization of PEA takes place on and/or in the polymerizing base particles after adsorption or migration of PEA to the base particles.

Figure 6.35 shows scanning electron micrographs of base particles and PEA-

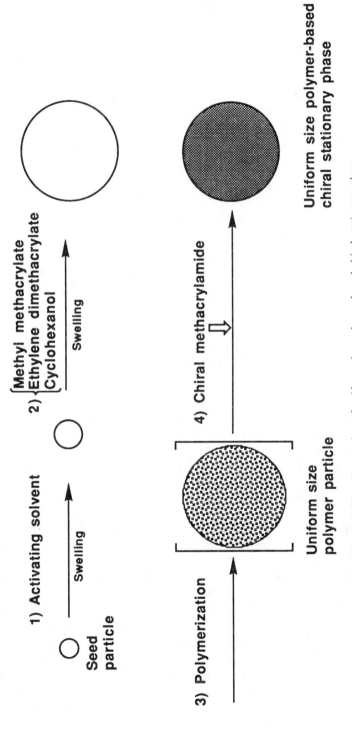

Figure 6.33. Preparation of uniform size polymer-based chiral stationary phase.

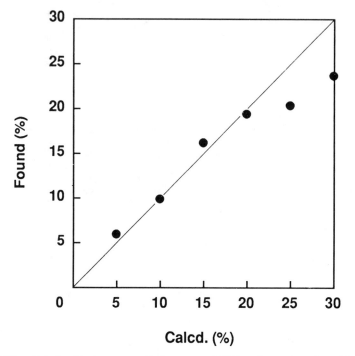

Figure 6.34. Relationship between added amount of PEA and quantity of combined PEA on base particle: solid curve is the line for $y = x$.

ADD. The good size uniformity of the base particles was not corrupted by the addition of PEA. Moreover, both particle types were relatively similar in outward appearance. These observations strongly suggest that the added PEA does not form any new homopolymeric particles (new generation) but is adsorbed and incorporated onto the surface of the base particles of uniform size.

Chromatographic properties of the prepared particles are summarized in Table 6.6. Pore volume of PEA-MIX was not changed from that of the base particle, but that of PEA-ADD was reduced. Since almost identical amounts of PEA were incorporated into both PEA-MIX and PEA-ADD, the reduction in pore volume of PEA-ADD suggests that the added PEA is incorporated not only onto the external surface but also onto the internal surface of the base particles. While in PEA-ADD, the PEA is thought to be incorporated as a copolymer unit.

The values for α_{CH_2} describing the hydrophobicity of the particles in the reversed-phase mode were decreased by the addition of PEA, which is an effect of relatively hydrophilic PEA onto the base particle. PEA-MIX was found to be more hydrophilic than PEA-ADD, which reminds us that PEA is incorporated more effectively because the steric selectivity toward plane aromatic compounds described by $\alpha_{T/O}$ and $\alpha_{P/TP}$ is higher with PEA-ADD than with PEA-MIX, which values are similar to those with the base particles. These findings imply that the

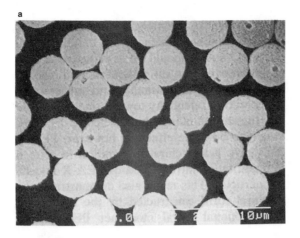

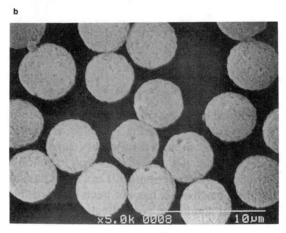

Figure 6.35. Scanning electron micrographs of (a) base particles and (b) PEA-ADD

Table 6.6   CHROMATOGRAPHIC PROPERTIES OF THE
PREPARED PARTICLES

Particle	Pore Volume (mL)[a]	Separation Factors		
		$\alpha_{\mathrm{CH_2}}$[b]	$\alpha_{\mathrm{T/O}}$[c]	$\alpha_{\mathrm{P/TP}}$[d]
Base	0.95	1.32	1.21	1.07
PEA-MIX	0.94	1.22	1.25	1.00
PEA-ADD	0.79	1.25	1.39	1.27

[a] Determined by size exclusion chromatography in tetrahydrofuran.

[b] $k'_{\mathrm{amylbenzene}}/k'_{\mathrm{butylbenzene}}$ in 60% aqueous acetonitrile.

[c] $k'_{\mathrm{triphenylene}}/k'_{o\text{-terphenyl}}$ in 60% aqueous acetonitrile.

[d] $k'_{\mathrm{anthracene}}/k'_{\mathrm{triptycene}}$ in 60% aqueous acetonitrile.

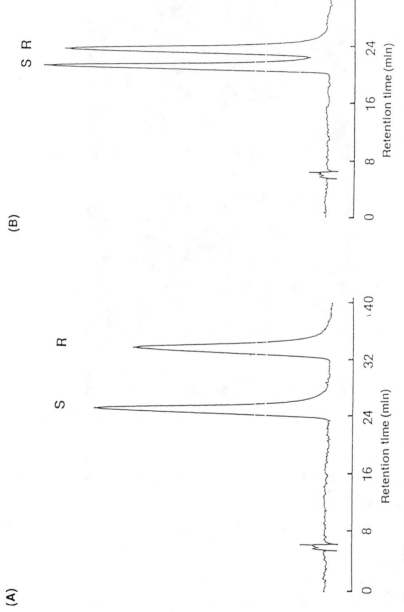

Figure 6.36. Chiral separation of 2,2′-dihydroxy-1,1′-binaphthyl with (A) PEA-ADD and (B) PEA-MIX: mobile phase, hexane/ethyl acetate 1:1 (v/v); flow rate, 1 mL/min; UV detection at 254 nm; column size, 4.6 mm i.d. × 300 mm.

Table 6.7   OPTICAL RESOLUTION WITH PCSPs

PCSP	Amide (%)[a]	k'	α	Rs
PEA-ADD	12.5	2.70 (R)	1.29	1.53
		2.10 (S)		
PEA-X-ADD	9.8	1.39 (R)	1.01	0.03
		1.37 (S)		

[a] Combined methacrylamide calculated based on elemental analyses to determine the percentage of nitrogen; mobile phase, 1:1 hexane/ethyl acetate; flow rate, 1 mL/min; column size, 4.6 mm i.d. × 150 mm.

form of incorporated PEA in the base particles differs somehow between PEA-MIX and PEA-ADD.

Chiral separations toward 2,2'-dihydroxy-1,1'-binaphthyl (DHB) examined with PEA-MIX and PEA-ADD are shown in Figure 6.36. The chiral resolutions toward DHB show significant differences. With PEA-ADD, baseline separation can be achieved with a longer retention time in a normal phase mode, while the resolution is much worse with PEA-MIX which indicates the added PEA works more effectively as a chiral selector with PEA-ADD than with PEA-MIX.

Table 6.7 shows chiral separation using PEA-ADD and PEA-X-ADD, which is prepared by means of the chiral cross-linking agent indicated in Figure 6.37. If only chirality derived from phenylethylamine influences the chiral recognition, this PEA-X should be as effective as PEA. Moreover, since on polymer-based packing mate-

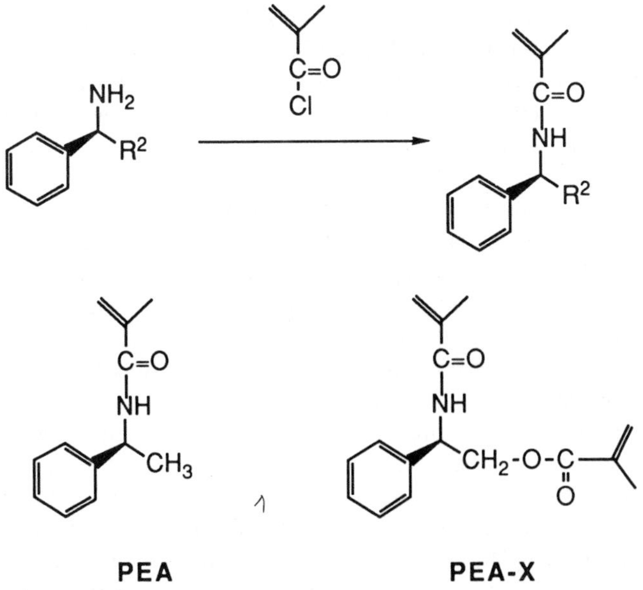

Figure 6.37.   Structures of methacrylamide monomers prepared for chiral selectors.

rials, micropores usually tend to determine separation selectivity toward small molecules [48], the cross-linking chiral selector has been thought more effective. Even so, however, PEA-X-ADD gave quite low chiral recognition in spite of the quantitative introduction of PEA-X into the base particle. This property also emphasizes the great importance of the chiral form constructed by the addition of the monofunctional chiral monomer PEA in this polymer-based chiral stationary phase, which agrees with the chiral recognition of polymeric chiral selectors reported by Blaschke [77].

A polymer-based chiral stationary phase of uniform size and having polymethacrylamide as a chiral selector can be prepared by a very simple one-pot procedure. All the preparation conditions have not yet been optimized, and the details of the separation mechanism are not clear, but this method should be very effective in preparing chemically stable polymer-based chiral stationary phases for HPLC with polymeric chiral selectors.

6.8 Molecularly Imprinted, Uniform-Size, Polymer-Based Stationary Phase for HPLC; Structural Contribution of Cross-Linked Polymer Network on Specific Molecular Recognition

The molecular imprinting technique is an effective strategy for preparing stationary phases having a specific capacity for molecular recognition [78–80]. In this technique, a template molecule is admixed in monomers to afford polymeric separation media after polymerization has occurred. The template molecule usually involves relatively polar functional groups [81,82] such as carboxylic acid, hydroxyl, amino, and/or aromatic groups, and therefore appropriate host monomers having functional groups that may interact with the functional groups of the template molecule (through, e.g., ionic interaction) are often utilized to compose a more effective imprinted site [83].

Usually, various nonaqueous bulk polymerization methods [81] including methods of preparing continuous rod-type separation media [84] have been utilized to obtain molecularly imprinted separation media, although molecularly imprinted polymers prepared by bulk polymerization methods should be grained and sieved to produce packing material [85]. This precaution is necessary because normal (O/W) suspension polymerization methods, which can afford spherical polymer beads, require an aqueous suspension medium, and because of the high polarity of the water medium adjacent to oil droplets of the monomer mixture, the water is thought to weaken the interaction between the template molecule and its host molecule. In 1994 a highly selective imprinted material was prepared in situ using aqueous porogen [86]. The author also mentioned that to produce the recognition effect, weak or moderate templates required low polymerization temperature and exclusion of water porogen during imprinting. Also in 1994, we reported molecularly imprinted, uniform size, polymer-based stationary phases for HPLC using an isomer of diaminonaphthalene as template [87]. Although we applied a typical two-step swelling and polymerization method [88] calling for water as the suspension medium, the prepared stationary phase showed a molecular recognition equivalent to that on the previously reported continuous rod-type polymeric separation medium, which was prepared by a kind of nonaqueous bulk polymerization [89]. This finding suggests that molecularly imprinted, spherical, polymer-based separation media can be prepared without loss of molecular recognition in a suspension polymerization system of the usual type.

In our 1994 report [87], methacrylic acid was utilized as a host molecule to the isomer of diaminonaphthalene [89] in expectation of possible interaction between the amino group and the host monomer's carboxylic acid. Actually the interaction played an important role for isomeric separation, while the relatively strong ionic interaction between the amino group and carboxylic acid produced severe peak tailing to amines. This interaction also affects the retention of other solutes having amino functionality, especially with uniform size polymeric separation media, because, again, this method requires an aqueous polymerization medium and thermal initiation of polymerization.

6.8.1 Diaminonaphthalene as Template

The author has prepared molecularly imprinted, uniform size, polymer-based stationary phases and has investigated their chromatographic properties in terms of the effects of a monofunctional host molecule on basic separation selectivity and of a chiral cross-linking agent on molecular recognition ability.

As reported in the paper cited above [87], the isomer of diaminonaphthalene utilized as template is eluted after the other isomer on the each imprinted stationary phase in 100% acetonitrile as the mobile phase, as shown in Figure 6.38. On these stationary phases, methacrylic acid serves as host monomer to the template, diaminonaphthalene. On both chromatograms, more hydrophobic naphthalene is found to be eluted, much faster, than both 1,5- and 1,8- diaminonaphthalenes, even in a reversed-phase mode. This finding supports the contention that combined methacrylic acid plays an important role in retaining the diaminonaphthalenes through interaction between the acid and the amino groups.

This contribution of the methacrylic acid functionality to favorable retention to amino group is also found for other aromatic amines as summarized in Table 6.8. Aniline is eluted after benzene with a broader peak shape; furthermore, pyridine affords the largest k' value in all the solutes tested.

Addition of 0.1% of triethylamine into the mobile phase does not change the k' value of benzene, while other amine compounds including diaminonaphthalenes are eluted much faster than those in 100% acetonitrile as the mobile phase. Since triethylamine would competitively block the methacrylic acid functionality on the stationary phase, these findings support the role of that functionality in retentivity, as mentioned before. The effect also confirms earlier results obtained with buffered mobile phases [90].

Methacrylic acid was found to be important in the molecular imprinting tech-

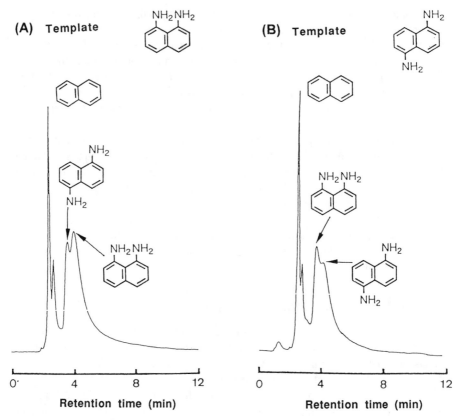

Figure 6.38. Separations on molecularly imprinted stationary phases of isomers of diaminonaphthalene: (A) 1,8- and (B) 1,5-. Mobile phase, 100% acetonitrile; flow rate, 1.0 mL/min; UV detection at 254 nm; column size, 4.6 mm i.d. × 150 mm. Solutes: 1,8-diaminonaphthalene (0.3 μg), 1,5-diaminonaphthalene (0.3 μg).

Table 6.8 EFFECT OF TRIETHYLAMINE ON SEPARATION AND RETENTIVITY IN CHROMATOGRAPHIC STUDY[a] USING DIAMINONAPHTHALENE[b] AS TEMPLATE

Template	Mobile Phase[c]	Separation, α $(k'_{1,5-}/k'_{1,5-})$	$k'_{1,5-}$	$k'_{1,8-}$	$k'_{pyridine}$	$k'_{aniline}$	$k'_{benzene}$
					Retentivity		
1,5-	AN	1.55	1.46	0.94	1.51	0.40	0.11
1,5-	AN + TEA	1.38	0.99	0.72	0.79	0.30	0.12
1,8-	AN	0.79	0.93	1.18	1.23	0.34	0.11
1,8-	AN + TEA	0.84	0.69	0.82	0.70	0.28	0.12

[a] Flow rate, 1 mL/min; UV detection at 254 nm.
[b] 1,5-, 1,5-diaminonaphthalene; 1,8-, 1,8-diaminonaphthalene.
[c] AN, 100% acetonitrile; AN + TEA, 100% acetonitrile, (0.1% triethylamine).

nique because it facilitates separation between the template and its isomer with larger retention. However, other solutes involving the same functionality as the template (e.g., pyridine) were also affected, resulting in larger retention with broader peak shape, which tends to cause peak overlapping with the template molecule in the analysis of mixed solutes. In other words, strong host functionalities often afford even worse selectivity when high retentivity is found between a solute having the same functionality and the solute used as template.

Although the retention contributed by the methacrylic acid functionality is blocked by the addition of triethylamine into the mobile phase, each diaminonaphthalene used as template was still retained longer than the isomer on the corresponding imprinted stationary phase. Since the base stationary phase prepared without the template molecule could not separate the two diaminonaphthalenes from each other, the observed moderate molecular recognition can be explained only on the basis of a contribution of the recognition site formed on the cross-linked polymer network.

To confirm the foregoing findings concerning the putative contribution of a cross-linked polymer network, we prepared an imprinted stationary phase but did not use methacrylic acid. The results on a stationary phase using 1,8-diaminonaphthalene as a template as well as a nonimprinted base stationary phase are summarized in Table 6.9. The retention time for both diaminonaphthalenes was found to be shorter on the stationary phase prepared without the host than on the stationary phase prepared with the host, while the k' values obtained were larger than those reported from the nonimprinted base stationary phase prepared without the host monomer.

Although the base stationary phase could not separate the isomers of diaminonaphthalene, the imprinted stationary phase without the host monomer (i.e., methacrylic acid) could separate the isomer where the template isomer was retained longer. Interestingly, pyridine was eluted much faster on both the base stationary phase and the imprinted stationary phase prepared without the host. These differences in the k' values of pyridine support the model according to which the methacrylic acid functionality contributes to the retentivity of the amino functionality.

Table 6.9 RETENTIVITY ON IMPRINTED PARTICLES PREPARED WITHOUT THE HOST MONOMER IN A CHROMATOGRAPHIC STUDY[a] USING DIAMINONAPHTHALENE[b] AS TEMPLATE

Template	Host	Separation, α ($k'_{1,8\text{-}}/k'_{1,5\text{-}}$)	Retentivity		
			$k'_{1,8\text{-}}$	$k'_{1,5\text{-}}$	$k'_{pyridine}$
None	None	1.00	0.46	0.46	0.15
1,8-	None	1.10	0.66	0.60	0.14
1,8-	Methacrylic acid	1.27	1.18	0.93	1.23
None	None	0.94	0.50[c]	0.47[c]	0.15[c]

[a] Mobile phase, 100% acetonitrile; flow rate, 1.0 mL/min; UV detection at 254 nm.
[b] 1,8-, 1.8-diaminonaphthalene; 1,5-, 1,5-diaminonaphthalene.
[c] Triethylamine (0.1%) added to mobile phase.

The addition of triethylamine to the mobile phase for the nonimprinted base stationary phase did not affect the k' values of all the solutes tested. This finding means that interaction between ethylene dimethacrylate and the amino functionality is relatively weak and that the observed molecular recognition on the imprinted stationary phase prepared without the host monomer is due to the existence of a recognition site for the shape of the template on the cross-linked polymer network.

Although the addition of triethylamine affords small recognition to diaminonaphthalene, this recognition is in inverse proportion to the observed molecular recognition on the imprinted stationary phase; thus this effect is almost negligible. Since the chemical composition of both stationary phases is absolutely the same, the observed molecular recognition suggests that a specific site for molecular recognition is formed on the cross-linked structure, as expected on the basis of the preceding paragraph.

6.8.2 (S)-$(+)$-N-(3,5-Dinitrobenzoyl)-α-methylbenzylamine as Template

Chiral resolution is a good target for the molecular imprinting technique. There are many reports of chiral solutes such as drugs being recognized or separated with molecularly imprinted separation media [91–93]. The template selected for study is (S)-$(+)$-N-(3,5-dinitrobenzoyl)-α-methylbenzylamine, (S)-DNB, as shown in Figure 6.39. Although nonderived α-methylbenzylamine can be used as template and should be separated by a molecularly imprinted stationary phase using methacrylic acid as a host molecule, here the amide derived from α-methylbenzylamine is employed to confirm the structural contribution of the cross-linked polymer network as discussed in Section 6.8.1 using diaminonaphthalene as the template.

As summarized in Table 6.10, a nonimprinted base stationary phase prepared using only ethylene dimethacrylate as cross-linking agent could not resolve enantiomers of three kinds of chiral amide employed as solutes. When (S)-DNB was utilized as template, a molecular recognition to the template molecule was found to afford chiral resolution with larger retention times. Although the k' value of the template and the antipode became larger by 73 and 34%, respectively, those of N-(4-nitrobenzoyl)-α-methylbenzylamine (NB) having similar structure to the template became larger by 11%, while those of N-benzoyl-α-methylbenzylamine (B) were only 7% larger without resolution.

As found in Section 6.8.1, the imprinted stationary phases prepared without methacrylic acid afforded larger k' values to the template molecule and its isomer only, compared with nonimprinted base stationary phase. Here, the imprinted stationary phase with the cross-linking agent, ethylene dimethacrylate, again affords chiral resolution with larger k' values to the template and the antipode, in comparison with those on NB or B. These results suggest that the cross-linked polymer network can memorize the shape of the chiral template specifically.

To enhance specific molecular recognition within the imprinted polymer network, a chiral cross-linking agent, N,O-dimethacryloyl phenylglycinol, (DPGL: Figure 6.39) is selected. We selected this chiral cross-linking agent because it is

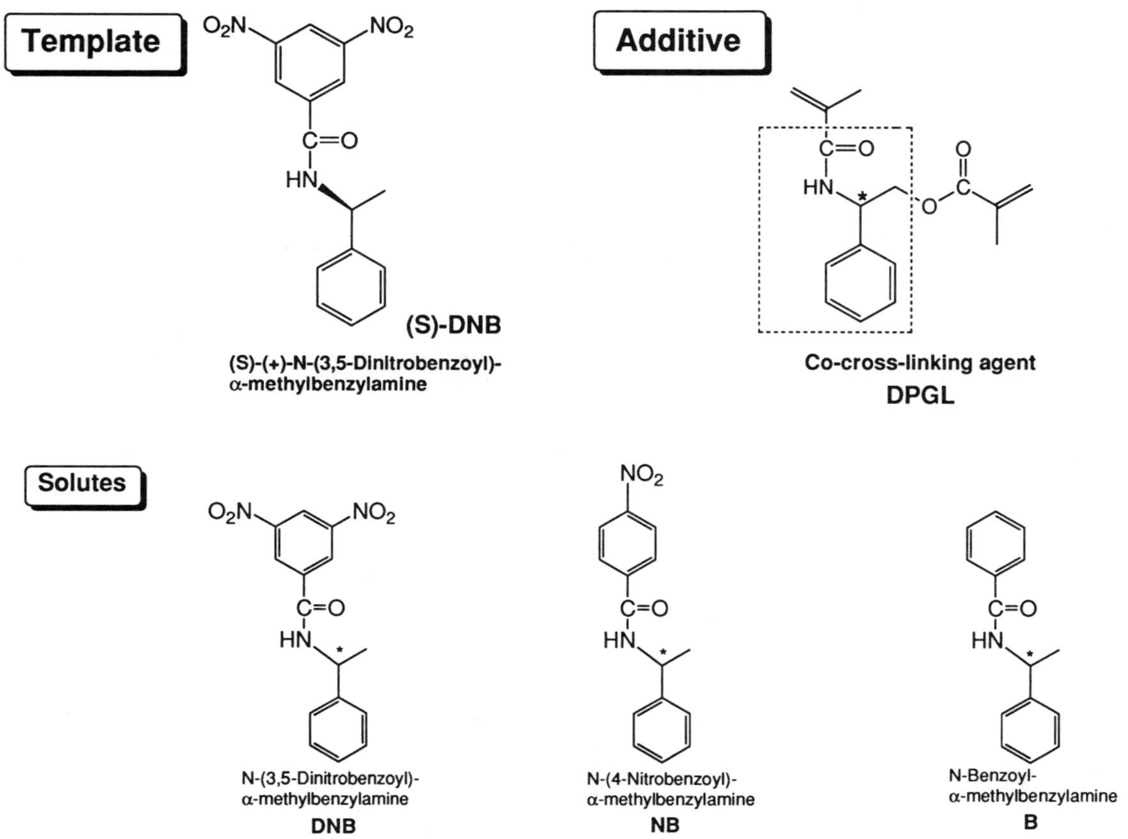

Figure 6.39. Chiral template, additive, and solutes, and their abbreviations.

Table 6.10 SEPARATION, α, OF CHIRAL AMIDE DERIVATIVES[a]

Template	Additive	DNB			NB			B		
		$k'_{(R)-}$	$k'_{(S)-}$	α	$k'_{(R)-}$	$k'_{(S)-}$	α	$k'_{(R)-}$	$k'_{(S)-}$	α
No	No	1.73	1.73	1.01	1.63	1.63	1.02	1.21	1.21	1.00
(S)-DNB	No	2.32	3.00	1.29	1.82	1.82	1.02	1.30	1.30	1.01
(S)-DNB	(R)-DPGL	2.26	4.88	1.91	1.77	1.94	1.09	1.25	1.31	1.05
(S)-DNB	(±)-DPGL	2.55	3.69	1.44	1.86	1.91	1.02	1.31	1.31	1.00
(S)-DNB	(S)-DPGL	2.41	3.00	1.24	1.75	1.70	0.97	1.19	1.18	0.99

[a] Abbreviations given in Figure 6.39.

easy to prepare and involves an amide functionality, which may make a favorable molecular interaction through possible amide–amide interaction to the template. In addition, as reported earlier with respect to uniform size, polymer-based chiral separation media with methacrylamide as the chiral selector [94], this chiral cross-linking agent did not show any effective chiral resolution when the usual copolymerization technique was utilized for preparation of chiral stationary phase with the cross-linking agent. Thus the chiral cross-linking agent DPGL does not work as a chiral selector on normal chiral stationary phases prepared by the copolymerization technique. Similar poor recognition was reported using a similar chiral cross-linking agent [95]. Therefore, we utilized co-cross-linking agents, namely, (R)-DPGL and its antipode (S)-DPGL, as well as racemic DPGL.

(R)-DPGL, which has the same chirality as the template, is found to enhance the specific molecular recognition, and the largest α value is afforded to the chiral template, as shown in Figure 6.40. Interestingly, the k' value of the solute used as the template becomes much larger than that on the imprinted stationary phase prepared without the chiral cross-linking agent, while k' values of the antipode as well as another solutes are almost equivalent (Table 6.10). Although the chiral resolution of DNB is drastically affected by the addition of (R)-DPGL, almost no separations are found for the similar solutes NB and B. This again is because (R)-DPGL itself affords low chiral recognition.

If the chiral selectivity of the added chiral cross-linking agent is the dominant factor in the enhancement of specific molecular recognition to the template molecule, the addition of racemic DPGL should result in the loss from the imprinted stationary phase of the enhancement of specific molecular recognition. However, the imprinted stationary phase shows intermediate molecular recognition between the stationary phases prepared with and without (R)-DPGL. In addition, the antipode of (R)-DPGL is found to afford molecular recognition quite similar to that of the imprinted stationary phase prepared without DPGL. This means that (S)-DPGL does not disturb the chiral recognition observed with the imprinted stationary phase prepared without any chiral cross-linking agents (DPGL).

The addition of racemic or (S)-DPGL does not change the k' values of other solutes as found with the imprinted stationary phase with (R)-DPGL, while a very small inversion of chiral recognition is observed with the stationary phase prepared with (S)-DPGL. The latter result probably is due to the inverse chirality of the cross-linking agent, but as mentioned before, chiral recognition of DPGL is not high [93]. Thus the inversion of chiral recognition is almost negligible.

These findings strongly suggest that only favorable interaction between solutes and combined chiral cross-linking agents within the imprinted site enhances specific molecular recognition toward the solute used as the template with specific enhancement of the retention to the template. It also appears that where the template molecule fits well, relatively weak intermolecular interaction can be enhanced within a specific recognition site.

This work still involves speculative explanations; however, a molecularly imprinted stationary phase prepared without a host monomer and having relatively strong interaction with the template molecule affords moderate molecular recogni-

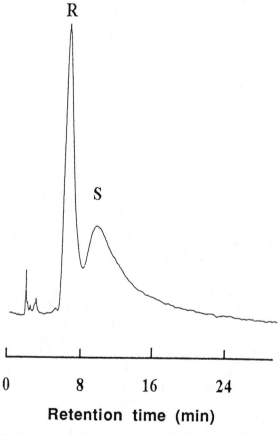

Figure 6.40. Resolution of DNB on the imprinted stationary phase prepared with (R)-DPGL; mobile phase, hexane/ethyl acetate, 1:1 (v/v); flow rate, 1.0 mL/min; UV detection at 254 nm; column size, 4.6 mm i.d. × 150 mm.

tion because the cross-linked polymer network seems to memorize the shape of the template utilized. The host monomer plays an important role in affording specific molecular recognition between isomers utilized as the template, but other solutes have functionality similar to the template are also affected by the host functionality, and the result is nonspecific enhancement of retention. On the other hand, because intermolecular interactions are relatively weak, a cross-linked, molecularly imprinted stationary phase prepared without a strong host monomer can enhance retention only of the solute used as the template; thus specificity within the solutes having similar functionality is also obtained. Addition of a favorable chiral cross-linking agent enhances this specific molecular recognition without the loss of solute specificity.

Acknowledgment

The author thanks Professor Jean M. J. Fréchet (Cornell University) and Professor Nobuo Tanaka (Kyoto Institute of Technology) for useful suggestions and helpful comments.

References

1. P.D.G. Dean, W.S. Johnson, and F.A. Middle, *Affinity Chromatography.* IRL Press, Oxford, 1985.

2. S.G. Allenmark, *Chromatographic Enantioseparations.* Ellis Horwood, Chichester, 1988.

3. R. Epton, *Chromatography of Synthetic and Biological Macromolecules,* Vols. 1 and 2. Ellis Horwood, Chichester, 1978.

4. T. Kremmer and L. Boross, *Gel Chromatography.* Wiley-Interscience, New York, 1979.

5. D. Ishii, *Introduction to High-Performance Liquid Chromatography.* Verlag Chemie, Weinheim, 1988.

6. D.C. Sherrington and P. Hodge, *Syntheses and Separations Using Functional Polymers.* Wiley, Chichester, 1988. R. Arshady, *J. Chromatogr.* **586,** 199 (1991).

7. N. Tanaka, T. Ebata, K. Hashizume, K. Hosoya, and M. Araki, *J. Chromatogr.* **475,** 195 (1989).

8. N. Tanaka, K. Kimata, Y. Mikawa, K. Hosoya, T. Araki, Y. Ohtsu, Y. Shiojima, R. Tsuboi, and H. Tsuchiya, *J. Chromatogr.* **535,** 13 (1990).

9. Y. Yamasaki, T. Kitamura, S. Nakatani, and Y. Kato, *J. Chromatogr.* **481,** 391 (1989).

10. N. Tanaka and M. Araki, *Adv. Chromatogr.* **30,** 81 (1989).

11. A. Nahum and C. Horvath, *J. Chromatogr.* **203,** 53 and 65 (1981).

12. P.C. Sadek, P.W. Carr, and L.W. Bowers, *J. Liquid Chromatogr.* **8,** 2369 (1985).

13. R. Arshady, *J. Chromatogr.* **586,** 181 (1991).

14. For example, J.C. Moore, *J. Polym. Sci., Part A-2,* 835 (1969). D. Horak, F. Svec, M. Ilavsky, M. Bleha, and J. Kalal, *Angew. Makromol. Chem.* **95,** 117 (1981).

15. W. Rolls, F. Svec, and J.M.J. Fréchet, *Polymer,* **31,** 165 (1991).

16. S.M. Ahmed, *Dispersion Sci. Technol.* **5,** 421 (1984).

17. T. Allen, *Particle Size Measurement,* 2nd ed. Chapman & Hall, London, 1975.

18. C.D. Scott, *Anal. Biochem.* **42**, 292 (1968).

19. J.V. Dawkins, T. Stone, and G. Yeadon, *Polymer*, **18**, 1179 (1977).

20. L.R. Snyder and J.J. Kirkland, *Introduction to Modern Liquid Chromatography*, 2nd ed. Wiley, New York, 1979.

21. T. Ellingsen, O. Aune, J. Ugelstad, and S. Hagen, *J. Chromatogr.* **535**, 147 (1990).

22. L.I. Kulin, P. Flodin, T. Ellingsen, and J. Ugelstad, *J. Chromatogr.* **514**, 1 (1990).

23. K. Hosoya and J.M.J. Fréchet, *J. Liquid Chromatogr.* **16**, 353 (1993).

24. R.M. Fich, *Science and Technology of Polymer Colloids*. Martinus Nijhoff, the Hague, 1983.

25. V. Smigol, F. Svec, K. Hosoya, Q. Wang, and J.M.J. Fréchet, *Angew. Makromol. Chem.* **195**, 151 (1992).

26. A. Guyot and M. Bartholin, *Prog. Polym. Sci.* **8**, 277 (1982).

27. Q. Wang, K. Hosoya, F. Svec, and J.M.J. Fréchet, *Anal. Chem.* **64**, 1232 (1992).

28. S. Matsumoto and Y. Takashima, *Kagaku-kogaku*, **47**, 540 (1983).

29. C.M. Cheng, F.I. Micale, J.W. Vanderhoff, and M.S. El-Aasser, *J. Polym. Sci., Part A*, **30**, 235 (1992). C.M. Cheng, J.W. Vanderhoff, and M.S. El-Aasser, *J. Polym. Sci., Part A*, **30**, 245 (1992).

30. J. Ugelstad, P.C. Mork, H.R. Mfutakamba, E. Soleimany, I. Nordhuus, R. Schmid, A. Berge, T. Ellingsen, O. Aune, and K. Nustad, *Science and Technology of Polymer Colloids*. Martinus Nijhoff, the Hague, 1983, pp. 51.

31. I.H. Hagestem and T.C. Pinkerton, *Anal. Chem.* **57**, 1757 (1985).

32. T.C. Pinkerton, T.D. Miller, S.E. Cook, J.A. Perry, J.D. Rateike, and T.J. Szczerba, *Biochromatography*, **1**, 96 (1986).

33. S.E. Cook and T.C. Pinkerton, *J. Chromatogr.* **368**, 233 (1986).

34. T. Nakagawa, A. Shibukawa, N. Shimono, T. Kawashima, H. Tanaka, and J. Haginaka, *J. Chromatogr.* **420**, 397 (1987).

35. T.C. Pinkerton and K.A. Koeplinger, *J. Chromatogr.* **458**, 129 (1988).

36. J.A.O. Meriluoto, K. Isaksson, H. Soini, S.E. Nygard, and J.E. Eriksson, *Chromatographia*, **30**, 301 (1990).

37. J. Haginaka, N. Yasuda, J. Wakai, H. Matsunaga, H. Yasuda, and Y. Kimura, *Anal. Chem.* **61**, 2445 (1989).

38. J. Haginaka, J. Wakai, N. Yasuda, H. Yasuda, and K. Kimura, *J. Chromatogr.* **515**, 59 (1990).

39. K. Kimata, R. Tsuboi, K. Hosoya, N. Tanaka, and T. Araki, *J. Chromatogr.* **515**, 73 (1990).

40. Y. Sudo, R. Miyagawa, and Y. Takahata, *Chromatography*, **9**, 179 (1988).

41. D.J. Gisch, R.T. Hunter, and B. Feibush, *J. Chromatogr.* **422**, 264 (1988).

42. D.J. Gisch, B. Feibush, B.T. Hunter, and T.L. Ascah, *Biochromatography*, **4**, 206 (1989).

43. J. Haginaka and J. Wakai, *Chromatographia*, **29**, 223 (1990).

44. J. Haginaka and J. Wakai, *Anal. Chem.* **62**, 997 (1990).

45. K. Kimata, K. Hosoya, N. Tanaka, T. Araki, R. Tsuboi, and J. Haginaka, *J. Chromatogr.* **558**, 19 (1991).

46. J. Haginaka, J. Wakai, and H. Yasuda, *J. Chromatogr.* **535**, 163 (1990).

47. M. Beth, K.K. Unger, M.P. Tsyurupa, and V.A. Davankov, *Chromatographia*, **36**, 351 (1993).

48. K. Hosoya, S. Maruya, K. Kimata, H. Kinoshita, T. Araki, and N. Tanaka, *J. Chromatogr.* **625**, 121 (1992).

49a. Asahipak '93–'94 Technical Data Compendium, pp. 83. (b) Asahipak Technical Note 1.

50. M. Gewehr, K. Nakamura, N. Ise, and H. Kitano, *Makromol. Chem.* **193**, 249 (1992).

51. S. Fujishige, *Polym. J.* **19**, 297 (1987).

52. S. Fujishige, K. Kubota, and I. Ando, *J. Phys. Chem.* **93**, 3311 (1989).

53. A. Guyot, in D.C. Sherrington and P. Hodge, Eds., *Syntheses and Separations Using Functional Polymers*. Wiley, New York, 1988.

54. J.M.J. Fréchet, and K. Hosoya, U.S. Patent 5,306,561 (1994).

55. J.A. Riddick, W.B. Bunger, and T.K. Sakano, *Organic Solvents,* 4th ed. Wiley, New York, 1986, pp. 138, 217.

56. S. Shinkai, H. Kinda, and O. Manabe, *J. Am. Chem. Soc.* **104**, 2933 (1982).

57. I. Karube, Y. Ishimori, S. Suzuki, and T. Sato, *Biotechnol. Bioeng.* **20**, 1775 (1978).

58. I.C. Kwon, *Makromol. Chem., Macromol. Symp.* **33**, 265 (1990).

59. H. Okuzaki and Y. Nagata, *Polym. Prepr. Japan*, **40**, 615 (1991).

60. I. Nozawa, Y. Suzuki, S. Sato, K. Sugibayashi, and Y. Morimoto, *J. Biomed. Mater. Res.* **25**, 577 (1991).

61. T. Okano, Y.H. Bae, H. Jacobs, and S.W. Kim. *Adv. Drug Delivery Syst.* **4**, 255 (1990).

62. N. Negishi, K. Ishihara, I. Shinohara, T. Okano, K. Kataoka, and Y. Sakurai, *Makromol. Chem., Rapid Commun.* **2**, 95 (1981).

63. N. Negishi, K. Ishihara, I. Shinohara, T. Okano, K. Kataoka, Y. Sakurai, and T. Akaike, *Chem. Lett.* 681 (1981).

64. K. Ishihara, I. Shinohara, and T. Okano, *Kagaku No Ryoiki*, **37**, 873 (1983).

65. K. Yamamoto, H. Kanazawa, Y. Matsushima, N. Takai, T. Okano, and Y. Sakurai, in 114th National Meeting of the Pharmaceutical Society of Japan, Tokyo, 1994, p. 160.

66. K. Hosoya, E. Sawada, K. Kimata, T. Araki, N. Tanaka, and J.M.J. Fréchet, *Macromolecules*, **27**, 3973 (1994).

67. B.H. Hofstee, *J. Prep. Biochem.* **5**, 7 (1975).

68. K. Hosoya, E. Sawada, K. Kimata, T. Araki, and N. Tanaka, *J. Chromatogr.* **662**, 37 (1994).

69. K. Hosoya, Y. Kishii, N. Tanaka, K. Kimata, T. Araki, and H. Kiniwa, *Chem. Lett.* 745 (1993).

70. G. Blaschke, H.P. Kraft, K. Fickentscher, and F. Köhler, *Arzheim.-Forsch.* **29**, 1640 (1979).

71. G. Blaschke and A.-D. Schwanghart, *Chem. Ber.* **109**, 1967 (1976).

72. G. Blaschke and F. Donow, *Chem. Ber.* **108**, 2792 (1975).

73. G. Blaschke, *J. Liquid Chromatogr.* **9**, 341 (1986).

74. Japan Chemical Society, Ed., *Kikan Kagaku Sosetsu, Vol. 6:* JCS, Tokyo, 1989.

75. T. Shibata, T. Sei, H. Nishimura, and K. Deguchi, *Chromatographia*, **24**, 552 (1987).

76. K. Kimata, R. Tsuboi, K. Hosoya, and N. Tanaka, *Anal. Method. Instrum.* **1**, 14 (1993).

77. G. Blaschke, W. Bröker, and W. Fraenkel, *Angew. Chem., Int. Ed. Engl.* **25**, 830 (1980).

78. B. Sellergren, B. Ekberg, and K. Mosbach, *J. Chromatogr.* **347,** 1 (1985).

79. K.J. Shea and D.Y. Sasaki, *J. Am. Chem. Soc.* **113,** 4109 (1991).

80. G. Wulff, *Am. Chem. Soc. Symp. Ser.* **308,** 186 (1986).

81. G. Wulff and A. Sarhan, *Angew. Chem.* **84,** 364 (1972).

82. L.I. Andersson, B. Sellergren, and K. Mosbach, *Tetrahedron Lett.* **25,** 5211 (1984).

83. O. Ramström, L.I. Andersson, and K. Mosbach, *J. Org. Chem.* **58,** 7562 (1993).

84. F. Svec and J.M.J. Fréchet, *Anal. Chem.* **64,** 820 (1992).

85. G. Wulff, A. Sharhan, and K. Zabrocki, *Tetrahedron Lett.* **14,** 4329 (1973).

86. B. Sellergren, *J. Chromatogr. A,* **673,** 133 (1994).

87. K. Hosoya, K. Yoshizako, N. Tanaka, K. Kimata, T. Araki, and J. Haginaka, *Chem. Lett.* 1437 (1994).

88. K. Hosoya and J.M.J. Fréchet, *J. Polym. Sci., Part A, Polym. Chem.* **31,** 2129 (1993).

89. J. Matsui, T. Kato, T. Takeuchi, M. Suzuki, K. Yokoyama, E. Tamiya, and I. Karube, *Anal. Chem.* **65,** 2223 (1993).

90. B. Sellergren and K.J. Shea, *J. Chromatogr.* **654,** 17 (1993).

91. G. Vlatakis, L.I. Andersson, R. Müller, and K. Mosbach, *Nature,* **361,** 645 (1993).

92. L. Fischer, R. Müller, B. Ekberg, and K. Mosbach, *J. Am. Chem. Soc.* **113,** 9358 (1991).

93. M. Kempe and K. Mosbach, *J. Chromatogr. A,* **664,** 276 (1994).

94. K. Hosoya, K. Yoshizako, N. Tanaka, K. Kimata, T. Araki, and J.M.J. Fréchet, *J. Chromatogr. A,* **666,** 449 (1994).

95. L. Andersson, B. Ekberg, and K. Mosbach, *Tetrahedron Lett.* **26,** 3623 (1985).

CHAPTER

7

Molecular Recognition in Complexation Gas Chromatography

Volker Schurig

7.1 Introduction

Complexation gas chromatography is defined as a method that utilizes the fast and reversible 1:1 equilibrium between a metal coordination compound A (selector) in a nonvolatile inert solvent S and a solute B (selectand) along a coated packed or open column. The principle is depicted in Figure 7.1.

In conventional gas chromatography, the solute B is distributed along the column between the mobile gas phase and the stationary liquid phase consisting of the nonvolatile solvent S. The elution of B is described by the *retention factor* (or capacity factor) k'° according to equation 7.1 (the superscript $^{\circ}$ refers to the left-hand column in Figure 7.1)

$$k'^{\circ} = \frac{t_{R}^{\circ} - t_{M}^{\circ}}{t_{M}^{\circ}} = \frac{t_{R}'^{\circ}}{t_{M}^{\circ}} = \frac{t_{\text{stationary phase}}}{t_{\text{mobile phase}}} = K_{L}^{\circ} \frac{1}{\beta^{\circ}} \qquad (7.1)$$

where t_{R}° = total retention time of B
$t_{R}'^{\circ}$ = adjusted retention time of B
t_{M}° = gas holdup (dead) time
K_{L}° = distribution constant
β° = phase ratio

and k'° is the time fraction of B spent in the stationary phase versus the mobile phase.

Chemical selectivity is introduced into the chromatographic distribution process

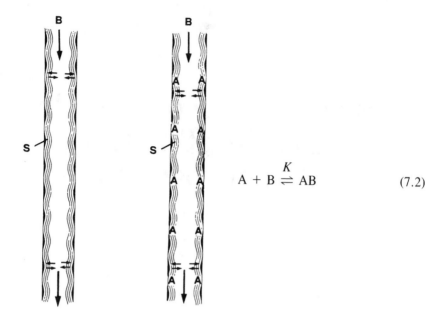

$$k'^{\circ} = \frac{t_{R}^{\circ} - t_{M}^{\circ}}{t_{M}^{\circ}} = K_{L}^{\circ} \frac{1}{\beta^{\circ}} \quad (7.1)$$

$$R' = \frac{k' - k'^{\circ}}{k'^{\circ}} = K\, a_{A} \quad (7.3)$$

Figure 7.1. The principle of complexation gas chromatography: *left*, reference column and *right*, complexation column.

when a selector A, displaying a chemical affinity toward the selectand B, is added to a nonvolatile *inert* solvent S according to the equilibrium shown in equation 7.2 (cf. Figure 7.1, right)

$$A + B \overset{K}{\rightleftharpoons} AB \quad (7.2)$$

where K is the thermodynamic complexation constant.

The fast and reversible 1:1 interaction between A and B in the nonvolatile inert solvent S leads to a *retention-increment* (or chemical capacity factor) R' of B, which is defined by equation 7.3 (note the formal similarity between equations 7.1 and 7.3):

$$R' = \frac{k' - k'^{\circ}}{k'^{\circ}} = \frac{t_{complexed}}{t_{noncomplexed}} = K\, a_{A} \quad (7.3)$$

where k'° = retention factor of B measured with a reference column without A (see Figure 7.1, left)

k' = retention factor of B measured with the complexation column containing A in S (see Figure 7.1, right)

K = thermodynamic complexation constant

a_A = activity of A in S

The retention-increment (retention-increase or chemical capacity factor) R' of B is equal to the product of the thermodynamic complexation constant K and the activity of A in S, a_A; R' is the time fraction of B spent in the complexed state $[R' / (R' + 1)]$ versus the noncomplexed state $[1 / (R' + 1)]$ in the stationary phase. Thus, according to equation 7.3, the thermodynamic complexation constant K or $-\Delta G$, and from the temperature dependence of $-\Delta G$, also $-\Delta H$ and ΔS, are accessible from chromatographic retention data, provided the activity of A in S, a_A, is known.

Complexation gas chromatography has two important merits. First, the introduction of chemical selectivity into the chromatographic distribution process can be utilized for particularly difficult separations (e.g., resolution of constitutional and configurational isomers and isotopomers X and Y). The method is highlighted by the separation of enantiomers (optical isomers) on stationary phases containing a chiral selector A via enantioselective complexation (*enantioselective complexation gas chromatography*). Second, without knowledge of the activity of A in S, a_A, the ratio of thermodynamic complexation constants K_X/K_Y or the free enthalpy (Gibbs energy) difference, $-\Delta_{X,Y} (\Delta G)$, can easily be determined from relative retention data employing a reference column (see Figure 7.1, left) and a complexation column (see Figure 7.1, right) at a given temperature. Using an inert noncomplexing reference compound B*, coinjected with B, relative retention data are obtained that are independent of chromatographic parameters such as kind of column, phase ratio, flow rate, and pressure drop.

7.2 Historical Perspective

The selective separation of alkene isomers (e.g., isomeric butenes) on packed columns containing silver nitrate in ethylene glycol described by Bradford, Harvey, and Chalkley [1] represented the first instance of complexation gas chromatography involving olefin metal coordination. *Argentation gas chromatography* has since been used to determine complexation constants K of olefin silver(I) coordination complexes from gas chromatographic retention data [2–5]. From the temperature dependence of K, the *Gibbs–Helmholtz* parameters $-\Delta H$ and ΔS also were calculated [6]. This approach was extended to selective olefin rhodium(I) [7–11] and olefin rhodium(II) [11,12] coordination.

Complex-forming inorganic selectors have widely been applied to analytical pursuits in gas chromatography [13–15] employing the metal complexes (e.g., carboxylates [12,16,17], salicylaldimines and glyoximes [18,19], β-ketoenolates [20–24], phthalocyanins [25], or phosphanes [26] as solids [16,19,20], polymers [23,26,27], dispersions [12,15], or solutions in coordinating [17] or noncoordinating [21,22,24] solvents. Not all these systems, however, are readily available for thermodynamic studies of the complexation equilibria in solution. To obtain complexation constants according to the GC approach advanced by Purnell [4], the following criteria for the selector A must be considered [28]:

- nonvolatility and thermal stability
- solubility in an nonvolatile inert solvent S suitable for GC (e.g., squalane, polysiloxanes)
- fast and reversible complexation with B of 1:1 stoichiometry at ambient temperatures and in high dilution
- absence of dissociation or polymerization equilibria

In a pioneering work, Feibush et al. studied the complexation of σ-donor molecules with dilute solutions of lanthanide(III) tris[3-(trifluoroacetyl)-(1R)-camphorates] **1** in squalane ($C_{30}H_{62}$) [21]. No linear relationship between the retention-increment R' and the selector concentration was observed, however, owing to a dissociation equilibrium of the oligomeric metal compounds involved in solution [21]. On the other hand, the dimeric manganese(II) cobalt(II) and nickel(II) bis[3-(trifluoroacetyl)-(1R)-camphorates] **2** essentially fulfilled the conditions just outlined and were used in a squalane ($C_{30}H_{62}$) solution for the determination of the Gibbs–Helmholtz parameters of molecular complexation with 30 σ-donor molecules in the apolar solvent squalane ($C_{30}H_{62}$) [28]. While these measurements were performed with packed columns containing the selector in squalane and coated on Chromosorb as the support, the use of refined, coated, high resolution open tubular columns allowed not only the resolution of isotopomers with a mass difference of one dalton—that is, the separation of α-deuterated tetrahydrofurans $C_4H_{4-n}^\alpha D_n^\alpha H_4^\beta O$, on cobalt(II) bis[3-(trifluoroacetyl)-(1R)-camphorate] **3b** [29] but also the enantiomer separation of chiral aliphatic oxiranes, thiiranes, and aziridines on nickel(II) bis[3-(trifluoroacetyl)-(1R)-camphorate] **2c** [30,31] and manganese(II) bis[3-(trifluoroacetyl)-(1R)-camphorate] **2a** [32]. This development was preceded by the separation of all deuterated ethenes $C_2H_{4-n}D_n$ [33] and by the first enantiomer separation of racemic 3-methylcyclopentene on dicarbonyl-rhodium(I) 3-(trifluoroacetyl)-(1R)-camphorate **4** by complexation gas chromatography [34].

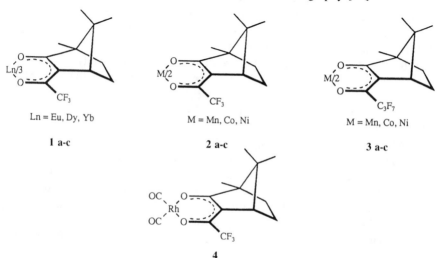

Ln = Eu, Dy, Yb

1 a-c

M = Mn, Co, Ni

2 a-c

M = Mn, Co, Ni

3 a-c

4

These results, emerging from the author's laboratory [35], provided the basis for further extensive studies of molecular recognition in complexation gas chromatography including enantioselectivity. The topic is treated in this chapter.

7.3 Comprehensive Theory

Consider a volatile solute B (selectand) migrating through a complexation column containing a dilute solution of the nonvolatile selector A (selector) in the nonvolatile inert solvent S (see Figure 7.1, right). When a 1:1 complex is formed rapidly and reversibly between A and B, two *separate* equilibria occur

$$B_{(g)} \overset{K_L^o}{\rightleftharpoons} B_{(l)} \tag{7.4}$$

and

$$A + B_{(l)} \overset{K}{\rightleftharpoons} AB \tag{7.2}$$

or

$$K_L^o = \frac{c_{B(l)}}{c_{B(g)}} \tag{7.5}$$

and

$$K = \frac{a_{AB}}{a_A a_{B(l)}} = \frac{c_{AB}}{c_A c_{B(l)}} \frac{\gamma_{AB}}{\gamma_A \gamma_{B(l)}} \tag{7.6}$$

where the subscripts g and l refer to the gas and the liquid phase, respectively; K_L^o is the distribution constant (partition coefficient) of B between the gas and the pure liquid phase S (= physical contribution to retention, neglecting the presence of A in S), K is the thermodynamic complexation constant of A and B in S (= chemical contribution to retention), and a_i is the activity of the species i, with the convention that $a_i \to c_i$ when $c \to 0$. Since both AB and $B_{(l)}$ usually are highly diluted in the gas chromatographic experiment equation 7.6 can be rewritten

$$K a_A = \frac{c_{AB}}{c_{B(l)}} \tag{7.7}$$

The total concentration of selectand B present in the liquid phase S is $c_{B(l)} + c_{AB}$. Hence, the *apparent* distribution constant K_L, assuming no volume change on dissolution of B, is [2]

$$K_L = \frac{c_{B(l)} + c_{AB}}{c_{B(g)}} \tag{7.8}$$

With equations 7.5 and 7.7 this can be rewritten

$$K_L = \frac{c_{B(l)}}{c_{B(g)}} + \frac{c_{AB}}{c_{B(g)}} = \frac{c_{B(l)}}{c_{B(g)}} + \frac{c_{B(l)}}{c_{B(g)}} \frac{c_{AB}}{c_{B(l)}} = K_L^\circ + K_L^\circ Ka_A \tag{7.9}$$

or

$$Ka_A = \frac{K_L}{K_L^\circ} - 1 \tag{7.10}$$

Thus, the thermodynamic complexation constant K can be obtained from the plot of K_L versus a_A (equation 7.9) [2,3].

From equation 7.10 a very useful relationship between K and chromatographic retention data can be developed when a reference column, containing the pure solvent S (see Figure 7.1, left, recalling that the reference column is designated by the superscript $^\circ$: e.g., K_L°, k'°) and an inert noncomplexing reference compound B*, coinjected with B, is employed [8,28,36,37] (symbols referring to B* are marked with an asterisk: e.g., $k'*$).

The distribution constant of the selectand B between the gas and the pure liquid phase S in the reference column, K_L°, is defined by

$$K_L^\circ = k'^\circ \beta^\circ \tag{7.11}$$

and the *apparent* distribution constant of the selectand B between the gas and the liquid phase S containing the selector A in the complexation column K_L is defined by

$$K_L = k'\beta \tag{7.12}$$

where k'° is the retention factor of B in the pure solvent S in the reference column, k' is the retention factor of B in the solution of A in S in the complexation column, β° is the phase ratio of the reference column, and β is the phase ratio of the complexation column. Equation 7.10 can now be rewritten as

$$Ka_A = \frac{k'}{k'^\circ} \frac{\beta}{\beta^\circ} - 1 \tag{7.13}$$

For the noncomplexing reference compound B* (e.g., a saturated hydrocarbon), it is required that $K^* = 0$. From equation 7.13, it follows for B* that

$$\frac{k'*}{k'^{\circ*}} \frac{\beta}{\beta^\circ} = 1 \tag{7.14}$$

The ratio β/β° is the same for B and for coinjected B*, and equations 7.13 and 7.14 can be combined to give

$$Ka_A = \frac{k'}{k'^\circ} \frac{k'^{\circ*}}{k'*} - 1 = R' \tag{7.15}$$

with

$$\frac{k'^{\circ}}{k'^{\circ *}} = \frac{t_R'^{\circ}}{t_R'^{\circ *}} = r^{\circ} \tag{7.16}$$

and

$$\frac{k'}{k'^{*}} = \frac{t_R'}{t_R'^{*}} = r \tag{7.17}$$

a relationship, reminiscent of equation 7.3 but referring to relative retention data r° and r, is obtained:

$$K a_A = \frac{r - r^{\circ}}{r^{\circ}} = R' \tag{7.18}$$

where r is the relative retention of the selectand B with respect to the inert reference compound B* in a complexation column containing the selector A with activity a_A in the inert solvent S (see Figure 7.1, right), and r° is the relative retention of the selectand B with respect to the same reference compound in the reference column containing the pure solvent S, devoid of the selector A (see Figure 7.1, left). Whereas it is experimentally difficult, if not impossible, to determine absolute retention factors k'° and k' of B according to equation 7.3 with strictly identical parameters for the reference and complexation column, the relative retention data r° and r in equation 7.18 are independent of column type and of all operating conditions except the temperature. Since r refers to retention factors k' or adjusted retention times t', respectively, the gas holdup (dead) time t_M of the columns must be determined precisely either as the time of the air peak, if present, or, as an approximation, as the time of the peak of coinjected methane [38].

According to the basic equations 7.3 and 7.18, the retention-increment (or chemical capacity factor) R' is linearly related to a_A at a given temperature when a 1:1 molecular complex is formed between A and B (as can be expected when minute amounts of B are injected into the column). Since equation 7.18 refers to relative retentions, it is not necessary to know gas chromatographic data such as the phase ratio (determined by such parameters as the volumes of the mobile and stationary phases or film thickness and internal diameter), flow rate of the mobile phase, column pressure drop, and column length. Equation 7.18 is also independent of the kind of column used (packed or open tubular).

A similar approach has been advanced in argentation GC [5,37]. Since the high concentration of silver nitrate in ethylene glycol is critical in view of a salting-out effect [3], lithium nitrate solutions in ethylene glycol have been used for the reference column. Again, the use of an inert reference compound greatly facilitated the method.

The validity of equation 7.18 has been scrutinized by careful and extensive experiments [28]. While the retention-increment R' can accurately be measured, an error in the *absolute* value of the thermodynamic complexation constant K according to equation 7.18 may arise as a result of the incertitude of the activity a_A of the selector A in the solvent S. The unknown activity a_A can be substituted by the molarity M_A or, preferably, by the molality m_A concentration scale, only in very dilute solutions. The unit molality is independent of the temperature, and, for practical reasons, it is advantageous to add A to the weighed amount of S [21,28]. Throughout this chapter, the molality concentration m is applied. In gas chromatographic experiments, the activity (or the molality, respectively) of A in S may change with time either by activation (removal of competing ligands from A such as water by the carrier gas) or deactivation (thermal decomposition of A). Furthermore, if A is volatile it may be depleted from the stationary phase at elevated temperatures. Loss of A or diffusion toward the end of the column may lead to concentration gradients in the stationary phase. To avoid these effects, polymeric selectors in which the metal compound is linked (e.g., to a polysiloxane backbone) may be used. However, the molality m_A of such polymeric stationary phases is difficult to assess.

The term selectivity S' is generally associated with the direct comparison of two (or more) selectands B_i (X, Y, Z, . . .) competing for the same selector A. In many instances it suffices to consider the ratio of thermodynamic complexation constants K_X/K_Y. According to equation 7.19, the selectivity S' is directly related by the ratio of the retention-increments R' for two (or more) competing species, and it is thus independent of a_A, the activity of A in S.

$$S'_{X,Y} = \frac{K_X}{K_Y} = \frac{R'_X}{R'_Y} = \frac{r_X - r^\circ_X}{r_Y - r^\circ_Y} \cdot \frac{r^\circ_Y}{r^\circ_X} \qquad (7.19)$$

From equation 7.19 and the Gibbs–Helmholtz equation, 7.20, where R is the gas constant,

$$-\Delta_{X,Y}(\Delta G) = -\Delta_{X,Y}(\Delta H) + T\Delta_{X,Y}(\Delta S) = RT \ln \frac{K_X}{K_Y} = RT \ln \frac{R'_X}{R'_Y} \qquad (7.20)$$

the thermodynamic selectivity parameters $-\Delta_{X,Y}(\Delta G)$, $-\Delta_{X,Y}(\Delta H)$, as well as $\Delta_{X,Y}(\Delta S)$ can be obtained by van't Hoff plots when measurements are performed at different temperatures T according to

$$R \ln \frac{K_X}{K_Y} = R \ln \frac{R'_X}{R'_Y} = \frac{-\Delta_{X,Y}(\Delta G)}{T} = \frac{-\Delta_{X,Y}(\Delta H)}{T} + \Delta_{X,Y}(\Delta S) \qquad (7.21)$$

Thus, only relative retention data r°_i and r_i have to be measured for the determination of thermodynamic data of competing selectands B_i using a reference column containing the pure solvent S, devoid of the selector A, and a complexation column containing A in S, independently of all chromatographic parameters except the temperature.

An interesting situation arises for enantiomers (optical isomers). They may be separated when the selector A is chiral and nonracemic. The enantioselectivity is customarily defined by the free enthalpy (Gibbs energy) difference $-\Delta_{D,L}(\Delta G)$ (the subscripts D, *dextro*, right, and L, *levo*, left, denote enantiomers, irrespective of their absolute configuration. Arbitrarily, D is eluted after L from the column below $T_{isoenant}$; see Section 7.8.3). Significantly, inasmuch as enantiomers cannot be separated on the achiral reference column, r° is identical for D and L. Hence, equation 7.19 simplifies in the case of enantioselectivity to

$$S'_{D,L} = \frac{K_D}{K_L} = \frac{R'_D}{R'_L} = \frac{r_D - r^\circ}{r_L - r^\circ} \tag{7.22}$$

Since the enantiomers compete for the same selector A in S, the ratio K_D/K_L is directly related to the ratio of their retention-increments R'_D/R'_L, which is accessible from relative retention data r_D, r_L, and r° and, significantly, this ratio is independent of the activity of A in S, and thus from the concentration of the selector. Since r° is identical for the enantiomers D and L, in principle, r° need not be determined separately but can be extrapolated from two sets of data for the relative retention r of the enantiomers (D and L) at two arbitrary concentrations (1 and 2) of A in S as a consequence of the general equation 7.18.

$$r_0 = \frac{r_D^{(1)} r_L^{(2)} - r_L^{(1)} r_D^{(2)}}{(r_D^{(1)} + r_L^{(2)}) - (r_L^{(1)} + r_D^{(2)})} \tag{7.23}$$

Equation 7.23 can be used to assess nonenantioselective contributions to retention, when r° cannot easily be obtained—for example, with polysiloxane-anchored chiral selectors (see Section 7.11). The validity of equation 7.23 has been verified [39].

Thermodynamic data of enantioselectivity are derived according to equations 7.20 and 7.23

$$-\Delta_{D,L}(\Delta G) = -\Delta_{D,L}(\Delta H) + T \Delta_{D,L}(\Delta S)$$

$$= RT \ln \frac{K_D}{K_L} = RT \ln \frac{R'_D}{R'_L} = RT \ln \frac{r_D - r^\circ}{r_L - r^\circ} \tag{7.24}$$

where $-\Delta_{D,L}(\Delta G)$ is independent of the activity a_A of A in S, hence also from its concentration (molality). Isothermal measurements at different activities (molalities) should yield the same value for $-\Delta_{D,L}(\Delta G)$. This requirement has been verified in complexation gas chromatography [39,40]. Even concentration gradients in the dilute stationary phase of A in S would not affect $-\Delta_{D,L}(\Delta G)$.

The ratio R'_D/R'_L of a chiral selectand B depends on the enantiomeric purity of the selector A. The maximum $-\Delta_{D,L}(\Delta G)$ refers to enantiomer discrimination obtained by an enantiomerically pure (ee = 100%) chiral selector A. When A is not enantiomerically pure (i.e., ee < 100%) the relative retention r_{ee} (measured with A of given ee) can be extrapolated to r_{100} (corresponding to ee = 100%) via equations 7.25 and 7.26 [40,41]:

$$r_{D_{100}} = \frac{1}{2}(r_{D_{ee}} + r_{L_{ee}}) + \frac{50}{ee}(r_{D_{ee}} - r_{L_{ee}}) \tag{7.25}$$

and

$$r_{L_{100}} = \frac{1}{2}(r_{D_{ee}} + r_{L_{ee}}) - \frac{50}{ee}(r_{D_{ee}} - r_{L_{ee}}) \tag{7.26}$$

with

$$ee\% = \frac{D - L}{D + L} \times 100 \tag{7.27}$$

Note that the right-hand sides of equations 7.25 and 7.26 require for ee → 0 (racemic selector), $r_D - r_L \to 0$ (no enantiomer separation). This derivation is based on the reasonable assumption that the enantiomeric excess of A is linearly related to $-\Delta_{D,L} (\Delta G)$. Nonlinear relationships as observed in enantioselective catalysis [42] are in principle also possible in enantioselective chromatography but have not been described yet.

To define selectivity S' in gas chromatography, it is customary to define the separation factor α.

$$\alpha = \frac{t'_D}{t'_L} = \frac{r_D}{r_L} \tag{7.28}$$

In complexation chromatography and related techniques, the separation factor α_{dil} is concentration-dependent when A is *diluted* in S [43]. By substituting r_D and r_L in equation 7.28 by equation 7.18, with $r°$ being equal for enantiomers, it follows that [43]

$$\alpha_{dil} = \frac{K_D a_A + 1}{K_L a_A + 1} = \frac{R'_D + 1}{R'_L + 1} \tag{7.29}$$

According to equation 7.29, the curves for α versus a_A level off at high values of a_A (high concentrations of A in S). The optimum is reached already at low concentrations if chemical complexation is strong (i.e., large K or $R' >> 1$). Figure 7.2 shows simulated plots of α versus a_A for $K_D = 5.26$ and 210.5 according to equation 7.29.

Thus, while $-\Delta_{D,L} (\Delta G)$ represents a true concentration-independent measure of enantioselectivity, the separation factor α has only practical importance. Its numerical value underestimates the underlying chiral discrimination, since the adjusted

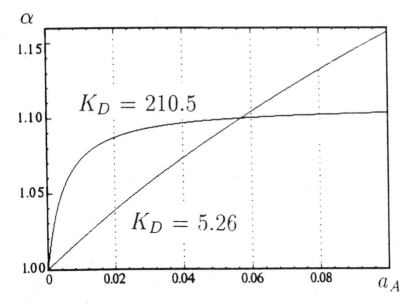

Figure 7.2. Simulated plots of α versus a_A for $K_D = 5.26$ and $K_D = 210.5$ (at $a_A = 0.057$, the same $\alpha = 1.10$ results when $R'_D = 0.3$ and $R'_D = 12$, respectively [43].

relative retention time from which it is calculated is the sum of the nonenantioselec-tive physical contribution to retention ($r°$) and the enantioselective chemical contri-bution to retention ($r - r°$). Only the ratio of the latter leads to enantiomer separa-tion according to equation 7.22: that is, $(r_D - r°)/(r_L - r°)$. Only in the rare case of $r°$ very low and complexation very strong (i.e., $r \gg r°$) may $-\Delta_{D,L} (\Delta G)$ be approximated from equation 7.30, previously employed for undiluted selectors A [41]

$$-\Delta_{D,L}(\Delta G) = RT \ln \alpha_{undil} \approx RT \ln \frac{t'_{R_D}}{t'_{R_L}} \approx RT \ln \frac{r_D}{r_L} \qquad \text{for} \quad r \gg r° \qquad (7.30)$$

A related argument for distinguishing nonselective from enantioselective interac-tions in enantioselective liquid chromatography on proteins, presented in 1993, arrives at the same expressions for the separation factor α [44].

7.4 Selective Rhodium Olefin Complexation

7.4.1 Rhodium(I) Olefin Complexation

Square planar d^8-rhodium(I) coordination compounds such as **5** are coordinatively and electronically unsaturated and are therefore prone to olefin exchange reactions.

As determined by IR spectroscopy, the relative thermodynamic complexation constants K_{rel} at 50°C decrease in the order ($K_{rel(ethene)}$ arbitrarily normalized to 1000) [45]:

> 1-butene (143) > *cis*-2-butene (30) > *trans*-2-butene (11) > isobutene (3)

The dicarbonyl-rhodium(I) coordination compounds **4** and **6–9** dissolved in squalane ($C_{30}H_{62}$) and coated on Chromosorb in packed columns [7,8] or coated on the inner surface of stainless steel open tubular columns undergo a fast and reversible complexation with olefins which can be observed by complexation gas chromatography. Figure 7.3 shows a representative complexation gas chromatogram featuring the high chemical selectivity observed.

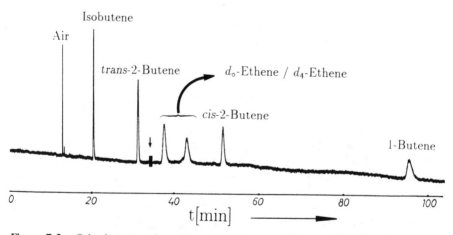

5

6

7 $R^1 = R^2 = CH_3$

8 $R^1 = CH_3 ; R^2 = CF_3$

9 $R^1 = R^2 = CF_3$

Figure 7.3. Selective separation of isomeric butenes and of d_0- and d_4-ethene isotopomers by complexation gas chromatography on 0.06 *m* (molality) **4** in squalane at 25°C: column, 100 m × 0.25 mm i.d. open tubular column made from nickel; carrier gas, dinitrogen; arrow, elution time of the reference compound *n*-pentane [9].

While 1-butene and isobutene cannot be resolved on pure squalane, a separation factor $\alpha = 11$ results for this isomeric pair when **4** is added to squalane (0.06 m, molality) [9]. This exceedingly high selectivity would allow a quantitative separation in a few seconds on a miniaturized open tubular column. The geometrical isomers cis- and trans-2-butene are separated with $\alpha \approx 2$. The highest retention-increment R' is observed for the parent olefin, ethene, which is eluted after n-pentane. The separation factor ethene/ethane is $\alpha \approx 50$. 1,3-Butadiene, which is eluted only after several hours, undergoes a very strong interaction with rhodium(I). As a result of an inverse secondary isotopic effect, d_4-ethene (tetradeuteroethene) is separated from d_0-ethene (ethene) by a separation factor $\alpha = 1.2$ (see Section 7.5).

Using equations 7.18 and 7.19, relative thermodynamic complexation constants K_{rel} have been determined which decrease in the order ($K_{rel(ethene)} = 1000$) at 25°C [46]:

1-butene (165) > cis-2-butene (43) > trans-2-butene (29) > isobutene (6)

Taking into account the different structures of **4** and **5** and the different methodologies in determining the thermodynamic complexation constants (IR vs. GC), the high degree of agreement between the data of rhodium(I) butene complexation is striking. Encouraged by this result, investigators determined the relative thermodynamic complexation constants K_{rel} of 40 acyclic and cyclic C-2–C-7 monoolefins with rhodium(I) (50°C) and rhodium(II) (100°C; see Section 7.4.3) by complexation gas chromatography based on equation 7.18 [11] and compared these with data already available for silver(I) [3] (see Table 7.1).

The advantages of the gas chromatographic method for the determination of relative thermodynamic complexation constants K_{rel} are [46] as follows.

- Complexation equilibria are measured in an apolar solvent and an inert gas (carrier gas helium).
- The method is reliable, simple, and reproducible and allows the simultaneous determination of different 1:1 equilibria in one run by the injection of mixtures of olefin vapors.
- The temperature is easily controllable over a wide range.
- The low concentration of B (selectand) precludes 2:1 molecular complexation with A in S.

The selectivity of complexation of olefins with rhodium(I) (see Table 7.1) is clearly mirrored by the rates of hydroformylation of the same olefins with cobalt(I) [47] and rhodium(I) catalysts [11], implying that olefin metal equilibration is the rate-determining step in the catalytic reaction. As is evident from Table 7.1, general trends in olefin rhodium(I) complexation can be recognized as follows [8,11] (see Sections 7.4.1.1–7.4.1.4).

Table 7.1   RELATIVE THERMODYNAMIC COMPLEXATION
CONSTANTS K_{rel} OF OLEFIN–METAL INTERACTION[a]

Olefin	Rh(I)[b]	Ag(I)[c]	Rh(II)[d]
Ethene	23.19	4.13	
Propene	3.40	1.68	
2-Methylpropene	0.15	0.72	0.94
1-Butene	3.62	1.43	0.26
trans-2-Butene	0.49	0.26	0.22
cis-2-Butene	**1.00**	**1.00**	**1.00**
1-Pentene	3.60	0.91	0.19
trans-2-Pentene	0.59	0.20	0.11
cis-2-Pentene	1.69	0.80	0.60
3-Methyl-1-butene	2.85	0.95	0.12
2-Methyl-1-butene	0.12	0.55	0.50
2-Methyl-2-butene	0.04	0.15	0.66
1-Hexene	3.85	0.80	0.19
trans-2-Hexene	0.60	0.15	0.13
cis-2-Hexene	1.38	0.57	0.53
trans-3-Hexene	0.73	0.18	0.08
cis-3-Hexene	2.60	0.72	0.42
4-Methyl-1-pentene	3.23	0.52	0.17
3-Methyl-1-pentene	2.07	0.63	0.07
2-Methyl-1-pentene	0.08	0.39	0.41
4-Methyl-2-pentene	0.06	0.11	0.38
3,3-Dimethyl-1-butene	0.13	0.67	0.03
2,3-Dimethyl-1-butene	0.05	0.44	0.13
2,3-Dimethyl-2-butene	0.00	0.02	0.57
2-Ethyl-1-butene	0.05	0.65	0.23
trans-4-Methyl-2-pentene	0.43	0.13	0.04
cis-4-Methyl-2-pentene	2.04	0.57	0.09
trans-3-Methyl-2-pentene	0.03	0.13	0.28
cis-3-Methyl-2-pentene	0.05	0.13	0.37
Cyclopentene	0.22	1.35	0.63
Cyclohexene	0.02	0.67	0.35
Cycloheptene	1.00	2.37	0.45
Methylenecyclobutane	22.2	1.07	0.88
Methylenecyclopentane	0.56	0.74	0.31
Methylenecyclohexane	0.79	1.11	0.40
1-Methylcyclopentene	0.01	0.35	0.24
3-Methylcyclopentene	0.22		0.37
1,2-Dimethylcyclopentene	0.00		0.10
1-Ethylcyclopentene	0.01	0.43	0.10
1-Methylcyclohexene	0.00	0.09	0.08
3-Methylcyclohexene	0.01	0.65	0.23
4-Methylcyclohexene	0.01	0.70	0.37

[a]   Determined by complexation gas chromatography according to equation 7.18 and arbitrarily normalized to cis-2-butene, $K_{rel} = 1.00$ [3,11].

[b]   Dicarbonyl-rhodium(I) 3-(trifluoroacetyl)-(1R)-camphorate **4** dissolved in squalane (50°C).

[c]   Silver nitrate dissolved in ethylene glycol (40°C).

[d]   Dirhodium(II)-tetrakis(heptafluorobutyrate) **10** suspended in chlorofluorohydrocarbon (Fluorolube) (100°C).

7.4.1.1 Alkyl Substitution at the Olefin Double Bond

Unsubstituted ethene undergoes the strongest interaction with rhodium(I), which is arbitrarily normalized to $K_{rel} = 1000$ for the sake of comparison (Figure 7.4). Monomethyl substitution of the olefin double bond reduces complexation by a factor of 7. Geminal dimethyl substitution reduces complexation considerably (cf. isobutene), while cis disubstitution is less severe than trans disubstitution. Trisubstitution reduces interaction by a factor of 700, whereas tetrasubstituted ethenes (cf. tetramethylethene) behave like saturated hydrocarbons and exhibit no measurable complexation.

A similar selectivity is observed for the rate of homogeneous hydrogenation of these olefins with $Rh(H)(CO)(PPh_3)_3$, which decreases in the order [48] 1-butene > cis-2-butene > trans-2-butene. Trimethylethene and tetramethylethene are not hydrogenated [48].

7.4.1.2 Influence of Alkyl Chain Length and Alkyl Branching

The chain length of the monoalkyl substituent at the olefin double bond has no marked influence on the complexation with rhodium(I) (Figure 7.5). Likewise the rate of catalytic hydrogenation of linear C-6–C-12 terminal olefins with $RhCl(PPh_3)_3$ is the same. Branching of the monoalkyl substituent at the olefin double bond greatly reduces complexation with rhodium(I) when a tert-butyl group is involved.

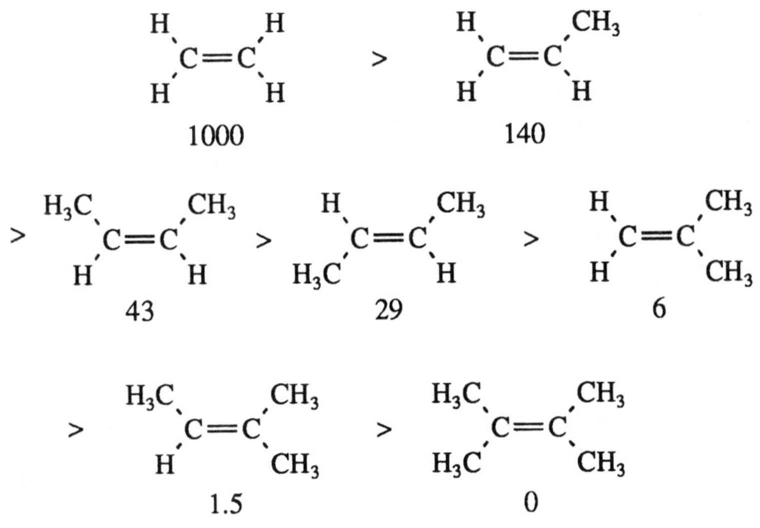

Figure 7.4. Steric effect of methyl substitution at the olefin double bond ($K_{rel(ethene)} = 1000$).

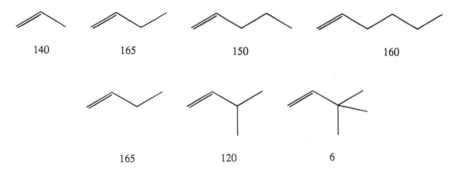

140   165   150   160

165   120   6

Figure 7.5. Influence of alkyl chain length and alkyl branching ($K_{rel(ethene)}$ = 1000).

7.4.1.3 Geometry at the Double Bond

Compared to the *trans*-olefins, the cis isomers always undergo a stronger complexation with rhodium. *cis*-Olefins are also hydrogenated at a faster rate than trans isomers with rhodium(I) catalysis. Thus, the rate of homogeneous hydrogenation of *cis*-4-methyl-2-pentene is greater by a factor of 5 than for *trans*-4-methyl-2-pentene with RhCl(PPh$_3$)$_3$ [49]. The stronger complexation tendency of *cis*-olefins is explained by relief of strain between cis substituents upon metal coordination. This effect is clearly corroborated by the observation that branching in the alkyl substituent increases the interaction with rhodium(I) (inverse steric effect) for the *cis*-, but not for the *trans*-olefins (Figure 7.6).

7.4.1.4 Influence of Ring Size

The complexation of *endo*-cycloolefins with rhodium markedly decreases in the order $C_7 > C_5 \gg C_6$ (Figure 7.7). This result is reminiscent of trends in preparative olefin rhodium(I) chemistry. In the synthesis of (cycloolefin)$_4$·Rh$_2$Cl$_2$, a stable

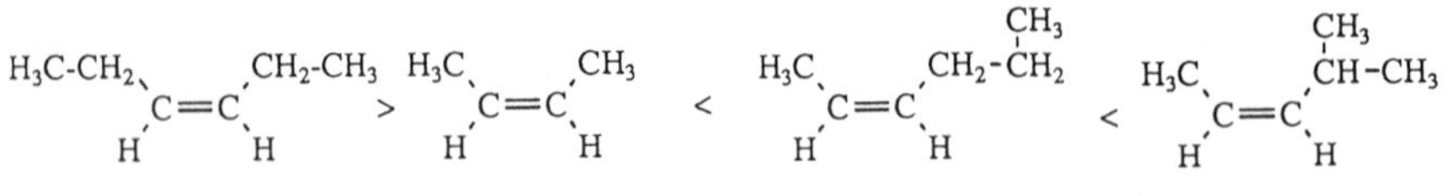

108   43   58   84

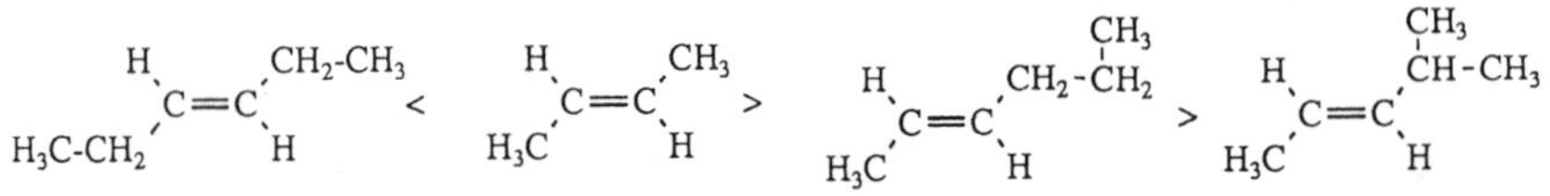

30   29   25   18

Figure 7.6. Inverse steric effect in *cis/trans*-olefins ($K_{rel(ethene)}$ = 1000).

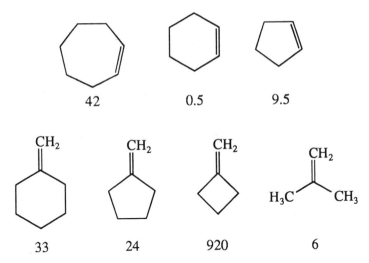

Figure 7.7. Relative stability of cycloolefins with rhodium(I) ($K_{rel(ethene)}$ = 1000).

complex has been isolated with cycloheptene, a reaction observed with cyclopentene; but no affinity of cyclohexene is noticed for rhodium(I) [50]. *exo*-Cycloolefins, when compared with the acyclic analogues, exhibit a very strong complexation with rhodium(I). This trend is especially pronounced for methylenecyclobutane, which complexes almost as strong as unsubstitued ethene ($K_{rel(ethene)}$ = 1000) with rhodium(I). The selectivity difference between cyclic methylenecyclobutene and acyclic isobutene is indeed remarkable.

7.4.2 Activation Phenomena in Complexation Gas Chromatography

Complexation of olefins with **4** and **6–9** involves an unusual activation phenomenon uncommon in gas chromatography [35,51]. The retention-increment R' steadily increases with time of operation until a stable threshold value is reached. This activation period is directly related to the speed of the carrier gas flow. Moreover, an activity gradient along the column during the activation period has been detected [35]. The column can be reversibly deactivated with carbon monoxide or dioxygen. These results are in agreement with loss of either carbon monoxide or dioxygen from an inactive species introduced into the column upon coating. This peculiar activation phenomenon is useful in its own right because it allows us to probe the validity of equation 7.18 with a single column. For 1-butene and isobutene, r_0, which is the same for both isomers, can accurately be determined via equation 7.23 without resorting to a reference column. As expected, the relative complexation constants K_{rel} of the isomeric butenes determined from the retention-increments R' according to equation 7.18 remain constant at a very high level of confidence over the whole period of activation.

7.4.3 Rhodium(II) Olefin Complexation

Diamagnetic dirhodium(II) tetracarboxylates **10** (e.g., rhodium benzoate) represent an interesting class of compounds in which the unusual electron configuration of rhodium(II), namely, d^7, is stabilized by metal–metal bonding in a dimer containing four carboxylate bridges. The coordinatively and electronically unsaturated clusters undergo a fast and reversible complexation with olefins which can be measured by complexation gas chromatography [11,12]. As shown in Figure 7.8, a striking difference in the selectivity for isomeric butenes is observed. On d^7-rhodium(II) the elution order is different from that of d^8-rhodium(I):

cis-2-butene > isobutene > $trans$-2-butene > 1-butene

A comprehensive set of data of relative olefin rhodium(II) complexation constants is contained in Table 7.1. Whereas with silver(I) and rhodium(I), alkyl substitution at the olefin double bond reduces complexation, an opposite trend is observed with rhodium(II), as shown in Figure 7.8. Also, for example, 2-methyl-2-butene and 2,3-dimethyl-2-butene interact quite strongly with rhodium(II) but not with silver(I) and rhodium(I). In some cases (e.g., for isomeric methylbutenes), interac-

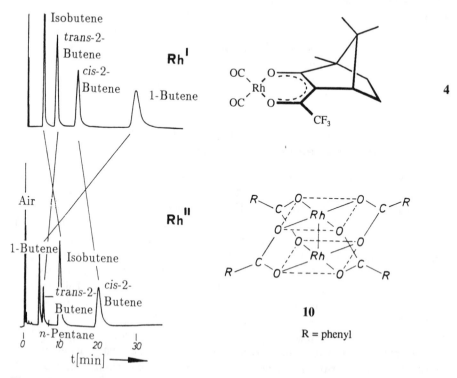

Figure 7.8. Different selectivity of rhodium(I) and rhodium(II) for isomeric butenes observed by complexation gas chromatography (for conditions, see Table 7.1) [46].

tion with rhodium(II) increases as alkyl substitution at the double bond is increased (see Tables 7.1 and 7.2).

Spectroscopic evidence also suggests that increasing alkyl substitution at the double bond favors complexation of olefins with rhodium(II) [52]. Obviously, the acceptor properties increase with increasing positive charge of the metal ion when going from d^8-rhodium(I) to d^7-rhodium(II). Thus, olefins act as strong σ-donor molecules, and the inductive effect of alkyl groups may overcome the negative steric effect toward complexation (inverse steric effect). The tendency of rhodium(II) carboxylates **10** to interact with olefins can also be demonstrated by NMR spectroscopy [11,53]. The use of **10** containing chiral carboxylate residues R* has been suggested for enantiomer discrimination of chiral olefins in complexation gas chromatography [11] and was recently demonstrated in NMR spectroscopy [53]. The D_4 symmetry of the chiral complexes should make the interpretation of enantioselectivities quite straightforward, since the chiral olefins will experience the same chiral environment when complexed to each of the four indistinguishable (homotopic) O—Rh—O-bonding axes in **10** [11].

7.5 Separation of Deuterated Ethenes: The Inverse Secondary Isotope Effect

The separation of ^{2}H-, or ^{13}C- and ^{3}H-doubly labeled olefins has been carried out by complexation gas chromatography on silver nitrate/ethylene glycol with a 75 m packed column [54]. The separation of isotopomers differing by only one dalton was only partially successful. The inverse (i.e., $K_D > K_H$) secondary isotopic effect in this system has been quantified and accounted for by the observation that steric and electronic parameters increase the donor strength of the double bond of deuterated ethenes [6,55].

As shown in Figure 7.9 all deuterated ethenes $C_2H_{4-n}D_n$ (n = number of

Table 7.2   COMPLEXATION SELECTIVITY OF METHYLBUTENES WITH RHODIUM(II), RHODIUM(I), AND SILVER(I)

Alkyl substitution at the double bond	Mono	Di	Tri
Rh(II)	0.12	0.50	0.66
Rh(I)	2.85	0.12	0.04
Ag(I)	0.95	0.55	0.15

Source: Data from Table 7.1, $K_{rel(cis\text{-}2\text{-}butene)}$ = 1.00 [46].

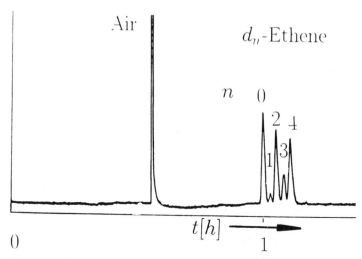

Figure 7.9. Separation of deuterated ethenes by complexation gas chromatography on 0.04 m (molality) **4** in squalane at 22°C: 200 m × 0.5 mm i.d., stainless steel column; carrier gas, dinitrogen [33].

deuterium atoms) are analytically separated on dicarbonyl-rhodium(I) 3-(trifluoro-acetyl)-(1R)-camphorate **4** in squalane in 75 minutes at 22°C.

The separation factors $\alpha = t'_D / t'_H$ are logarithmically additive; that is, $\alpha = (1.055)^n$ at 22°C, $\alpha = (1.050)^n$ at 42°C, $\alpha = (1.042)^n$ at 63°C, and they are the highest values measured thus far. The chemical isotopic effect $K_D / K_H > 1$ is much larger, and opposite in sign, compared to the physical isotope effect $r^\circ_D < r^\circ_H$. It is therefore called inverse, and the true chemical isotopic effect or chemical isotopic selectivity $S'_{D/H} = K_D / K_H$ is thus greater, as is evident from the separation factor α, which does not distinguish between the opposite physical and chemical contributions to isotopic discrimination. Equations 7.18 and 7.19 can be used in the measurement of S'. Because of the exceedingly high complexation tendency of ethene with **4** (cf. Table 7.1 and Figure 7.3), r is much greater than r°, and equation 7.19 can be simplified to

$$S'_{D/H} = \frac{K_D}{K_H} = \frac{R'_D}{R'_H} \approx \frac{r_D}{r_H}\frac{r^\circ_H}{r^\circ_D} \tag{7.31}$$

and

$$-\Delta_{D,H}(\Delta G) = -\Delta_{D,H}(\Delta H) + T\,\Delta_{D,H}(\Delta S)$$

$$= RT \ln \frac{K_D}{K_H} = RT \ln \frac{R'_D}{R'_H} \approx RT \ln \frac{r_D}{r_H}\frac{r^\circ_H}{r^\circ_D} \tag{7.32}$$

From the known vapor pressures p of d_4- and d_0-ethene [56] and the reasonable assumption that $\ln(p_D/p_H) \approx \ln(r_H^\circ/r_D^\circ)$, the thermodynamic parameters of the inverse secondary isotope effect for d_4- and d_0-ethene have been calculated from $K_D/K_H = 1.280$ (22°C), $K_D/K_H = 1.242$ (42°C), and $K_D/K_H = 1.215$ (63°C) [33,46]:

d_4- and d_0-ethene isotopomers:
$-\Delta_{D,H} (\Delta H) = 0.24 \pm 0.01$ kcal/mol
$\Delta_{D,H} (\Delta S) = -0.34 \pm 0.04$ cal/K·mol

More recently, the chemical secondary isotope effect of rhodium(I)/d_0-ethene and rhodium(I)/d_1-ethene measured by complexation gas chromatography with **4** in squalane on a 30 m packed column was compared with the statistical isotope effect derived from IR measurements [57]. It is noted that instead of r and r°, adjusted retention times t_R' AND THE GAS HOLDUP (DEAD) TIME t_M, respectively, were used in equation 7.19 and that the thermodynamic data derived vary from those cited earlier.

7.6 Enantiomer Separation of a Chiral Olefin on a Chiral Rhodium(I) Complex

When 3-methylcyclopentene is eluted from a open tubular column coated with squalane, its retention relative to the reference compound methylcyclohexane is $r^\circ = 0.26$ at 22°C. When the chiral olefin is eluted on a corresponding column coated with 0.04m (molality) of chiral dicarbonyl-rhodium(I) 3-(trifluoroacetyl)-(1R)-camphorate, (1R)-**4**, in squalane, the enantiomers elute after the reference compound methylcyclohexane with $r_L = 1.069$ and $r_D = 1.097$ at the same temperature (see Figure 7.10) [34].

According to equation 7.24, the difference of the free enthalpy (Gibbs energy) of the enantioselective complexation of 3-methylcyclopentene with (1R)-**4** is only $-\Delta_{D,L} (\Delta G) = 0.02$ kcal/mol (295 K) as calculated from $K_D/K_L = 1.035$. It is noted that this small free enthalpy difference leads to a separation factor $\alpha = 1.025$, which is sufficient for a quantitative enantiomer separation by high resolution open tubular column gas chromatography.

Figure 7.10 illustrates the first enantiomer separation achieved by means of complexation gas chromatography [34–36]. To prove this important result, two control experiments were performed. Racemic (1R,1S)-**4** was employed as a stationary phase on which only one peak was observed for the enantiomers (*peak coalescence of the first kind;* see Section 7.8.8) [35] and (1S)-**4** of opposite configuration was used as a stationary phase on which a reversal of the order of elution was observed for a sample of the olefin enriched in the (R)-enantiomer (*peak inversion of the first kind*) [36]. The occurrence of peak coalescence and peak inversion represents unequivocal evidence for the enantiomer separation of a racemic selectand on a chiral selector.

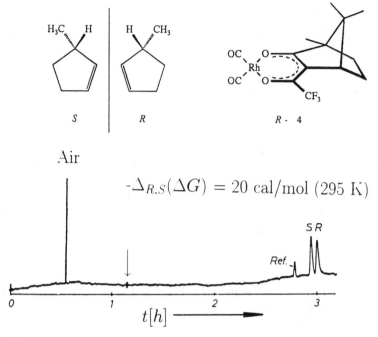

Figure 7.10. Enantiomer separation of 3-methylcyclopentene by complexation gas chromatography on 0.04 m (1R)-**4** in squalane at 22°C, $\alpha = 1.025$: 200 m × 0.5 mm i.d. stainless steel column; carrier gas, dinitrogen; reference compound, methylcyclohexane; arrow, elution position of the olefin on pure squalane [34].

Based on the gas chromatographic separation of enantiomers, the specific rotation of 3-methylcyclopentene has been determined and corrected to more than double that reported in literature [36]. Furthermore, a small kinetic resolution could be detected for 3-methylcyclopentene during the enantioselective rhodium(I)/phosphane-catalyzed hydrogenation (cf. Figure 7 in ref. 58). The monitoring of the enantiomeric ratio (er) or enantiomeric excess (ee), respectively, can easily be carried out from the reaction mixture by head space analysis [58].

7.7 Selective Transition Metal(II) Acceptor/σ-Donor Equilibria

The thermodynamic properties of the 1:1 complexation equilibria of group V and VI σ-donor selectands, B, with dimeric transition metal(II) bis[3-(trifluoroacetyl)-(1R)-camphorates] (metal = manganese, cobalt, nickel) **2a**, in apolar squalane have been determined by complexation gas chromatography using equations 7.18 and 7.21 [28]. A strictly linear (zero intercept) relationship between the retention-increment R' of the selectands B and the molal concentration m of the selectors A in squalane S was observed. Thus, the following conclusions were drawn [28].

- A fast and reversible 1:1 complexation between A and B takes place.
- The dimers of the metal coordination compounds A retain their structural integrity in dilute squalane solution ($\leq 0.05\ m$).
- The validity of equation 7.18 is confirmed, and the activity of A in S, a_A can be replaced by the molality m_A.

Relative thermodynamic complexation constants K_{rel} measured at 100°C decrease in the order alcohols > cyclic ethers > ketones > esters > aldehydes > aliphatic ethers. The overall complexation strength is drastically reduced in the order nickel(II) >> cobalt(II) > manganese(II) >> zinc. Table 7.3 gives some representative data, demonstrating the high selectivities observed.

Gibbs–Helmholtz parameters $-\Delta H$ and ΔS have also been measured for 30 selectands with dimeric nickel(II) bis[3-(trifluoroacetyl)-(1R)-camphorate] **2c** between 75 and 125°C according to equation 7.21 [28]. All complexation equilibria are exothermic and are accompanied by loss of entropy. Comparison of $-\Delta H$ values for oxygen and sulfur donor molecules revealed that the nickel(II) compound is a borderline acid with regard to the HSAB (hard–soft/acid–base) concept of Pearson [59]. While a linear relationship between R' and m_A was observed for the dimeric metal(II) bis[3-(trifluoroacetyl)-(1R)-camphorates] **2** (i.e., $R' = Ka_A$, cf. equation 7.18), only a linear relationship between R' and the square root of a_A, [i.e., $K(a_A)^{1/2}$] was noted with lanthanide(II) tris[3-(trifluoroacetyl)-(1R)-camphorates] **1** (21,22). This behavior has been attributed to a dimer–monomer dissociation equilibrium of the selector, with only the monomer being capable of complexing with the selectand [21]. This result should be kept in mind in view of the importance of, for example, praseodymium(III), europium(III), and ytterbium(III) tris [3-(trifluoroacetyl)-(1R)-camphorates] **1** as versatile paramagnetic chiral shift reagents in NMR spectroscopy [60,61], indicating that the complexation chemistry involved in solution is complex.

**Table 7.3   RELATIVE THERMODYNAMIC
COMPLEXATION CONSTANTS**[a]

Selectand	Mn(II)	Co(II)	Ni(II)
Tetrahydrofuran	4.6	18	80
1,4-Dioxane	2.6	12	106
Methyloxirane	**1.00**	2.2	7.2
Methylthiirane	0.13	0.90	6.9
Diethyl ether	0.39	0.86	2.8
Diethyl sulfide	0.20	2.3	17

[a] Arbitrarily normalized to $K_{rel} = 1.00$ for methyloxirane on Mn(II),

of ethers and thio ethers and dimeric **2** (0.05 m in squalane) at 22 °C.

Source: Ref. 28.

7.8 Enantiomer Separation on Metal(II) bis[α-(Perfluoroacyl)terpene Ketonates

7.8.1 Analytical Enantiomer Separation

It had long been recognized that the complexation equilibria discussed thus far have important analytical implications. The chemical selectivity in complexation gas chromatography is ideally suited to solve particularly difficult analytical separation problems. Since the selectors are chiral and nonracemic, they may be used for enantiomeric separation of chiral selectands, which was the original impetus of this work [7,9,21,28].

While the scope of enantiomer separation of olefins on dicarbonyl-rhodium(I) 3-(trifluoroacetyl)-(1R)-camphorate **4** was limited (see Section 7.6), the use of metal(II) bis[3-(trifluoroacetyl)-(1R)-camphorates] **2** and metal(II) bis[3-(heptafluorobutanoyl)-(1R)-camphorate] **3** (metal = Mn, Co, Ni) as chiral selectors for the enantiomer separation of chiral group V and VI σ-donor selectands with nitrogen-, oxygen-, and sulfur-containing functional groups soon emerged as a routine method for gas chromatographic enantiomer separations. Typical enantiomer separations on **3c** are shown in Figure 7.11 [62]. The occurrence of enantiomer separation has been proved by peak inversion of the enantiomers of (E)- and (Z)-2-ethyl-1,6-dioxaspiro[4.4]nonane (chalcogran; principal aggregation pheromone of *Pityogenes chalcographus* L.) when the chirality of the selector was inverted [(1R)-**3c** vs. (1S)-**3c**; cf. Figure 7.11, left].

Thus, metal(II) bis[3-(heptafluorobutanoyl)-(R)-camphorates] **3** proved to be versatile stationary phases for the separation of underivatized cyclic ethers, ketones, alcohols, and related classes of compounds [30,63,64]. In principle, any volatile chiral selectand containing a complexing function closely juxtaposed to the chiral center should be prone to enantiomer separation by complexation gas chromatography on a suitable chiral, nonracemic, metal-containing selector. The application of high resolution glass or fused-silica open tubular column technology in complexation gas chromatography [32,62,63] has aided many applications of enantiomer analysis in enzymatic reactions [65–68], in pheromone [69] and flavor chemistry [70].

Compounds **2** and **3** were originally used as solutions in squalane. Later on, polysiloxanes were employed as solvents. To improve the thermal stability of the selector, **2** and **3** were eventually bonded to polysiloxanes yielding chirasil–metal stationary phases (see Section 7.11).

7.8.2 Thermodynamic Data of Enantioselectivity

In enantiomer separation by complexation gas chromatography, useful information can be obtained from the peak parameters listed in Scheme 7.1.

It has been explained that thermodynamic data describing chemical selectivity and enantioselectivity (items 1 and 2 in Scheme 7.1) may readily be obtained from relative retention data r and $r°$ by complexation gas chromatography according to

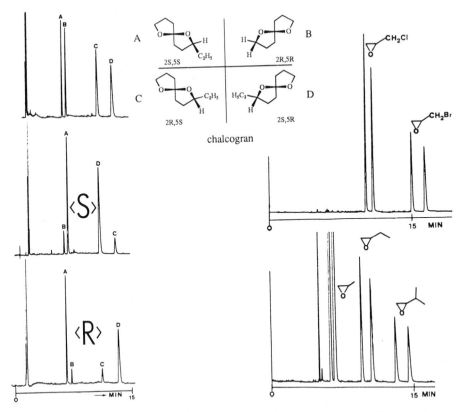

Figure 7.11. *Left:* Enantiomer separation and peak inversion of (Z)- and (E)-2-ethyl-1,6-dioxaspiro[4.4]nonane (chalcogran) at 93°C (25 m × 0.25 mm i.d. fused-silica column coated with 0.1 *m* nickel(II) bis[3-heptafluorobutanoyl)-(1R)-camphorate] (1R)-**3c** and nickel(II) bis[3-heptafluorobutanoyl)-(S)-camphorate] (1S)-**3c**, respectively, in SE-54, 0.2 μm, inlet pressure, 1 bar dinitrogen). *Top:* racemic chalcogran on (1R)-**3c**. *Center:* chalcogran enriched in the 2S-enantiomers on (1S)-**3c**. *Bottom:* chalcogran enriched in the 2S-enantiomers on 1R-**3c**. Peak assignment: A = 2S,5S; B = 2R,5R; C = 2R,5S; D = 2S,5R [62]. *Right:* Enantiomer separation of simple aliphatic oxiranes (25 m × 0.25 mm i.d. fused-silica column coated with 0.1 *m* cobalt(II) bis[3-(heptafluorobutanoyl)-(1R)-camphorate] **3b** in SE-54, 0.2 μm; oven temperature, 55°C; inlet pressure, 1 bar dinitrogen). *Top:* (chloromethyl)oxirane and (bromomethyl)oxirane; *bottom:* methyl-, ethyl-, and isopropyloxirane [62].

equations 7.18 and 7.22 [31,71]. The complexation equilibria between 15 chiral mono-, di-, and trialkyl-substituted cyclic ethers (mainly oxiranes) and 11 non-racemic nickel(II) bis-[α-(heptafluorobutanoyl)terpene ketonates] **3c** and **11–20** in squalane ($C_{30}H_{62}$) have been measured at 60°C, and interesting insights into trends of chiral recognition were obtained by considering the free enthalpy (Gibbs energy) differences, $-\Delta_{D,L}$ (ΔG), observed [40]. Surprisingly, selectors **3c** and **11** showed similar enantioselectivities, implying that structural rigidity is not a prerequisite for

1. *Peak retention*	chemoselectivity	$-\Delta G$
2. *Peak separation*	enantioselectivity	$-\Delta_{D,L}(\Delta G)$
3. *Peak coalescence* (fourth kind)	enantiomerization barrier	$\Delta G^{\ddagger}$
4. *Peak ratio*	enantiomeric ratio	er
5. *Peak assignment*	enantiomeric configuration	D, L

Scheme 7.1 Five peak parameters in chromatographic enantiomer separation [30,71].

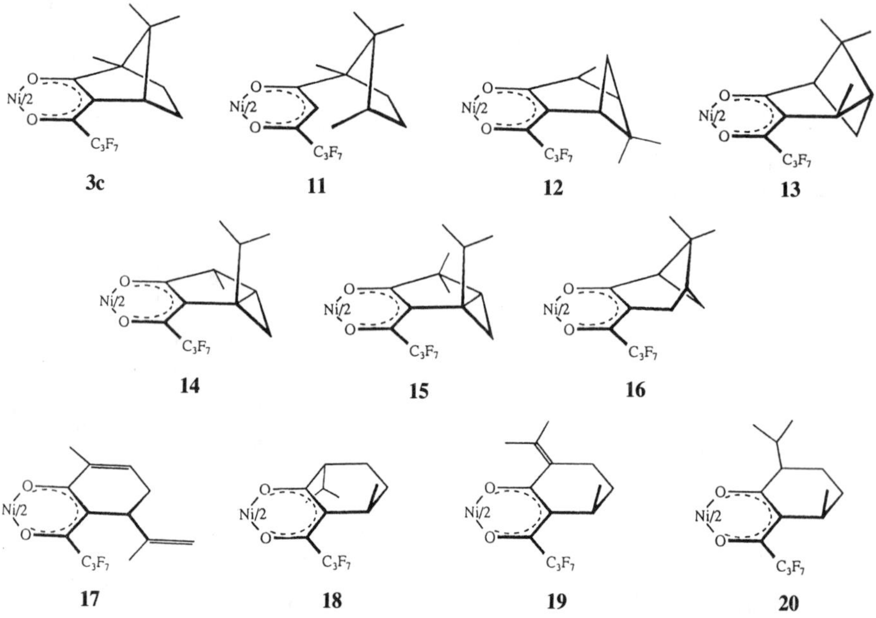

efficient chiral recognition. The highest enantiomeric bias was obtained with selector **13**. Not only selectors containing bicyclic molecular architectures but also those with monocyclic structures (e.g., differing by the geometric relation of the 1,4-dialkyl substituents **18–20**) are efficient for enantiomer separation [40]. Although **17** consists of a planar system except for the chiral center carrying the isopropenyl group, this selector is highly enantioselective and can even be used for the efficient semipreparative enantiomer separation of spiroketals such as 2-ethyl-1,6-dioxaspiro[4.4]nonane, chalcogran [72]. Some selectors were not readily available in an enantiomerically pure form. Therefore, the maximum $-\Delta_{D,L}(\Delta G)$ was extrapolated according to equations 7.25–7.27.

All complexation selectivities between alkyl-substituted oxiranes and selectors **3c** and **11–20** followed a common trend. It was interpreted in terms of opposing electronic and steric effects of the oxiranes [40]. No clear-cut relationship between the strength of the molecular complexation of the two enantiomers $-\Delta G_{mean} = 1/2 (\Delta G_L + \Delta G_D)$ and the magnitude of enantiomer discrimination, $\Delta_{D,L}(\Delta G)$ defined as the chiral recognition factor χ, was found [40]

$$\chi = \frac{-\Delta_{D,L}(\Delta G)}{-\Delta G_{\text{mean}}} \tag{7.33}$$

This trend is vividly demonstrated in Figure 7.12 for chiral 2-methyl-substituted cyclic ethers separated on manganese(II) bis[3-(heptafluorobutanoyl)-(1R)-camphorate] **3a** [32]. A very high complexation selectivity K_{rel}, differing by a factor of 65, is observed as the result of changing the ring size of the cyclic ethers (i.e., oxetane $\gg$ oxolane $\geq$ oxirane $>$ oxane). The exceedingly strong complexation of oxetanes and metal(II) coordination compounds **2** and **3**, the reason for which is not yet well understood, may give rise to extreme retention times, notably for the strong acceptor nickel(II) [40]. As outlined in Section 7.3, the separation factor α (see equation 7.28) does not represent a reliable criterion for enantioselectivity because it ignores the different volatilities (i.e., as evident from the different values of r°) for the cyclic ethers. Thus, it is not readily obvious from the chromatogram of Figure 7.12 that the true enantioselectivity, as correctly expressed by $-\Delta_{D,L}(\Delta G)$, decreases in the order oxirane $>$ oxetane $>$ oxolane $>$ oxane. Thus, neither a strong complexation (as for the oxetane) nor a very weak complexation (as for the oxane) is essential for an efficient enantioselectivity.

The highest degree of chiral recognition is observed when the complexation

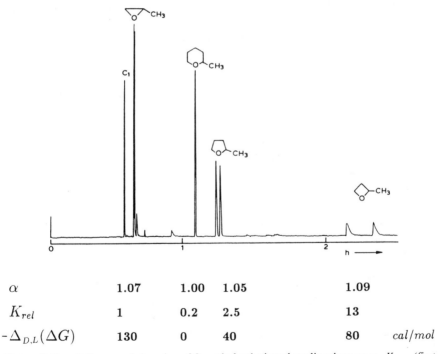

α	1.07	1.00	1.05	1.09	
K_{rel}	1	0.2	2.5	13	
$-\Delta_{D,L}(\Delta G)$	130	0	40	80	cal/mol

Figure 7.12. Influence of ring size of 2-methyl-substituted cyclic ethers on α, $K_{\text{rel,1}}$ (first eluted enantiomer) and $-\Delta_{D,L}$ (ΔG) at 60°C: 160 m × 0.4 mm i.d. stainless steel column coated with 0.05 m **3a** in squalane; carrier gas, dinitrogen [32].

strength $-\Delta G_{mean}$ is intermediate, (i.e., for oxiranes). Therefore, the intuitive assumption that strong molecular complexation is a prerequisite for a high enantioselectivity is not verified by experiment. This argument applies also when the different acceptor properties of the metal(II)-containing selector are considered. Although, according to Table 7.3 manganese(II) [and even more so zinc(II)] are very weak acceptors compared to nickel(II), high enantioselectivities are often observed with the former despite the low retention-increment R', or $-\Delta G_{mean}$, respectively [32]. Thus, the chiral recognition factor χ (see equation 7.33) decreases in the order manganese(II) > cobalt(II) > nickel(II).

For chiral alkyl-substituted oxiranes, a consistent relationship between the enantiomer configuration and the order of elution (cf. item 5, Scheme 7.1) was observed for all oxiranes on **3c** and **11–20** [32,40] in accordance with a *quadrant rule* proposed earlier [73]. However, the magnitude and the sign of enantioselectivity, $\pm\Delta_{D,L}\,(\Delta G)$, between alkyl-substituted oxiranes and **3c** and **11–20** could be neither predicted nor rationalized by simple molecular models [32,40]. This situation is further complicated by the temperature dependence of the enantioselective complexation equilibria. Thus, it had been noted [32] that the enantiomers of isopropyloxirane are not separated on nickel(II) but are well separated on manganese(II) and cobalt(II) (see Figure 7.11, right) at elevated temperatures. Indeed, a temperature-dependent reversal of the enantioselectivity [i.e., of the sign of $\Delta_{D,L}\,(\Delta G)$] for isopropyloxirane resolved on nickel(II) bis[3-(heptafluorobutanoyl)-8-(methylene)-(1R)-camphorate] **21a** was discovered in 1992 [71]. This phenomenon arises from the temperature term in the Gibbs–Helmholtz equation as the result of enthalpy/entropy compensation [39,74–76], as discussed next.

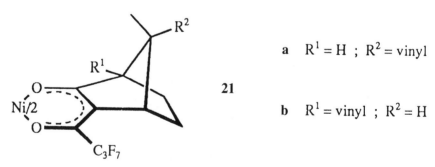

a $R^1 = H$; $R^2 = $ vinyl

21

b $R^1 = $ vinyl ; $R^2 = H$

7.8.3 The Temperature Dependence of Enantioselectivity: The Isoenantioselective Temperature

According to equation 7.24, enantiomer separation by complexation gas chromatography represents a thermodynamic chiral recognition process, that is,

$$RT \ln \frac{K_D}{K_L} = -\Delta_{D,L}\,(\Delta G) = -\Delta_{D,L}\,(\Delta H) + T\,\Delta_{D,L}\,(\Delta S) \qquad (7.24)$$

Enantioselectivity, as expressed by the free enthalpy (Gibbs energy) difference of molecular association $-\Delta_{D,L}$ (ΔG) is thus governed by an enthalpy term $-\Delta_{D,L}$ (ΔH) and an entropy term $\Delta_{D,L}$ (ΔS). For a 1:1 complexation process, these terms oppose each other in determining $-\Delta_{D,L}$ (ΔG). This enthalpy/entropy compensation arises because the more tightly bonded associate ($-\Delta H_D > -\Delta H_L$) is at the same time more ordered ($\Delta S_D < \Delta S_L$). Since the entropy term increases with temperature T according to equation 7.24, $-\Delta_{D,L}$ (ΔG) and thus enantioselectivity is temperature dependent. At a coalescence temperature, or *isoenantioselective temperature* T_{isoenant}, $\Delta_{D,L}$ (ΔG) becomes nil ($K_D = K_L$; no enantioselectivity), that is,

$$T_{\text{isoenant}} = \frac{\Delta_{D,L}(\Delta H)}{\Delta_{D,L}(\Delta S)} \tag{7.34}$$

At T_{isoenant}, *peak coalescence of the third kind* (see Section 7.8.8) is observed and the enantiomers cannot be separated. Below it, the sign of enantioselectivity $\Delta_{D,L}$ (ΔG) is governed by $-\Delta_{D,L}(\Delta H)$ and above it, by $\Delta_{D,L}(\Delta S)$. At T_{isoenant} (no enantiomer separation), the sign of enantioselectivity will change (*peak inversion of the second kind*).

A reversal of enantioselectivity in complexation gas chromatography above 200°C has been postulated for (Z)- and (E)-2-ethyl-1,6-dioxaspiro[4.4]nonane, chalcogran, on **3c** based on the following thermodynamic parameters determined below T_{isoenant} [39]:

Z-chalcogran enantiomers:
$-\Delta_{D,L}$ (ΔG) = 0.25 kcal/mol (353.16 K)
$-\Delta_{D,L}$ (ΔH) = 0.92 kcal/mol
$\Delta_{D,L}$ (ΔS) = −1.90 cal/K·mol
T_{isoenant} = 484 K ($\approx$ 210°C)

E-chalcogran enantiomers:
$-\Delta_{D,L}$ (ΔG) = 0.20 kcal/mol (353.16 K)
$-\Delta_{D,L}$ (ΔH) = 0.81 kcal/mol
$\Delta_{D,L}$ (ΔS) = −1.75 cal/K·mol
T_{isoenant} = 463 K ($\approx$ 190°C)

A lower value for T_{isoenant} for (E)-chalcogran is predicted from the thermodynamic data determined on selector **21b** obtained by substituting the 10-methyl group of camphor in **3** by a vinyl group.

E-chalcogran enantiomers:
$-\Delta_{D,L}$ (ΔH) = 0.62 kcal/mol
$\Delta_{D,L}$ (ΔS) = −1.75 cal/K·mol
T_{isoenant} = 354 K ($\approx$ 80°C)

Indeed, the low coalescence temperature T_{isoenant} = 80°C was confirmed, and the predicted inversion of the elution order above T_{isoenant} was later verified by the experiment [76].

This intriguing result was subsequently confirmed with a different selectand, that is, for isopropyloxirane on **21a** (see Figure 7.13) [71]. Thus, the coordination interaction between the R- and S-enantiomers of isopropyloxirane and nickel(II) bis [3-(heptafluorobutanoyl)-8-(methylene)-(1R)-camphorate] **21a** steadily decreases between 55 and 110°C, as seen from the drop of the retention time from 50 minutes to 2.5 minutes, respectively (Figure 7.13). While no separation of the enantiomers is observed between 70 and 90°C, the R-enantiomer is eluted after the S-enantiomer at 55°C. Above the coalescence region at T_{isoenant}, $+\Delta_{\text{D,L}}$ (ΔG) increases again and enantiomer separation resumes at 110°C, whereby the S-enantiomer is now eluted after the R-enantiomer (*peak inversion of the second kind*). As expected from the thermodynamic origin of the phenomenon, the van't Hoff plot of $R \ln(K_{\text{D}}/K_{\text{L}})$ versus $1/T$ in 5 C° intervals between 55 and 110°C is strictly linear, as is evident from Figure 7.14, crossing the zero line at the coalescence temperature observed experimentally, in agreement with the following data (the subscripts D and L merely denote enantiomers, irrespective of the absolute configuration of the oxirane):

Isopropyloxirane enantiomers:
$$-\Delta_{\text{D,L}}\ (\Delta H) = 0.297\ \text{kcal/mol}$$
$$\Delta_{\text{D,L}}\ (\Delta S) = -0.82\ \text{cal/K·mol}$$
$$T_{\text{isoenant}} = 362\ \text{K}\ (\approx 89°\text{C})$$

Quite unexpectedly, minor structural modifications of the metal compound in a position opposite to the coordination site (at C-8 of camphor) remarkably influence T_{isoenant} for isopropyloxirane (Table 7.4, left). Minor changes of the oxirane structure drastically influence T_{isoenant} (Table 7.4, right). Thus, for the diastereomeric

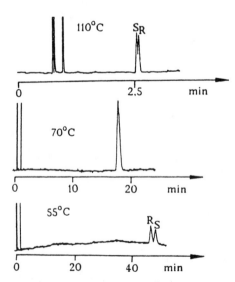

Figure 7.13. Temperature-dependent reversal of enantioselectivity for the enantiomers of isopropyloxirane by complexation gas chromatography on nickel(II) bis-[3-(heptafluorobutanoyl)-(1R)-8-(methylene)-camphorate] **21a** [71].

$-\Delta_{D,L}(\Delta G)/T \ [10^{-4} kJ \ mol^{-1} K^{-1}]$

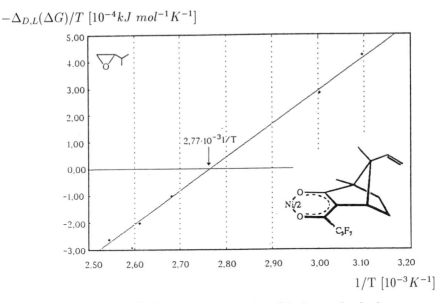

$1/T \ [10^{-3} K^{-1}]$

Figure 7.14. Linear van't Hoff plot and determination of the isoenantioselective temperature $T_{isoenant}$ [71].

sec-butyloxiranes only one of the two enantiomeric pairs (i.e., *R,S / S,R—unlike* CIP descriptors) shows a low $T_{isoenant}$ similar to that observed for isopropyloxirane, while for the other pair (i.e., *R,R / S,S—like* CIP-descriptors), $T_{isoenant}$ is more than 500 K higher (Table 7.4, right) [71]. It should be noted that a surprisingly low $T_{isoenant}$ has been observed only for isopropyloxirane and for one of the diastereomers of *sec*-butyloxirane, but not for other alkyl-substituted oxiranes. The reason for this observation is unknown at present.

Enthalpy/entropy compensation is of importance for the understanding of mechanistic aspects of enantiomer recognition in chiral environments. Above $T_{isoenant}$, preferential recognition of one enantiomer is caused by an entropy effect due to a higher degree of disorder in the resulting chemical interaction, while enantiomer recognition below $T_{isoenant}$ is caused by the enthalpy effect of chemical bonding [76]. Thus, entropic contributions to chiral recognition represent an important parameter not to be neglected in mechanistic considerations (e.g., by molecular modeling) of enantioselectivity. Temperatures below $T_{isoenant}$ will always be favorable for chromatographic enantiomer separation. Enantiomer separation should become independent of the separation temperature when $-\Delta_{D,L}(\Delta H)$ is zero. Such a situation may apply to chiral recognition due to entropy changes $-\Delta_{D,L}(\Delta S)$ only.

7.8.4 Enantiomerization

Complexation gas chromatography not only represents a useful quantitative method to measure *thermodynamic* data of enantioselectivity, it can also be used to obtain

Table 7.4 THERMODYNAMIC ENANTIOSELECTIVITY DATA

	For Isopropyloxirane and Structurally Modified Nickel(II) Selectors 21				For Isopropyloxirane and sec-Butyloxirane and 3c		
	$-\Delta_{D,1}(\Delta H)$ (cal/mol)	$-\Delta_{D,1}(\Delta S)$ (cal/K·mol)	T_{isoenant} (°C)		$-\Delta_{D,1}(\Delta H)$ (cal/mol)	$-\Delta_{D,1}(\Delta S)$ (cal/K·mol)	T_{isoenant} (°C)
	300	−0.8	90		270	−0.8	64
	270	−0.8	60		280 / 180	−0.8 / −0.2	94 / 622
	150	−0.5	20				

Source: Ref. 71.

kinetic data of enantiomerization (dynamic gas chromatography: see item 3 in Scheme 7.1) [31,77–79]. Heretofore, only enantiomers retaining their stereochemical integrity during separation, have been considered. Elution patterns that arise when configurationally or conformationally labile enantiomers are chromatographed on a chiral selector have been predicted [80] and afterward described both experimentally [30] and theoretically [31]. The term "enantiomerization" implies a process in which the individual stereoisomers of a racemic or enriched mixture of enantiomers undergo inversion of chirality during their chromatographic separation. The dynamic behavior of interconverting enantiomers results in intrinsic distortions of the elution curves, and these anomalies are amenable to the acquisition of kinetic data by peak form analysis of interconversion profiles [30]. The equilibrium of enantiomerization (i.e., D $\rightleftharpoons$ L), which occurs independently in the mobile and stationary phases, is appealing because it is governed by entropic changes only and represents a simple case of a reversible unimolecular reaction. A typical interconversion profile is characterized by the appearance of an overlapping zone (plateau), caused by inverted molecules, between the terminal peaks of the noninverted molecules (Figure 7.15). If enantiomerization is rapid, only one peak will be observed (*peak coalescence of the fourth kind:* see Section 7.8.8).

The observed interconversion profiles represent an important diagnostic tool for the recognition of inversion of chirality. Moreover, interconversion profiles can be calculated based on a cyclic process, represented in Scheme 7.2.

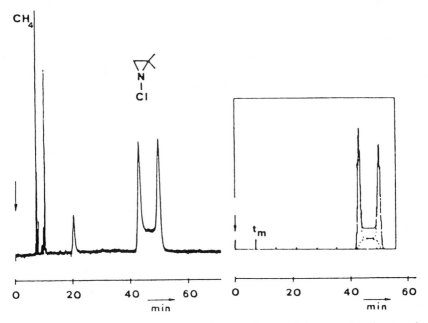

Figure 7.15. Peak profiles of interconverting enantiomers during complexation gas chromatography of 1-chloro-2,2-dimethylaziridine on **3c:** experimental (left) and simulated (right) chromatograms [77].

The interconverting enantiomers have a different energy in the stationary phase because there is a different apparent distribution constant $K_{L(D)} > K_{L(L)}$ (cf. equation 7.9 and Scheme 7.2) leading to enantiomer separation. Hence, the interconversion rates are rendered enantioselective in the stationary phase because $k^s_{L \to D}$ and $k^s_{D \to L}$ become *different* (Scheme 7.2) [31].

$$\frac{K_{L(D)}}{K_{L(L)}} = \frac{k^s_{L \to D}}{k^s_{D \to L}} \tag{7.35}$$

Although the first-eluted enantiomer L is inverted at a higher rate into the second eluted enantiomer D, the former is not depleted but is re-formed because the second eluted enantiomer D has a longer residence time in the column ($K_{L(D)} > K_{L(L)}$). As a consequence, depletion and enrichment of L (and, *mutatis mutandis*, enrichment and depletion of D) cancel each other and no overall deracemization is observed.

A more complicated situation arises when the chiral selector is present in a dissolved state (i.e., A in S). In this case three independent equilibria may be distinguished (Scheme 7.3), which are interrelated by the apparent distribution constant K_L according to equation 7.9 and the retention-increment R' according to equation 7.18 [78].

Examples featuring enantiomerization have been described in complexation gas chromatography for 1-chloro-2,2-dimethylaziridine [31,77], 1,6-dioxa-spiro[4.4] nonane [31], phenyloxirane [71], isopropenyloxirane [71], and homofuran [79 (see Figure 7.16).

For homofuran a perfect agreement between experimental and calculated chromatograms was obtained. Moreover, the enantiomerization barrier calculated by complexation gas chromatography corresponded to that measured by polarimetry for an optically enriched sample. It should be noted that only minute amounts of the *racemic* selectand suffice for the determination of kinetic data of inversion by complexation gas chromatography.

Unlike homofuran, examples have been found in which the presence of the metal-containing selector A considerably changes the barrier of enantiomerization in the stationary phase. Thus, a determination was made for 1-chloro-2,2-dimethyl-

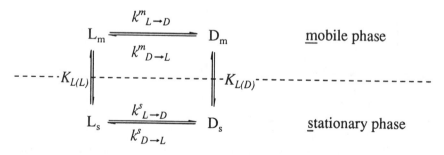

Scheme 7.2 Cyclic process involving enantiomerization in mobile and stationary phases [77,78].

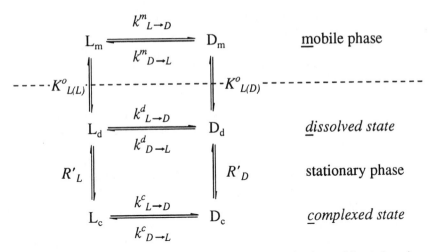

Scheme 7.3 Cyclic process involving enantiomerization in the mobile and stationary phases containing A in S [78].

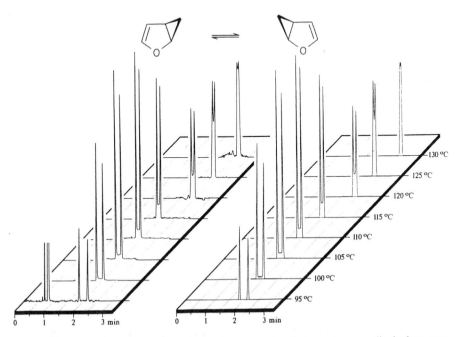

Figure 7.16. Comparison of experimental and calculated chromatograms displaying enantiomerization of homofuran by dynamic complexation gas chromatography on chirasil–nickel **22** [79].

aziridine, $\Delta G\ddagger \approx 25$ kcal/mol at 60°C by dynamic complexation gas chromatography [77]. Since the chiral aziridine could also be separated by complexation gas chromatography in a semipreparative fashion, it was possible to study the racemization kinetics of the isolated enantiomers in the gas phase at 65°C by analytical complexation gas chromatography. Under the second set of conditions, a much higher value ($\Delta G\ddagger \approx 28$ kcal/mol) was determined [81]. While the barrier of the enantiomerization of *trans*-2,3-dimethyloxirane via conrotatory thermal ring opening is $\Delta G\ddagger > 50$ kcal/mol, phenyloxirane and isopropenyloxirane undergo inversion of configuration during enantiomer separation on **3c** already at elevated temperatures [71]. Since saturated alkyl-substituted oxiranes do not show this phenomenon, it is probable that the metal-mediated enantiomerization is assisted by the presence of unsaturated substituents at the oxirane ring.

A very complex interconversion profile has been observed for the epimerization of 2-ethyl-1,6-dioxaspiro[4.4]nonane, chalcogran (structure see Figure 7.11) (A $\rightleftharpoons$ D and B $\rightleftharpoons$ C) on **3c.** The selector is apparently present in a catalytically active state after prolonged time of use, (cf. comparison of Figures 7.11, left, and 7.17).

7.8.5 Kinetic Resolution During Enantiomer Separation

Quantitative gas chromatographic enantiomer separation is prerequisite for the precise determination of enantiomeric ratios er (see item 4 in Scheme 7.1) [63]. A

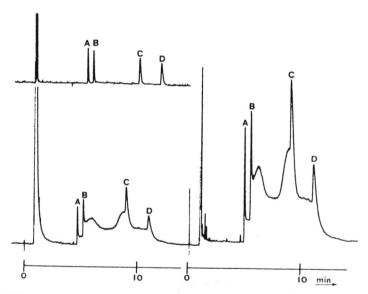

Figure 7.17. Interconversion profile with enhanced peak distortion due to epimerization of (Z)- and (E)-2-ethyl-1,6-dioxaspiro[4.4]nonane, chalcogran, at 93°C: 25 m × 0.25 mm i.d. fused-silica column coated with 0.1 *m* nickel(II) bis[3-(heptafluorobutanoyl)-(1R)-camphorate] **3c** in OV-101, 0.2 μm; inlet pressure, 0.4 bar dinitrogen [62].

serious error in the determination of enantiomeric ratios may be caused by preferential loss of one enantiomer during the gas chromatographic separation process. Nonenantioselective decomposition of a chiral selectand leads to a preferential depletion of the second eluted enantiomer, which spends a longer time in the column. Enantioselective decomposition mediated by the chiral metal coordination compound may lead to depletion of the first or the second eluted enantiomer. Both phenomena have been observed with strained oxiranes [62,71]. A deviation from the expected 1:1 ratio for racemic *n*-decyloxirane separated at 150°C on chirasil–nickel **22** (see Section 7.11) is shown in Figure 7.18 [71].

Apart from the preferential depletion of the first eluted enantiomer, a new broad elution zone, likely created by an unidentified reaction product, appears. Reversing the chirality of the selector does not alter the observed peak profiles because two enantioselective processes, chromatographic resolution and kinetic resolution, are involved.

7.8.6 Unusual Peak-Broadening Phenomena

Still another metal-mediated enantioselective phenomenon involving racemic *trans*-2,3-diethyloxirane has been discovered in chromatographic enantiomer separation [71]. It is thought to be related to different rates of coordination. As shown in Figure 7.19, the first eluted enantiomer shows a considerably broader peak width, also recognized by the reduced peak height compared to the second eluted

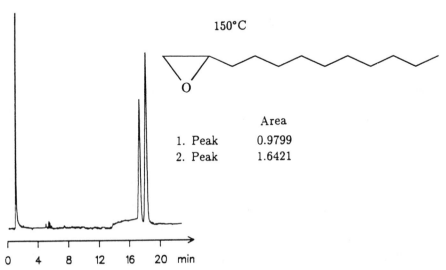

Figure 7.18. Deviation from the expected 1:1 ratio (*first* peak diminished) upon enantiomer separation of racemic *n*-decycloxirane on chirasil–nickel **22** by complexation gas chromatography at 150°C [71].

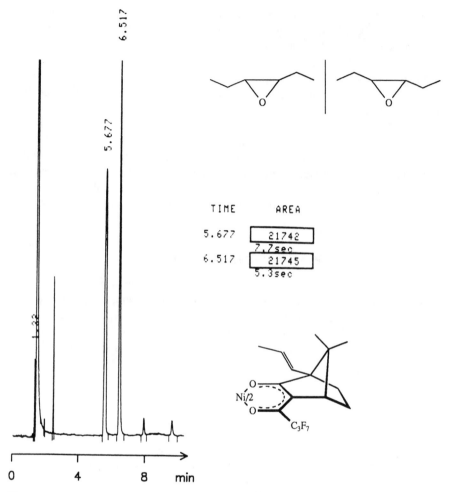

Figure 7.19. Deviation from the expected peak width for the first eluted enantiomer of racemic *trans*-2,3-diethyloxirane separated on nickel(II) bis[3-(heptafluorobutanoyl)-10-(ethylidene)-(1S)-camphorate] **21a′** [71].

enantiomer, although the peak areas are strictly identical. Since in chromatography peak broadening is related to the rate of the complexation process, the two enantiomers obviously differ in their kinetics of coordination. This result may imply that two different coordination sites of the (planar) metal compound are involved in the complexation of each individual enantiomer.

Another peak-broadening phenomenon, not yet understood, was observed when the enantiomers of (*E*)-2-ethyl-1,6-dioxaspiro[4.4]nonane, chalcogran, separated on **3c,** coalesced at T_{isoenant} (the reader is referred to note 6 in ref. 76).

7.8.7 Distinguishing Four Enantioselective Processes in Complexation Gas Chromatography

The complexation of aliphatic chiral oxiranes with nickel(II) bis[3-(heptafluoro-butanoyl)-(1R)-camphorate] (**3c**) or derivatives thereof (**21**) may give rise to four different enantioselective processes:

- *Enantioselective coordination* (via distinct thermodynamics): peak ratio 1:1 (chromatographic resolution mediated by the chiral metal compound; see Figure 7.11, right).
- *Enantioselective enantiomerization:* terminal peak ratio 1:1 and internal plateau formation (different rate of enantiomerization mediated by the chiral metal compound; see Figure 7.15).
- *Enantioselective reaction:* peak ratio $1 \neq 1$ (kinetic resolution mediated by the chiral metal compound; see Figure 7.18).
- *Enantioselective coordination* (via distinct kinetics): peak ratio 1:1, unusual peak width (different complexation sites at the metal compound; see Figure 7.19).

These enantioselective processes are easily detected in the chromatographic complexation column. The observations will guide further investigations in static reaction devices (e.g., kinetic resolution).

7.8.8 Distinguishing Six Coalescence Phenomena in Enantioselective Chromatography

It is useful to distinguish different coalescence phenomena in enantioselective complexation gas chromatography and related methods. We mention six kinds.

- *Peak coalescence of the first kind:* only one observable peak due to negligible enantioselectivity of the selector ($-\Delta_{D,L} (\Delta G) < 10$ cal/mol).
- *Peak coalescence of the second kind:* only one peak due to a racemic selector.
- *Peak coalescence of the third kind:* only one peak at the isoenantioselective temperature $T_{isoenant}$ (see Section 7.8.3).
- *Peak coalescence of the fourth kind:* only one peak due to fast enantiomerization of the selectand at the chromatographic time scale (see Section 7.8.4).
- *Peak coalescence of the fifth kind:* only one peak due to compensation of enantioselectivity of competing selectors present in the stationary phase, in the mobile phase, or in both [82].
- *Peak coalescence of the sixth kind:* only one peak due to slow instrumental response factors in very fast enantiomer separations.

7.9 Isotopic Separation of α-Deuterated Tetrahydrofurans

Another instance of the quantitative discrimination of isotopomers with a mass difference of one dalton (Section 7.5) is provided by the separation of α-deuterated tetrahydrofurans $C_4H_{4-n}^{\alpha}D_n^{\alpha}H_4^{\beta}O$, on cobalt(II) bis[3-(heptafluorobutanoyl)-(1R)-camphorate] **3b** in less than 30 minutes at 70°C [29]; see Figure 7.20.

The separation factors $\alpha = t_D'/t_H'$ are nearly logarithmically additive; that is, $\alpha = (1.03)^n$ at 70°C (n = number of deuterium atoms). The separation is due to a marked inverse secondary isotope effect $K_D/K_H > 1$. The thermodynamic data for α-d_4- and d_0-tetrahydrofuran between 50 and 80°C are with $\ln(p_D/p_H) \approx \ln(r_H^{\circ}/r_D^{\circ}) = 0.00015$:

α-d_4- and d_0-tetrahydrofuran isotopomers:
$-\Delta_{D,H}(\Delta H) = 0.2$ kcal/mol
$\Delta_{D,H}(\Delta S) = -0.4$ cal/K·mol

As expected, the inverse isotopic effect rapidly decreases with distance from the complexing oxygen atom. Thus, d_8-tetrahydrofuran and α-d_4-tetrahydrofuran could not be separated from each other. The choice of cobalt(II) in **3** proved to be optimal because the thermodynamic complexation constant K is too strong with nickel(II) and too weak with manganese(II) (cf. Table 7.3). Attempts to resolve [^{18}O]tetrahydrofuran from [^{16}O]tetrahydrofuran by complexation gas chromatography have failed thus far.

7.10 Attempted Separation of Protodeutero-Substituted Enantiomers

One of the challenges in separation science is to resolve stereoisomers owing their chirality to isotopically substituted compounds. Thus, $\alpha\alpha'$-d_2-tetrahydrofuran con-

Figure 7.20. Isotopic separation of α-deuterated tetrahydrofurans $C_4H_{4-n}^{\alpha}D_n^{\alpha}H_4^{\beta}O$ by complexation gas chromatography: 60 m × 0.25 mm i.d. glass open tubular column coated with 0.1 m cobalt(II) bis[3-(heptafluorobutanoyl)-(1R)-camphorate] **3b** in dimethylpolysiloxane (OV101, 0.2 μm); oven temperature, 70°C; inlet pressure, 1.9 bar dinitrogen; order of elution d_0-THF, α-d_1-THF, $\alpha\alpha'$-d_2-THF (two diastereomers), $\alpha\alpha\alpha'$-d_3-THF, $\alpha\alpha\alpha'\alpha'$-$d_4$-THF; peak assignment by GC-MS [29].

tains two stereogenic centers —OC*(H,D)CH$_2$— giving rise to the occurrence of three stereoisomers. As seen in Figure 7.20, separation of the meso/rac diastereomeric pair and resolution of the rac form on the chiral selector (1R)-**3b** is not observed. Chiral α-d_1-tetrahydrofuran and ααα'-d_3-tetrahydrofuran are also not resolved.

Further attempts to resolve by complexation gas chromatography chiral isotopically labeled compounds such as d_1-2,2-dimethyl-1-propanol, cis-(d_3-methyl)methyloxirane, and (d_9-tert-butyl)-tert-butylmethylcarbinol, and their derivatives have failed thus far. Interestingly, the enantiomers of d_1-2,2-dimethyl-1-propanol [83] and of (d_9-tert-butyl)-tert-butylmethylcarbinol [84] can be distinguished by ^{1}H or NMR spectroscopy in the presence of chiral lanthanide shift reagents such as europium(III) tris[3-(heptafluorobutanoyl)-(1R)-camphorate] **1**. This result provides evidence that isotopic enantiomer discrimination by NMR in chiral solvating agents is due to a chemical shift anisochrony of diastereotopic nuclei rather than to a difference of diastereomeric complexation constants K_D/K_L [84].

7.11 Chirasil–Metal as Polymeric Selector

The development of selectors as stable surface-bonded stationary phases by cross-linking or chemical binding ("immobilization") has important merits in complexation gas chromatography notably for analytical purposes [85].

Chiral polysiloxanes linked to *metal* compounds, that is, chirasil–metal, can be obtained by total synthesis [39] or, preferentially, by polymer-analogous reaction [85]. The synthesis of chirasil–nickel **22** is depicted in Figure 7.21. The fast

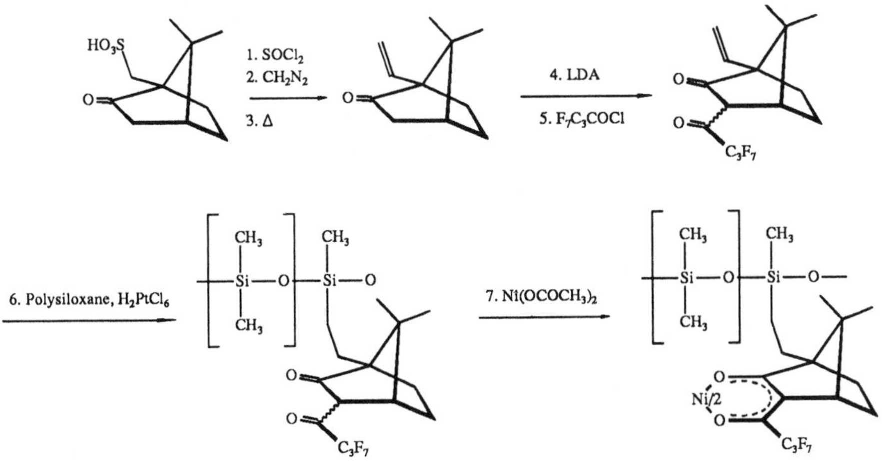

Figure 7.21. Synthesis of the chemically modified chiral polysiloxane chirasil–nickel **22** (LDA = lithium diisopropylamide). The random ratio of methylhydro-/dimethylsiloxane monomers in the polymer is about 1:9. The nickel content is about 10 mg/g polymer (≈ 0.2 m) [85,86].

separation of 2-methyl-substituted cyclic ethers on thermally immobilized chirasil–nickel **22** is shown in Figure 7.22.

Immobilized chirasil–nickel **22** is compatible with highly solvating mobile phases such as supercritical carbon dioxide. Hence, open tubular columns coated with chirasil–nickel **22** can be utilized in supercritical fluid chromatography (SFC) [84–86]. Quite unexpectedly, the dicarbonyl compound, carbon dioxide, used at high pressure, does not interfere with the complexation equilibria of nickel(II) and σ-donor selectands such as ethers, ketones, and alcohols [85,86]. Peak-broadening phenomena as discussed in Section 7.8.6 are very pronounced with chirasil–nickel **22** employed in the SFC mode, as demonstrated in Figure 7.23 [87].

While thermodynamic data have not yet been determined on chirasil–metal selectors, it is anticipated that the same results will be obtained as in systems containing physical mixtures of the selector A in the polysiloxane solvent S. Fortunately, according to equation 7.24, enantioselectivities are independent of the activity concentration of the selector in the solvent S, which is difficult to assess for polymeric selectors. Moreover, no reference column is necessary when equation 7.23 is used.

Europium(III) tris[3-(trifluoroacetyl)-(1R)-camphorate] **1** has also been linked to a polysiloxane. The resulting chiral polymer, chirasil–europium, can be used for enantiomer separation in complexation gas chromatography, as a recoverable polymeric chiral NMR shift reagent, and as a recyclable (after precipitation) Lewis acid catalyst in enantioselective synthesis [88].

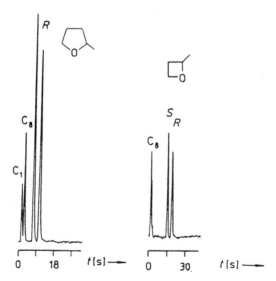

Figure 7.22. Fast enantiomer separation by complexation gas chromatography on chirasil–nickel **22** (miniaturized 1.5 m × 0.05 mm i.d. fused-silica column coated with chirasil–nickel). *Left:* 2-Methyltetrahydrofuran, 115°C, 2 bar dinitrogen. *Right:* 2-Methyloxetane, 140°C, 2 bar dinitrogen [85].

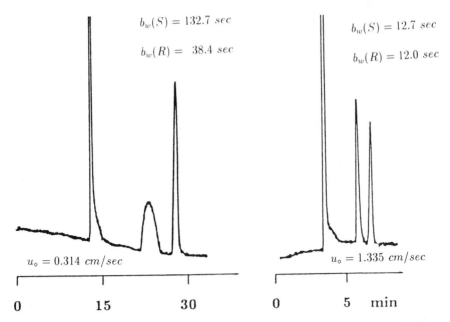

$b_w(S) = 132.7 \; sec$

$b_w(R) = \;\; 38.4 \; sec$

$b_w(S) = 12.7 \; sec$

$b_w(R) = 12.0 \; sec$

$u_o = 0.314 \; cm/sec$

$u_o = 1.335 \; cm/sec$

0   15   30   0   5   min

Figure 7.23.   Enantiomer separation of racemic camphor at different flow rates u_0 of super-critical carbon dioxide (15 MPa, 60°C) by supercritical fluid chromatography: 2 m × 0.05 mm i.d. fused-silica column coated with immobilized chirasil–nickel **22**, 0.25 μm; b_w is the peak width at half-height [87].

7.12 Applications in Inclusion Gas Chromatography

As outlined earlier, the concept of the retention-increment R', expressed by relative retention in a reference column $r°$ and a complexation column r permits the investigation of thermodynamic parameters of molecular complexation in diluted systems by complexation gas chromatography. The method may also be applied to other types of molecular association: hydrogen bonding using amino acid derivatives [89], charge-transfer interaction using fullerenes [90], or inclusion using cyclodextrins [43] and calixarenes [91] as selectors. Whereas it can safely be anticipated that there is virtually no interaction of saturated hydrocarbons, used as reference compound B*, with metal compounds A in S, employed as selectors in complexation gas chromatography, the method may not readily be applicable to inclusion-type selectors. Thus, with cyclodextrins, no truly inert reference compound B* can be conceived. Even unfunctionalized hydrocarbons form weak inclusion complexes as judged from the observation that the enantiomers of saturated hydrocarbons can be separated on cyclodextrin derivatives [92]. If the reference compound B* itself possesses a retention-increment R'_{B*}, however small, with respect to a hypothetical totally inert compound B**, a relationship between the measured retention-increment R'

$$R' = \frac{r - r^\circ}{r^\circ} \tag{7.18}$$

and the true retention-increment R'_{true}

$$R'_{true} = \frac{r_{true} - r^\circ}{r^\circ} \tag{7.36}$$

can be derived [43]. The r values (but not the r° values) must be corrected in analogy with equation 7.18.

$$R'_{B*} = \frac{r_{true} - r}{r} \tag{7.37}$$

or

$$R'_{true} = \frac{r_{true} - r^\circ}{r^\circ} = (1 + R'_{B*})\frac{r}{r^\circ} - (1 + R'_{B*}) + R'_{B*}$$

$$= (1 + R'_{B*})R' + R'_{B*} \tag{7.38}$$

where R'_{B*} is not an inaccessible quantity but can be estimated by comparison of calculated and measured α versus molality curves (see Figure 7.2) [43]. As demonstrated for enantioselective inclusion gas chromatography [43], values for $R'_{B*} = 0.1$–0.2 have typically been found for hydrocarbons on permethylated cyclodextrins.

Recently, thermodynamic data of the enantioselective complexation of the inhalation anesthetics enflurane, isoflurane, and desflurane with octakis(3-O-butanoyl-2,6-di-O-n-pentyl)-γ-cyclodextrin dissolved in a polysiloxane by gas liquid chromatography have been determined using equation 7.18, and reliable data could be obtained by control measurements with four different reference compounds B* [93].

Polysiloxanes may themselves interact with derivatized cyclodextrins. Equation 7.18 has therefore been modified to account for the nonideal behavior between A and S [94,95]

$$R' = K[(a_A - K_1 f(a_A)f(a_S)] \tag{7.39}$$

where K_1 = thermodynamic complexation constant of the selector A and the non-inert solvent S
$f(a_A)$ and $f(a_S)$ = activity functions.
However the experimental work, which revealed a nonlinearity between R' and the concentration of the cyclodextrin derivative in polysiloxane [94], has not taken into account that equation 7.18 is valid only at high dilution of A in S.

7.13 Conclusion

The introduction of chemical selectivity into the gas chromatographic separation process permits the solution of particularly difficult separation problems involving isomers, isotopomers, and enantiomers. In complexation gas chromatography, the underlying selective equilibria have been quantified. Thus, equations have been developed that permit the calculation of thermodynamic data of molecular association between the selector A and competing selectands B_i in the inert solvent S in a true inert environment (helium, dinitrogen) at a wide temperature range. The concept of the retention-increment (chemical capacity factor) R' of B on A, determined from relative retention data $r°$, and r, is used to calculate isomer-, isotopo-, and enantioselectivities. From the temperature dependence of $-\Delta_{x,y}(\Delta G)$, the Gibbs–Helmholtz parameters $-\Delta_{x,y}(\Delta H)$ and $\Delta_{x,y}(\Delta S)$ of the competing equilibria are also accessible.

Acknowledgments

The author is pleased to thank the late Professor Emanuel Gil-Av, Rehovoth, Israel, for his continuous interest in this work, and his co-workers, cited in the references, for their invaluable contributions to complexation gas chromatography. Generous support by the Deutsche Forschungsgemeinschaft and the Fonds der Chemischen Industrie is gratefully acknowledged.

References

1. B.W. Bradford, D. Harvey, and D.E. Chalkley, *J. Inst. Petrol., London*, **41**, 80 (1955).

2. E. Gil-Av and J. Herling, *J. Phys. Chem.* **66**, 1208 (1962).

3. M.A. Muhs and F.T. Weiss, *J. Am. Chem. Soc.* **84**, 4697 (1962).

4. J.H. Purnell, in A.B. Littlewood, Ed., *Gas Chromatography 1966*. Institute of Petroleum, London, 1967, p. 3.

5. C.L. de Ligny, T. van't Verlaat, and F. Karthaus, *J. Chromatogr.* **76**, 115 (1973).

6. R.J. Cvetanovič, F.J. Duncan, W.F. Falconer, and R.S. Irwin, *J. Am. Chem. Soc.* **87**, 1827 (1965).

7. V. Schurig and E. Gil-Av, *J. Chem. Soc., Chem. Commun.* 650 (1971).

8. E. Gil-Av and V. Schurig, *Anal. Chem.* **43**, 2030 (1971).

9. V. Schurig, R.C. Chang, A. Zlatkis, E. Gil-Av, and F. Mikeš, *Chromatographia*, **6**, 115 (1973).

10. A.N. Genkin and N.A. Petrova, *J. Chromatogr.* **105**, 25 (1975).

11. V. Schurig, *Inorg. Chem.* **25**, 945 (1986).

12. V. Schurig, J.L. Bear, and A. Zlatkis, *Chromatographia*, **5**, 301 (1972).

13. O.K. Guha and J. Janák, *J. Chromatogr.* **68**, 325 (1972).

14. R.J. Laub and R.L. Pecsok, *J. Chromatogr.* **113**, 47 (1975) (refs. 11–22 therein).

15. W. Szcepaniak, J. Nawrocki, and W. Wasiak, *Chromatographia*, **12**, 485, 559 (1979).

16. D.W. Barber, C.S.G. Phillips, G.F. Tusa, and A. Verdin, *J. Chem. Soc.* 18 (1959).

17. R.C. Castells and J.A. Catoggio, *Anal. Chem.* **42**, 1268 (1970).

18. R.T. Pflaum and L.E. Cook, *J. Chromatogr.* **50**, 120 (1970).

19. G.P. Cartoni, R.S. Lowrie, C.S.G. Phillips, and L.M. Venanzi, in R.P.W. Scott, Ed., *Gas Chromatography 1960*. Butterworth, London, 1960, p. 273.

20. G.P. Cartoni, A. Liberti, and T. Palombari, *J. Chromatogr.* **20,** 278 (1965).

21. B. Feibush, M.F. Richardson, R.E. Sievers, and C.S. Springer, Jr., *J. Am. Chem. Soc.* **94,** 6717 (1972).

22. J.J. Brooks and R.E. Sievers, *J. Chromatogr. Sci.* **11,** 303 (1972).

23. J.E. Picker and R.E. Sievers, *J. Chromatogr.* **203,** 29 (1972); **217,** 275 (1981).

24. W.J. Kowalski, *J. Chromatogr.* **349,** 457 (1985).

25. R.L. Pecsok and E.M. Vary, *Anal. Chem.* **39,** 289 (1967).

26. W. Wasiak, *Int. Lab.* **16**(7), 96 (September 1986).

27. J. Delventhal, H. Keck, and W. Kuchen, *Angew. Chem., Int. Ed. Engl.* **11,** 830 (1972).

28. V. Schurig, R.C. Chang, A. Zlatkis, and B. Feibush, *J. Chromatogr.* **99,** 147 (1974).

29. V. Schurig and D. Wistuba, *Angew. Chem., Int. Ed. Engl.* **22,** 772 (1983).

30. V. Schurig and W. Bürkle, *Angew. Chem., Int. Ed. Engl.* **17,** 132 (1978).

31. V. Schurig and W. Bürkle, *J. Am. Chem. Soc.* **104,** 7573 (1982).

32. V. Schurig and R. Weber, *J. Chromatogr.* **217,** 51 (1981).

33. V. Schurig, *Angew. Chem., Int. Ed. Engl.* **15,** 304 (1976).

34. V. Schurig, *Angew. Chem., Int. Ed. Engl.* **16,** 110 (1977).

35. V. Schurig, *Chromatographia*, **13,** 263 (1980).

36. V. Schurig and E. Gil-Av, Israel *J. Chem., Suppl. Proc. Israel Chem. Soc.* **9,** 220 (1971).

37. C.L. de Ligny, *J. Chromatogr.* **69,** 243 (1972).

38. B. Koppenhoefer, G. Laupp, and M. Hummel, *J. Chromatogr.* **547,** 239 (1991).

39. V. Schurig and R. Link, in D. Stevenson and I.D. Wilson, Eds., *Chiral Separations*. Proceedings of the International Meeting on Chromatography, University of Surrey, September 3–4, 1987. Plenum Press, New York and London, 1988, p. 91.

40. V. Schurig, W. Bürkle, K. Hintzer, and R. Weber, *J. Chromatogr.* **475,** 23 (1989).

41. U. Beitler and B. Feibush, *J. Chromatogr.* **123,** 149 (1976).

42. R. Noyori and M. Kitamura, *Angew. Chem., Int. Ed. Engl.* **30,** 49 (1991).

43. M. Jung, D. Schmalzing, and V. Schurig, *J. Chromatogr.* **552,** 43 (1991).

44. S. Allenmark, *Chirality,* **5,** 295 (1993).

45. R. Cramer, *J. Am. Chem. Soc.* **89,** 4621 (1967).

46. V. Schurig, *Chem. Ztg.* **101,** 173 (1977).

47. I. Wender, S. Metlin, S. Ergun, H.W. Sternberg, and H. Greenfield, *J. Am. Chem. Soc.* **78,** 5401 (1956).

48. J.P. Candlin and A.R. Oldham, *Discuss. Faraday Soc.* **46,** 60 (1968).

49. F.H. Jardine, J.A. Osborne, and G. Wilkinson, *J. Chem. Soc. A,* 1574 (1967).

50. G. Winkhaus and H. Singer, *Chem. Ber.* **99,** 3602 (1966).

51. A.N. Genkin and N.A. Petrova, *Rhodium Express,* **2,** 12 (1993); **3,** 11 (1994); **6,** 19 (1994).

52. M.P. Doyle, M.R. Colsman, and M.S. Chinn, *Inorg. Chem.* **23,** 3684 (1984).

53. K. Wypchlo and H. Duddeck, *Tetrahedron Asymmetry,* **5,** 27 (1994).

54. J.G. Atkinson, A.A. Russell, and R.S. Stuart, *Can. J. Chem.* **45,** 1963 (1967).

55. R.J. Cvetanovič, F.J. Duncan, and W.E. Falconer, *Can. J. Chem.* **41,** 2095 (1963).

56. R.L. Arnett and B.L. Crawford, *J. Chem. Phys.* **18,** 118 (1950).

57. D.Y. Youn, K.B. Hong, K.-H. Jung, D. Kim, and K.-R. Kim, *J. Chromatogr.* **607,** 69 (1992).

58. V. Schurig, in J.D. Morrison, Ed., *Asymmetric Synthesis,* Vol. 1. Academic Press, New York, 1983, p. 59.

59. R.G. Pearson, *J. Chem. Educ.* **45,** 581, 643 (1968).

60. H.L. Goering, J.N. Eikenberry, and G.S. Koermer, *J. Am. Chem. Soc.* **93,** 5913 (1971); **96,** 1493 (1974).

61. V. Schurig, *Tetrahedron Lett.* **16,** 1269 (1976).

62. V. Schurig, *J. Chromatogr.* **441,** 135 (1988).

63. V. Schurig, *Kontakte (Darmstadt),* 3 (1986–1).

64. V. Schurig and R. Weber, *J. Chromatogr.* **289,** 321 (1984).

65. D. Wistuba and V. Schurig, *Angew. Chem., Int. Ed. Engl.* **25,** 1032 (1986).

66. C.A.G.M. Weijers and J.A.M. de Bont, *Microbiol. Sci.* **5,** 156 (1988).

67. D. Wistuba, O. Träger, and V. Schurig, *Chirality,* **4,** 185 (1992).

68. N. Kasai, K. Tsujimura, K. Unoura, and T. Suzuki, *J. Ind. Microbiol.* **9,** 97 (1992).

69. R. Weber and V. Schurig, *Naturwissenschaften,* **71,** 408 (1984).

70. V. Schurig, in P. Schreier, Ed., *Bioflavour '87.* de Gruyter, Berlin, 1988, p. 35.

71. V. Schurig and F. Betschinger, *Chem. Rev.* **92,** 873 (1992).

72. V. Schurig, *Naturwissenschaften,* **74,** 190 (1987).

73. V. Schurig, B. Koppenhöfer, and W. Bürkle, *Angew. Chem., Int. Ed. Engl.* **17,** 937 (1978).

74. B. Koppenhoefer and E. Bayer, *Chromatographia,* **19,** 123 (1984).

75. K. Watabe, R. Charles, and E. Gil-Av, *Angew. Chem., Int. Ed. Engl.* **28,** 192 (1989).

76. V. Schurig, J. Ossig, and R. Link, *Angew. Chem., Int. Ed. Engl.* **28,** 194 (1989).

77. W. Bürkle, H. Karfunkel, and V. Schurig, *J. Chromatogr.* **288,** 1 (1984).

78. M. Jung and V. Schurig, *J. Am. Chem. Soc.* **114,** 529 (1992).

79. V. Schurig, M. Jung, M. Schleimer, and F.-G. Klärner, *Chem. Ber.* **125,** 1301 (1992).

80. V. Schurig, W. Bürkle, A. Zlatkis, and C.F. Poole, *Naturwissenschaften,* **66,** 423 (1979).

81. V. Schurig and U. Leyrer, *Tetrahedron Asymmetry,* **1,** 865 (1990).

82. S. Mayer, M. Schleimer, and V. Schurig, *J. Microcolumn Sep.* **6,** 43 (1994).

83. C.J. Reich, G.R. Sullivan, and H.S. Mosher, *Tetrahedron Lett.* **17,** 1505 (1973).

84. V. Schurig, in *Stereoselective Synthesis,* Vol. E21a, Houben-Weyl, Methods of Organic Chemistry. Thieme, Stuttgart, New York, 1995, cf. Figure 10, p. 167.

85. V. Schurig, D. Schmalzing, and M. Schleimer, *Angew. Chem., Int. Ed. Engl.* **30,** 987 (1991).

86. M. Schleimer and V. Schurig, *J. Chromatogr.* **638,** 85 (1993).

87. M. Schleimer, M. Fluck, and V. Schurig, *Anal. Chem.* **66,** 2893 (1994).

88. H. Weinmann, Diploma thesis, University of Tübingen, 1993.

89. T. Hobo, S. Susuki, W. Watabe, and E. Gil-Av, *Anal. Chem.* **57,** 364 (1985).

90. A. Hirsch and V. Schurig, unpublished results.

91. J. Jauch, B. Schulz, and V. Schurig, unpublished results.

92. V. Schurig and H.-P. Nowotny, *Angew. Chem., Int. Ed. Engl.* **29,** 939 (1990).

93. V. Schurig, M. Juza, and H. Grosenick, *Recl. Trav. Chim. Pays-Bas,* **114,** 211 (1995).

94. W.M. Buda, K. Jaques, A. Venema, and P. Sandra, 16th International Symposium on Capillary Chromatography, Riva del Garda, September 27–30, 1994, Vol. I. Hüthig, p. 164.

95. W.M. Buda, K. Jaques, A. Venema, and P. Sandra, *Fresenius Z. Anal. Chem.* **352,** 679 (1995).

Index